BARRON'S
E–Z
CALCULUS

Douglas Downing, Ph.D.
Seattle Pacific University

BARRON'S

Acknowledgments
The inspiration for this book came from the Shoreline High School 1975 calculus class and its teacher, Mr. Clint Charlson. He demonstrated how much it means to have an inspiring teacher, and I hope many other math teachers will feel encouraged to be an inspiration for their students as they teach this difficult subject. I am also indebted to my teachers in physics, astronomy, economics, and mathematics at Yale University. My thanks also go to my mother, my sister, Marlys, and Mark Yoshimi for their help in reviewing the manuscript, to Ruth Flohn, Mickey Wagner, and Michael Covington at Barron's for their work editing the book, to Mary Falcon, editor of the third edition, and to Linda Turner, editor of the fourth and Dave Rodman, editor of this edition. Special thanks also to Susan Detrich for the marvelous illustrations that have done so much to make the book come alive.

All inquiries should be addressed to:
Barron's Educational Series, Inc.
250 Wireless Boulevard
Hauppauge, NY 11788
www.barronseduc.com

ISBN-13: 978-0-7641-4461-5
ISBN-10: 0-7641-4461-8

Library of Congress Catalogue Card No.: 2010005225

Library of Congress Cataloging-in-Publication Data
Downing, Douglas.
    Calculus : the easy way / Douglas Downing, Ph.D. — 5th ed.
        p. cm. — (Ez calculus)
    Includes bibliographical references and index.
    ISBN-13: 978-0-7641-4461-5 (alk. paper)
    ISBN-10: 0-7641-4461-8 (alk. paper)
    1. Calculus. I. Title
    QA303.2.D69 2010
    515—dc22   2010005225

Printed in the United States of America
9 8 7 6 5 4 3

CONTENTS

LIST OF SYMBOLS

$\sqrt{}$	square root
$\|\ \|$	absolute value
$\pm$	plus or minus
$<$	less than
$>$	greater than
$\simeq$	approximately equal to
∞	infinity
$\int$	integral sign
$\dfrac{dy}{dx}$ or y'	derivative of y with respect to x
$\dfrac{d^2y}{dx^2}$ or y''	second derivative
$\dfrac{\partial y}{\partial x}$	partial derivative

Greek Letters

Δ	capital delta (used for "change in")
Σ	capital sigma (used for summation)
ρ	rho (represents density)
π	pi (= 3.14159 . . .)
Ω	capital omega (represents angular frequency)
ω	lowercase omega (represents angular frequency)
θ	theta

INTRODUCTION

This book tells of adventures that took place in the land of Carmorra. The story is told here because, by following these adventures, you can learn differential and integral calculus. This book includes material suitable for a first-year calculus course. It is designed to be used in a classroom, but it can also be used by someone wishing to learn calculus on his or her own, or as a supplement to a course. This book is unlike regular math books, though. You are invited to read the book as you would read a fantasy novel.

The subject of calculus stands at the gateway to much of higher mathematics, and to applications in many different fields such as physics, biology, chemistry, economics, business, and statistics. In arithmetic, operations are carried out on numbers; in algebra, operations are carried out on symbols that stand for numbers; whereas, in calculus, operations are carried out on functions that represent the relationship between two variable quantities. Some integration techniques date back to the time of the ancient Greek world, but what we now know as calculus was developed independently by Isaac Newton in 1666 and Gottfried Wilhelm Leibniz in 1675. Newton called his invention the method of *fluxions*, which he developed at the same time that he was developing the foundations of the branch of physics known as mechanics.

You will best appreciate this book if you have about the same mathematical background as the people of Carmorra had at the beginning of the story. The material in this book is designed to follow high school courses in algebra, trigonometry, and geometry. You should know basic algebra terminology and methods, such as how to solve an equation with the quadratic formula. Experience in factoring second-degree polynomials will also be beneficial. Calculus depends heavily on analytic geometry, so it helps if you are familiar with Cartesian coordinates, the slopes of lines, and the equation of figures such as circles, ellipses, and parabolas. Function notation, as in $f(x) = x^2$, is also used extensively throughout the book. The book *E-Z Algebra* contains an account of how the people in Carmorra discovered these topics.

You should be familiar with basic trigonometric functions and know some of their properties. See the book *E-Z Trigonometry* for more information. A review list of trigonometric identities is included in Chapter 11. Some familiarity with logarithmic and exponential functions will help, although it is not essential to understand the book. Imaginary numbers play a small role in Chapter 15, but familiarity with imaginary numbers is not needed anywhere else in the book.

This book is designed to let you solve applied problems as quickly as possible. Many of the results presented here are demonstrations rather than formal proofs. If you are planning further study in calculus, you should become familiar with some of the rigorous background theory, such as the meaning of continuity and of limit.

The people in Carmorra use a very bizarre system of measurement, so I have translated numerical measurements into the metric system or else left measurements in terms of general units. If you are a science student, in particular, you will have to learn to be rigorous in your treatment of units.

At the end of each chapter are exercises to provide practice in applying the concepts developed in that chapter. Understanding any mathematical material requires work. The answers are provided at the back of the book so you can tell for yourself how well you have mastered the problems. The problems in Chapter 18 provide a comprehensive test of material from throughout the book. The exercises came from a wide variety of sources; some were supplied by the gremlin and some were dreamed up by Professor Stanislavsky. The final test in Chapter 18 is presented here exactly as the gremlin presented it to us.

Many of the chapters contain worksheets with additional practice exercises. These often provide repetitive practice with basic steps. Once you become familiar with the routine of solving these exercises you will be equipped to take on more challenging exercises. The answers to the worksheet problems are included in Appendix 1.

Computers and calculators provide essential tools for working with calculus problems. They can perform a variety of calculations that would be too tedious for you to do yourself. If you have a graphing calculator you can use it to create the many graphs that appear in the book. The graphing calculator will allow you to increase your understanding of a diagram by zooming in or zooming out to see the graph from different perspectives. Another valuable tool is a computer spreadsheet. A spreadsheet can be used to perform many of the calculations and create many of the graphs. Some of the exercises in the book are designed to be done with a spreadsheet, and Chapter 17 illustrates different examples of using spreadsheets for solving calculus problems numerically.

You may also find it useful to use software that will do symbolic manipulations for you. For example, such software could tell you that the derivative of $y = x^2$ is $y' = 2x$. However, you shouldn't use such software as a replacement for learning how to do these operations yourself.

Here are some of the reasons calculus is a difficult subject. First, you need to be able to think in a very abstract, general way. For example, you should be able to picture that the graph of the equation $y = 17$ is a horizontal line. The challenge is to realize that in general, the graph of $y = c$, where c can be any constant number, will also be a straight line. When stated this way, the rule is completely general, so you don't need separate rules for $y = 18$, $y = 19$, etc. When you have made it over the hurdle of thinking generally in this fashion, then the work we do in this book becomes clearer.

Second, calculus is hard because there are some things to memorize. I hope the rather odd adventures in the book will create vivid pictures that will stick in your mind and will help a bit with the task of memorization. Third, calculus is hard because some derivations require a lot of steps, and it is very hard to complete a lengthy derivation without getting one of the steps wrong along the way. There is no way to avoid this work, although I hope you will appreciate that some calculus derivations are very short and elegant (for example, showing that the derivative of the function giving the area under a curve is actually the function represented by that curve; see Chapter 8).

Fourth, calculus might seem hard if you're not motivated to learn the subject. In fact, there are many applications of calculus, although many of the really interesting applications are too complicated to include in an introductory text. This book tries to give you a feel for some these applications. You may feel that you personally will not need to understand these applications which is fine if you're content to let other people understand how the world works, but I hope you will see the benefit of increasing your own understanding. Even if you never need to pilot an orbiting spaceship to rendezvous with another spaceship, you might find it very interesting to know how this is done (see Chapter 16, Exercise 49).

Three general categories of applications of calculus include the study of motion; the search for maximum and minimum points; and the summation of many small parts to find areas, lengths, and volumes. Some specific applications include:

How long will it take for a thrown ball to return to the ground? What angle should the ball be thrown at to maximize the distance it travels?

What is the optimal level of inventory?

What quantity should be produced to maximize the profit? How do you find the volume of a swimming pool?

How do you find the area of a circle or an ellipse? What shape for a cylindrical can maximize the volume for a fixed amount of material used in the can?

Where does the graph of a curve change direction?

How does the rate of a chemical reaction change if it depends on the amount of reactants available at a particular time?

How is a radio station tuned so it receives signals of the desired frequency? How do you estimate the number of people whose heights are between two specified values?

Why does the force of gravity coming from a spherical planet act as if the mass were all at the center of the planet? Why do planets move along elliptical orbits? How does the energy change as an object moves?

The first edition of the book was published in 1982 under its previous title *Calculus the Easy Way*. Since then the strange adventures encountered here have helped many readers learn this fascinating subject.

Look for more information on the book's web page: *http://myhome.spu.edu/ddowning/easycalc.html*. Here you will also find some spreadsheet files that you can download.

If you have never seen a calculus problem before, you are in the same position that Recordis, the professor, and the others are at the beginning of the story. You are now about to embark on the discovery of calculus.

The Slope of the Tangent Line

MAIN CONFERENCE ROOM

The storm struck my ship with devastating suddenness. Something hit me on the head, and my memory was completely knocked out. The next thing I remember I was being washed ashore on a strange land called Carmorra. The farmer who first met me, Mr. Floran, decided to take me to the capital city. He took me to the train station. We settled comfortably onto the train, but I was puzzled because there was no engine.

"All aboard!" the conductor cried, and then a giant appeared and started pushing on the back of the train. Amazed, I could feel the giant pushing as the train started moving faster and faster.

"That's Pal," Mr. Floran explained. "He is a very friendly giant who likes to help the people by doing things like pushing the train. He is very careful to push the train with a constant force, which means that the train keeps going faster and faster."

The countryside was speeding by. Soon we were in the hills and began racing around a sharp curve next to a giant canyon.

"Let go!" the conductor shouted to the giant.

"Sometimes Pal has so much fun that he forgets to let go of the train," Mr. Floran gasped as we desperately clutched our seats as the train spun around the curve.

Soon we arrived at the Capital, where Mr. Floran took me to the Royal Palace. A heated debate was going on in a room labeled "Main Conference Room."

"Have you made any more progress?" Mr. Floran asked after he had introduced me.

"No," a pleasant woman with intense, piercing eyes said sadly. ("That's Professor Stanislavsky," Floran whispered to me.)

"Yes, we have!" contradicted a middle-aged man with an elaborately carved pen in his hand and three more pens behind his ear. "It has been proved to be impossible to solve the problem." ("That's Marcus Recordis, the Royal Keeper of the Records," Floran whispered.")

"We have indeed seemingly reached an impasse," a man in a glittering robe said. ("That's the king," Floran informed me. "You had better bow to him.") After the necessary formalities were over, Floran introduced me to the other people in the room: Alexanderman Trigonometeris, the Royal Keeper of the Triangles, and Gerard Macinius Builder, the Royal Construction Engineer.

"You mustn't forget Igor," Recordis said.

"Who is Igor?" I asked, seeing no one else in the room.

"This is Igor," Recordis said, slapping a large object on the wall that looked like a combination television screen and blackboard. "This is the only Visiomatic Picture Chalkboard Machine in the world."

"Now we can explain the problem," the professor stated. "Up to now we have not been able to use Pal to his full capacity, because we are worried that if he keeps pushing the train it will go too fast. The problem happens when the train reaches Lurch Curve, overlooking Steepdrop Canyon."

"I remember," I shuddered.

"If the train travels too fast around the curve, it will fall off the track and into the canyon. The trouble is that we don't know how fast the train is going at a given time."

"Don't you know where the train is at a particular time?" I asked.

"Yes, it is a simple matter to tell the position of the train at any time. Draw the picture, Igor. (See Figure 1–1.) Recordis boards the train with his watch. All along the track we have markers telling how far it is from the start of the track. Every time one minute goes by, Recordis shouts 'Now!' and Trigonometeris quickly looks outside and writes down how far the train has gone. They made a table of their results." (See Table 1–1.)

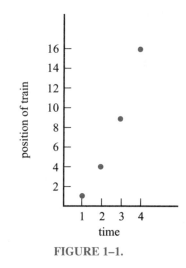

FIGURE 1–1.

Table 1–1

Time	Position
0	0
1	1
2	4
3	9
4	16
5	25

"With these numbers it is easy to make a position-time graph, like the one the professor just showed you," the king said. "Igor draws two perpendicular lines, marks time on the horizontal line and distance on the vertical line, and then puts a dot at every point where the time number directly under the point is equal to the time when the train is at the position number directly to the left of it.

"The next thing we can do with the graph is figure out exactly where the train was at some time in the middle, such as 3.5 minutes. Obviously, the train must be somewhere on the track, and it must be somewhere between the place where it was at 3 minutes and the place where it was at 4 minutes (because it always goes in the same direction). The only way to figure out exactly where it is is to draw a line or a curve that connects all the points on the graph. The most logical connecting line is a smooth curve." (See Figure 1–2.)

"This way we can represent the entire curve mathematically," Recordis said. "It just so happens that the distance that the train has moved from the starting point is equal to the amount of time it has been traveling multiplied by itself."

(distance train has moved) = (time in minutes) × (time in minutes)

"We can abbreviate this equation by denoting the distance traveled by some letter, such as d."

"Why?" the king asked.

"All right, call it y if you prefer," Recordis said. "Then I suppose we should call the time the train has been traveling x.

$$y = (x) \times (x)$$

"That is the same thing as writing x squared, or x^2," the king pointed out.

$$y = x^2$$

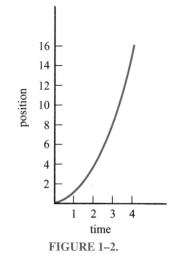

FIGURE 1–2.

"We can also write that as a function machine," the professor said. "We decided that a function was a machine that turned one number into another number according to some rule. If the number we put in was called x, then the number that came out of the machine was called $f(x)$. For example, one simple function is $f(x) = 2x$." (Figure 1–3.) [The expression $f(x)$ is read "f of x."]

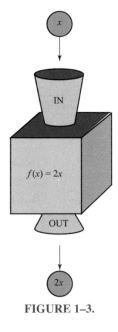

"In this case it is easy to figure out what number will come out," Recordis said. "If we put in 2, we'll get 4; if we put in 10, we'll get 20, etc."

"Also, we can call the in-number anything we like. We don't have to call it x," the king added. "If $f(x) = 2x$, then we can also say that $f(q) = 2q$, $f(a + b) = 2(a + b)$, and $f(3x^2 + 4x + 5) = 6x^2 + 8x + 10$. We can say that f of (in-number) = 2 times (in-number)."

"We also figured out the height of the train at different times if it veers off the track and plummets into the canyon," Builder said. "Let y represent the height of the train at a given moment in time (t). If it falls off the cliff at time $t = 0$, and the height of the cliff is

FIGURE 1–3.

h, then $y = h - \dfrac{1}{2}gt^2$ where g is a constant that measures the acceleration of gravity. But we don't know how fast it will be going at the moment it crashes to the ground at the bottom of the canyon."

Recordis shuddered. "If I'm on the train that's plunging downward, I'm actually glad we don't yet know how to calculate its speed at the instant of impact."

"We still need to know how to do it," the professor chided him. "So you see the problem: We can tell the position of the train at any given time, but we need to know the speed."

"It is easy to calculate the speed if an object is traveling at a constant speed," the king said. "For example, if Recordis walks at a constant speed of 4 miles per hour, then his position function is given by $f(x) = 4x$. If you make a graph of his position, it looks like a straight line." (Figure 1–4.) "In order to calculate the speed, we use this formula:

$$(\text{speed}) = \frac{(\text{distance traveled})}{(\text{time elapsed})}$$

"Since Recordis walks 8 miles in 2 hours, his speed is 8/2, which equals 4."

"Sometimes I walk faster," Recordis said proudly. "If my speed is 6, then the graph of my position function is different." (See Figure 1–4).

"It appears the line is steeper if you walk faster," I noted.

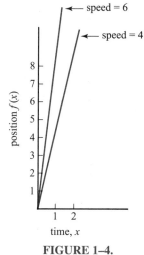

FIGURE 1–4.

"That's it!" the professor jumped into the air. "The speed of the object is (distance) divided by (time), which on the graph is represented as (vertical distance) divided by (horizontal distance). I knew I had seen the formula for speed before, and now I realize where: That is the formula for the slope of the line! So the speed of the object is the same as the slope of the line representing its position!" She quickly demonstrated her idea. "We can figure out the average speed of the train between any two times. At time $x = 1$, the position of the train is $1^2 = 1$. At time $x = 4$, the position of the train is $4^2 = 16$; so its average speed between those two times is

$$\frac{16-1}{4-1} = 15/3 = 5$$

which is the same as the slope of the line between those two points."

"There is a huge difference between the average speed of the train and its speed at any given moment!" Recordis protested vehemently. "My stomach can tell the difference between the early part of the ride, when the train is going very slow, and the later part of the ride, when the train is going very fast. My stomach doesn't care about the average speed for the whole trip."

"Right," the professor agreed. "When the speed of an object is changing, we need to calculate its instantaneous speed—that is, the speed it is traveling at a given instant."

"We can come close to the instantaneous speed by calculating the average speed over a very short time interval," the king suggested. "For example, at time 3.5, the position of the train is $3.5^2 = 12.25$. At time 3.7, the position of the train is $3.7^2 = 13.69$. Therefore, the average speed of the train during this interval is

$$\frac{(\text{distance traveled})}{(\text{time elapsed})} = \frac{13.69-12.25}{3.7-3.5} = \frac{1.44}{0.2} = 7.2$$

"That still doesn't give us a formula to calculate the speed of the train at a particular instant," Recordis said sadly. "What makes it even more frustrating is that we can come tantalizingly close. For example, we can make the time interval smaller and smaller, which lets us come closer and closer to the instantaneous speed." Recordis displayed a table that showed how the average speed changed as the time interval became smaller and smaller (Table 1–2).

Table 1–2

Start Time	Start Position	Ending Time	Ending Position	Length of Time Interval	Distance Traveled	Average Speed
3.5	12.25	3.7	13.69	0.2	1.44	7.2
3.5	12.25	3.6	12.96	0.1	0.71	7.1
3.5	12.25	3.51	12.3201	0.01	0.0701	7.01
3.5	12.25	3.501	12.257001	0.001	0.007001	7.001
3.5	12.25	3.5001	12.25070001	0.0001	0.00070001	7.0001

The king whispered to me, "When I stare at the table, the pattern seems to be so clear that I am almost willing to guess what the instantaneous speed must be at time 3.5 minutes, but I am afraid to say anything unless I am absolutely certain I am right."

"We do know one important clue," the professor said. "Remember that the speed of an object moving with constant speed is equal to the slope of the line representing the position of that object as a function of time. Therefore, in order to find the instantaneous speed of an object with variable speed, we need to find the slope of the curve representing the position of that object."

"How can you figure out the slope of a curve, like $f(x) = x^2$?" I asked. "At some points the curve has a very slight slope, but at other points it is sloped very steeply."

"My point exactly," the professor said. "The slope of the curve is changing because the speed of the train is changing. We can't determine the slope of a curve directly, but we can draw a line right next to the curve that has the same slope as the curve does at that point." (Figure 1–5.)

"We call that line the *tangent line* for the curve," the king told us. "Notice that this curve has lots of different tangent lines." (Figure 1–6.)

"You should also notice that a tangent line touches the curve at one and only one point," Recordis said.

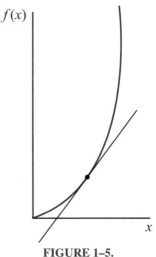

FIGURE 1–5.

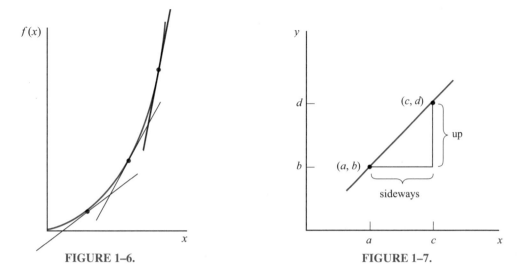

FIGURE 1–6.   FIGURE 1–7.

"That means that we can define the slope of the curve at a given point to be equal to the slope of the tangent line at that point," the professor added.

"Therefore, to find the speed of the train, all we need to do is find the slope of the tangent line to the curve," the king noted.

"That is what we have proved to be impossible," Recordis said heatedly. "We know very well how to draw a line. First, we start out with two points. We can call them anything we like, say (a, b) and (c, d). Now we can easily compute the slope of the line between these two points. A long time ago we defined the slope as equal to the distance the line goes up divided by the distance the line goes sideways. (See Figure 1–7.)

$$\text{slope} = \frac{(\text{up})}{(\text{sideways})}$$

"We know that (up) is equal to $(d - b)$, and that (sideways) is equal to $(c - a)$. This lets us say that the slope of the line is equal to (slope) $= (d - b)/(c - a)$. This method works for any line in the world, as long as we know *two* points on it. There is absolutely no way to find the slope of the tangent line, though, because we know only one point! We know lots of other points that are not on the line, but we don't know one other single point that *is* on the line."

There was a long silence as we contemplated what he had said.

"We must find the answer to this problem," the king stated. "I don't care how we have to do it."

"There has to be some solution," the professor said. "In fact, we placed a large wager with the gremlin that we would be able to reach a solution in the next few days."

"Who is the gremlin?" I asked.

"He is our arch-enemy," the king told me. "It is his sole purpose to disrupt our entire learning process and take over the kingdom of Carmorra. We have already defeated him several times concerning matters of algebra."

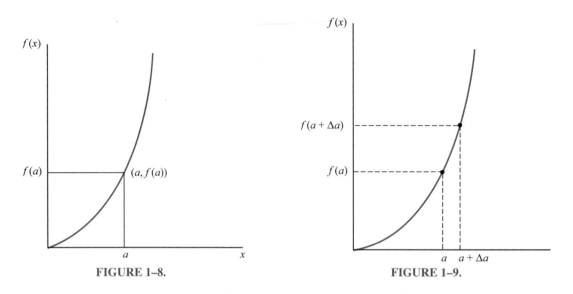

FIGURE 1–8.   FIGURE 1–9.

I looked at the drawing of the graph for several minutes. Finally I said, "It appears that our problem is that we need to find another point someplace. Recordis is certainly right when he says that we need two points to determine the slope of the line. Igor, draw a graph showing the point where we want to find the slope of the tangent line. (See Figure 1–8.)

"Let's say that this point represents the train at some time called a. This means that the distance the train has traveled equals a^2. The coordinates of this point are (a, a^2). We still have our position-locating function machine, so we can also write the coordinates as $(a, f(a))$. In order to find the slope of the tangent line, we need another point. Since the only points that we know very much about are the other points on the curve, we will have to use one of those. Igor, show me another point on the curve that is close to the first point. (See Figure 1–9.)

"It really doesn't make much difference how far away the second point is from the first point, so we can make up some distance and call it Δa. (The little triangle Δ is the fourth letter—capital form—of the Greek alphabet. It is known as *delta*. The symbol Δa is pronounced 'delta-a.') Then we know that the x coordinate, or the time coordinate, of the second point is equal to $(a + \Delta a)$."

"The y coordinate is still $y = f(x)$, so we can plug that into the machine and say that the y coordinate is equal to $f(a + \Delta a)$," the professor noted.

"I know the slope of the line between those two points," Recordis said.

$$\text{(slope of line between these two points)} = \frac{f(a + \Delta a) - f(a)}{a + \Delta a - a}$$

$$= \frac{f(a + \Delta a) - f(a)}{\Delta a}$$

"We can call that line a *secant* line," the professor stated. "We use that name for a line that intersects a curve in two points, rather than one point as the tangent line does."

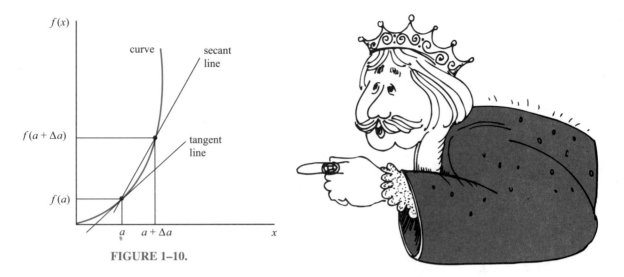

FIGURE 1–10.

"The secant line isn't very close to the tangent line," Recordis protested.

"Maybe we can make the secant line move closer to the tangent line," I said. "Igor, make a sketch of where you think the tangent line should be." (Figure 1–10.)

The king looked closely at the picture. "Couldn't we make the second point move closer to the first point? Wouldn't that make the secant line approach the tangent line?"

"That's it!" I said. "We can make the two points move closer together by making Δa smaller and smaller. Igor, draw a series of pictures showing what happens when the two points move closer together." (Figure 1–11.)

"What happens to the expression for the slope?" Recordis asked. "Remember what we have."

$$\text{(slope)} = \frac{f(a + \Delta a) - f(a)}{\Delta a}$$

"If we let Δa become too small, some weird things will happen to this fraction."

"No, they won't," the professor said. "Remember that when $a + \Delta a$ moves close to a we will also have $f(a + \Delta a)$ move close to $f(a)$. We will end up with the ratio of two very small numbers, and there is nothing wrong with that."

"But to get the slope of the tangent line we would have to let Δa become zero!" Recordis protested. "Then we would end up with a slope of 0/0, which doesn't tell us anything."

"This means that we cannot ever let Δa actually equal zero," I said. "The closer it gets to zero, though, the closer the slope of the secant line will come to the slope of the tangent line. Let's make the following definition."

$$\text{(slope of tangent line)} = \lim_{\Delta a \to 0} \frac{f(a + \Delta a) - f(a)}{\Delta a}$$

(The expression $\lim_{\Delta a \to 0}$ is read "The limit as delta-a goes to zero.")

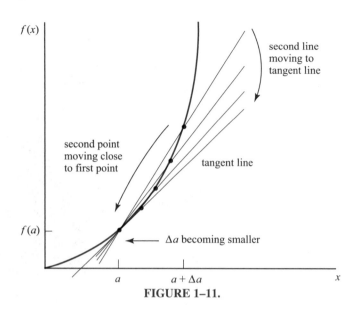

FIGURE 1–11.

"What does that mean?" Recordis protested. "What do you mean by that 'limit' thing?"

"I understand," the professor said. "We will let Δa move very, very, very, very close to zero, but we will put up a little fence that presents it from ever actually equaling zero."

"That's all very nice theoretically," the king remarked. "It still doesn't tell us how to find a number that represents the slope of the tangent line, though."

"We know what $f(x)$ is," I said. "We can rewrite the equation, only this time we will use $f(x) = x^2$."

$$(\text{slope}) = \lim_{\Delta a \to 0} \frac{(a + \Delta a)^2 - a^2}{\Delta a}$$

"I know what $(a + \Delta a)^2$ is," Recordis said. "That's algebra."

$$(\text{slope}) = \lim_{\Delta a \to 0} \frac{\boldsymbol{a^2} + 2a\Delta a + \Delta a^2 - \boldsymbol{a^2}}{\Delta a}$$

"The a^2 and the $-a^2$ will cancel out," the king pointed out helpfully.

$$(\text{slope}) = \lim_{\Delta a \to 0} \frac{2a\,\Delta a + \Delta a^2}{\Delta a}$$

We factored out Δa on top:

$$(\text{slope}) = \lim_{\Delta a \to 0} \frac{\Delta a(2a + \Delta a)}{\Delta a}$$

"We can divide both the top and the bottom of the fraction by Δa," the professor said. "After all, we never let Δa become zero."

$$(\text{slope}) = \lim_{\Delta a \to 0} \frac{2a + \Delta a}{1}$$

"Now it is very clear," I said. "As Δa approaches zero, the quantity $(2a + \Delta a)$ will approach $2a$."

"That's the answer!" the king shouted.

"That is the slope of the tangent line," the professor said reverently.

"That's amazing!" Trigonometeris said.

"That's too simple," Recordis protested suspiciously.

"We should make sure that it makes sense," I said cautiously. "Let's look at the graph. (See Figure 1–2.) At the place where $a = 0$, we know that the train has not yet started to move, so its speed must be zero. According to our formula for the slope, the slope should equal $2a$, which is $2 \cdot 0$, which is zero, so it looks right."

"And if we draw the tangent line at the point where $a = 0$, it is parallel to the axis," the king added. "For another example, if $a = 3$, the slope at that point will be $2 \times 3 = 6$."

The professor said excitedly, "The formula for the slope of the tangent line should work for any function, so we should record this result as our first definition."

The slope of the tangent line to the curve representing the function $f(x)$ at the point $(a, f(a))$ is given by this formula:

$$\lim_{\Delta a \to 0} \frac{f(a + \Delta a) - f(a)}{\Delta a}$$

"We have to think of a name for the subject we are getting into now," Recordis said. "We must do this systematically. I think I will have to start a new page in my record book."

Everybody thought of a name, but nobody came up with one that was satisfactory to all. The main problem was jealousy. Each person in the room wanted the subject named after him- or herself. The others finally turned to me and asked me to make up a name. I had vague memories of doing this sort of problem before, although I could not remember any details. For some reason the word "calculus" popped into my head, so I suggested that we call the subject *calculus*. Everyone agreed to this suggestion because the name sounded impressive.

"We will have to come back to this tomorrow," the professor said. "We can try other functions and see how they work out. First we should record what we found today."

$$\text{slope of tangent line for } f(x) = x^2 \text{ is } 2x$$

The group adjourned amidst great excitement, and they ran to the train to tell everyone that they knew how fast it went. Farmer Floran decided that he had better return to his home, so he boarded the train and Pal pushed him back to Coast City.

Later in the evening it was decided that in gratitude for my services I would be provided with lodging in the palace. I told the king that I want to go home as soon as I remembered where I had came from, but that I would be glad to stay and help for a while. In turn, the king promised to help me when I was ready to return home. "I think you arrived at the beginning of an exciting period," the king told me. "I wonder what we will discover tomorrow."

WORKSHEET

Let $y = x^2$. Each question below gives two values of x (call them x_1 and x_2). Calculate the value of y associated with each value, then calculate Δx, Δy, and the slope between the two points.

	x_1	y_1	x_2	y_2	$\Delta x = x_2 - x_1$	$\Delta y = y_2 - y_1$	slope $= \Delta y / \Delta x$
1.	10		15.0000				
2.	10		11.0000				
3.	10		10.1000				
4.	10		10.0100				
5.	10		10.0010				
6.	10		10.0001				

Exercises

1. The following pairs of points all define secant lines to the curve $y = x^2$ through the point $(2, 4)$. Find the slope of each secant line: $(2, 4)$ and $(3, 9)$; $(2, 4)$ and $(2.5, 6.25)$; $(2, 4)$ and $(2.3, 5.29)$; $(2, 4)$ and $(2.1, 4.41)$; $(2, 4)$ and $(2.05, 4.20)$.

2. Find the slope of the secant line defined by each of these pairs of points: $(1, 1)$ and $(2, 4)$; $(1.5, 2.25)$ and $(2, 4)$; $(1.7, 2.89)$ and $(2, 4)$; $(1.8, 3.24)$ and $(2, 4)$; $(1.9, 3.61)$ and $(2, 4)$; $(1.95, 3.80)$ and $(2, 4)$.

3. Find the equation of the tangent line to the curve $y = x^2$ at the point $(2, 4)$.

4. Find the equation of the tangent line to the curve $y = x^2$ through the point $(7, 49)$. Use this tangent line to estimate $\sqrt{50}$.

5. Draw a graph of the function

$$y = f(x) = \frac{8x + x^2}{x}$$

What is $f(0)$? What is $\lim\limits_{x \to 0} f(x)$?

6. Show that the formula for the slope of the tangent line is the same as the formula for the average speed of an object over a time interval that becomes very small.

7. Use a computer spreadsheet or graphing calculator to draw a graph of the curve $y = x^2$ between $x = a$ and $x = b$. What happens as you make a and b closer together while you increase the magnification to zoom in for a closer look at the curve? (Hint: you should see the curve become more like a straight line.) Next, include the secant line on your graph, and see what happens when a and b move closer together.

8. Use a spreadsheet to complete this table for $y = x^2$:

x	y	slope of tangent
−6		
−4		
−2		
0		
2		
4		
6		

Calculating Derivatives

Everybody gathered around Igor the next morning. The professor said that she had a whole series of new ideas to try out. Recordis started the meeting with an anguished complaint.

"This will never do!" he cried. "We must think of a shorter name for this whatever-it-is we've discovered. I can't write 'slope of the tangent line' all day. Already my wrist is developing a terrible cramp."

"Then we shall think of a name," the professor said matter-of-factly. "Let's look at our definition again."

$$(\text{slope of tangent line}) = \lim_{\Delta x \to 0} \frac{f(x + \Delta x) - f(x)}{\Delta x}$$

(We had decided to write "lim" as an abbreviation for "limit.")

"Now, what does that look like?" the professor asked.

"It looks to me like the time Pal spilled his letter blocks and we never could figure out all the words he had made," Trigonometeris said.

"We must take this seriously," the king rebuked. "We must think of a real name."

Everybody suggested names, but nothing sounded satisfactory. Finally the professor turned to me and said, "You were able to think of the name 'calculus.' Maybe you can think of a name for the slope of the tangent line."

I thought for a couple of minutes and came up with another name. "We could call it a *derivative*," I suggested.

"That sounds as good as any," Recordis said. "It surely was frustrating to derive it."

function $y = f(x)$

$$\text{derivative} = \text{slope of tangent line} = \lim_{\Delta x \to 0} \frac{f(x + \Delta x) - f(x)}{\Delta x}$$

"We still need to think of a symbol to stand for derivative," Recordis complained. "I can't write the word 'derivative' all the time."

Everyone looked at the board for a few minutes. "I have an ingenious idea," the professor said as modestly as she could. "Since a derivative is a slope, it should have units of y/x. For example, we wrote delta y over delta x ($\Delta y/\Delta x$) to stand for a tiny increment of y divided by a tiny increment of x. Why don't we say that dy/dx is the slope of the tangent line. In general, we could say that writing d/dx in front of a function means, 'Take the derivative of the function with respect to x.'"

Recordis was reluctant to agree to any symbol that required him to write four letters.

"I always liked the little prime ($'$) symbol," the king said. "I liked it when we wrote y' and called it 'why prime?' We could call the derivative y' or $f'(x)$?" (The symbol $f'(x)$ is read "f-prime of x.")

The professor looked hurt, but Recordis was happy. "I really like that!" he said.

"But you could get confused!" the professor protested. "What if the variable in the function isn't x? What if you have $y = f(t)$, $y = f(w)$, or $y = f(q)$? In my system you could write dy/dt, dy/dw, or dy/dq."

"But look at all that writing!' Recordis said.

"I don't think we need to have an argument here," I told them. "We'll use both systems. At any particular time we'll use whichever one seems to be the more convenient."

function $y = f(x)$

derivative $y' = f'(x) = \dfrac{dy}{dx} = \lim\limits_{\Delta x \to 0} \dfrac{f(x + \Delta x) - f(x)}{\Delta x}$

"We wasted too much time thinking of names," the professor said. "We must start with the important part. We must be systematic about this, and make a list of different kinds of functions and their derivatives."

"It seems to me that the simplest function is one that has the same value all the time," the king stated.

"I remember when we made an $f(x) = 2$ function," Recordis said. "Pal got tired of 2's, but no matter what number he put into the function machine he always got a 2 out."

"Let's see what our formula says if $f(x) = 2$," the professor said.

$$f(x) = 2$$

$$f'(x) = \lim\limits_{\Delta x \to 0} \frac{f(x + \Delta x) - f(x)}{\Delta x}$$

$$= \lim\limits_{\Delta x \to 0} \frac{2 - 2}{\Delta x}$$

$$= \lim\limits_{\Delta x \to 0} \frac{0}{\Delta x}$$

$$f'(x) = 0$$

"We should have a slope of zero," the professor said.

"Let's draw a graph of the function $y = 2$, just to make sure," Recordis said. (See Figure 2–1.)

"That's just a straight line with no slope," Trigonometeris pointed out. "I could have told you the slope of that before we developed all this hocus-pocus."

"But we know that zero must be the right answer," the king said. "Let's pretend that the graph represents the position of the train. Then this means that the train is 2 units from the ocean and is just staying in one place. If it is stopped, then of course its speed is zero."

"Like the time Pal was scared by that parakeet and just stopped in the middle of the track, still holding onto the train," Recordis reminded us.

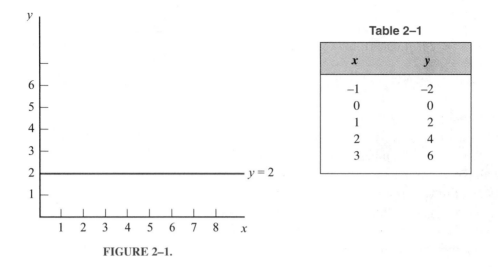

FIGURE 2–1.

Table 2–1

x	y
−1	−2
0	0
1	2
2	4
3	6

"The same thing should happen if y is any constant number, and not just 2," the professor said. "We could have y equals 2 or 7.6 or 819½ or anything else, just so long as it doesn't change. We can write that as our next rule: The derivative of a constant function is zero."

Function	Derivative
$y = c$	$y' = dy/dx = 0$ (when c is a constant)

"What if we have a tilted line?" Trigonometeris asked. (Figure 2–2.)

"We've had functions like this before. Igor, show us a table of values." (Table 2–1.)

"That function is easy to recognize," the professor said. "We have $f(x) = 2x$. Let's plug that into our formula."

$$\frac{dy}{dx} = y' = \lim_{\Delta x \to 0} \frac{2(x + \Delta x) - 2(x)}{\Delta x}$$

$$= \lim_{\Delta x \to 0} \frac{2x + 2\,\Delta x - 2x}{\Delta x}$$

$$= \lim_{\Delta x \to 0} \frac{2\,\Delta x}{\Delta x}$$

$$\frac{dy}{dx} = 2$$

"But I could have told you the slope was 2," Trigonometeris said. "We could have figured out the slope using the old method." (Figure 2–3.)

$$\text{slope} = \frac{\text{(up)}}{\text{(sideways)}} = 2$$

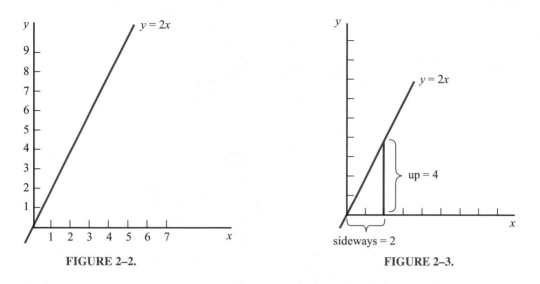

FIGURE 2–2.   FIGURE 2–3.

"I don't think you really need calculus methods whenever you have any function that is a straight line," the professor stated. "However, it is a good thing that the calculus methods give us the same answer for the slope as the regular methods did. If calculus turned out to be inconsistent with algebra and geometry, we'd be in real trouble."

"I bet that we can generalize this rule," the king said. "Suppose that $f(x)$ equals x times any constant number, say c."

$$f(x) = cx$$

$$f'(x) = \lim_{\Delta x \to 0} \frac{c(x + \Delta x) - cx}{\Delta x}$$

$$= \lim_{\Delta x \to 0} \frac{cx + c\,\Delta x - cx}{\Delta x}$$

$$= \lim_{\Delta x \to 0} \frac{c\,\Delta x}{\Delta x}$$

$$f'(x) = c$$

"That looks like a good rule," the professor agreed.

Function	Derivative
$y = cx$	$y' = dy/dx = c$ (when c is a constant)

"I know something that has a position function that looks like that," Recordis said. "Remember when we took Pal to Ice Skating Lake and gave him a push? We made a table of his position at different times." (Table 2–2.)

"What happened after that?" I asked.

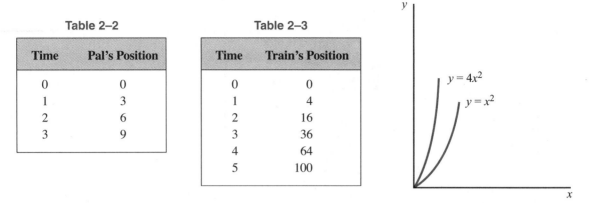

Table 2–2

Time	Pal's Position
0	0
1	3
2	6
3	9

Table 2–3

Time	Train's Position
0	0
1	4
2	16
3	36
4	64
5	100

FIGURE 2–4.

"The ice cracked and he almost fell through," the professor answered. "I remember that we established that he had been traveling with a constant speed of 3 units per second. Apparently whenever something travels with a constant speed the derivative of its position function is a constant number."

"I remember a complicated situation," Recordis said. "Remember the time we fed Pal some Extra-Strength Tablets before we let him play with the train? He started running very fast, and we kept a table of values." (Table 2–3.)

Igor drew a graph of the train's motion (Figure 2–4). We could see that the slope was much steeper than the slope of the graph we had looked at the day before.

"I recognize that function," the king said. "One squared times 4 is 4, 2 squared times 4 is 16, and 3 squared times 4 is 36."

"That's it," the professor said. "The function is $f(x) = 4x^2$."

We put that in the formula for the derivative:

$$f(x) = 4x^2$$

$$f'(x) = \lim_{\Delta x \to 0} \frac{4(x + \Delta x)^2 - 4x^2}{\Delta x}$$

Multiply out $(x + \Delta x)^2$:

$$= \lim_{\Delta x \to 0} \frac{4(x^2 + 2x\,\Delta x + \Delta x^2) - 4x^2}{\Delta x}$$

Multiply the 4 across:

$$= \lim_{\Delta x \to 0} \frac{\mathbf{4x^2} + 8x\,\Delta x + 4\,\Delta x^2 - \mathbf{4x^2}}{\Delta x}$$

Cancel $4x^2$ and $-4x^2$:

$$= \lim_{\Delta x \to 0} \frac{8x\,\Delta x + 4\,\Delta x^2}{\Delta x}$$

Factor out Δx:

$$= \lim_{\Delta x \to 0} \frac{(8x + 4\,\Delta x)\Delta x}{\Delta x}$$

Cancel Δx from the top and bottom of the fraction:

$$= \lim_{\Delta x \to 0} 8x + 4\,\Delta x$$

The $4\,\Delta x$ vanishes when you take the limit. This means that as Δx gets very close to zero, so does $4\,\Delta x$. It gets so close to zero that we can ignore the impact it has on the expression.

$$f'(x) = 8x$$

"That means the train's speed is $8x$!" the professor said. "It was going four times faster than usual."

"In other words, 5 minutes after the train left the station it was doing 5×8, or 40, units per minute," Recordis added.

"Maybe we can generalize this rule to see what happens when $f(x) = cx^2$, where c is any constant number," the professor said.

$$f(x) = cx^2$$

$$f'(x) = \lim_{\Delta x \to 0} \frac{c(x + \Delta x)^2 - cx^2}{\Delta x}$$

$$= \lim_{\Delta x \to 0} \frac{cx^2 + 2cx\,\Delta x + c\,\Delta x^2 - cx^2}{\Delta x}$$

$$= \lim_{\Delta x \to 0} \frac{2cx\,\Delta x + c\,\Delta x^2}{\Delta x}$$

$$f'(x) = 2cx$$

"That works when $c = 1$, because then we get $f'(x) = 2x$, which is what we did yesterday," the professor reminded us.

"Or it works when $c = 4$, because then we get $f'(x) = 8x$, which is what we just did," Recordis said.

"Or it even works when $c = 0$," the king noted. "Then we get $f(x) = 0$, which is a constant number so $f'(x) = 0$."

We generalized this rule to the case where any function $f(x)$ is multiplied by any constant c:

$$y = cf(x)$$

$$y' = \frac{dy}{dx} = \lim_{\Delta x \to 0} \frac{cf(x + \Delta x) - cf(x)}{\Delta x}$$

$$= c \lim_{\Delta x \to 0} \frac{f(x + \Delta x) - f(x)}{\Delta x}$$

$$y' = \frac{dy}{dx} = cf'(x)$$

"I remember once when the train didn't start from the ocean," Recordis said. "Pal started pushing when we were 5 units away from the zero point. We made a table of values." (Table 2–4.)

Igor drew a graph (Figure 2–5).

"I think the function is $f(x) = x^2 + 5$," the king suggested. "Let's try plugging that function into our formula for the derivative."

$$f(x) = x^2 + 5$$

$$f'(x) = \lim_{\Delta x \to 0} \frac{(x + \Delta x)^2 + 5 - (x^2 + 5)}{\Delta x}$$

$$= \lim_{\Delta x \to 0} \frac{x^2 + 2x\,\Delta x + \Delta x^2 + 5 - x^2 - 5}{\Delta x}$$

Table 2–4

Time	Train's Position
0	5
1	6
2	9
3	14
4	21

Table 2–5

Time	Trig's Position
0	5.00
1	6.01
2	9.02
3	14.03
4	21.04

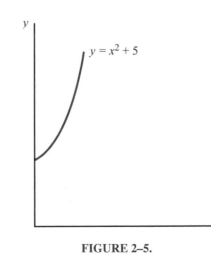

FIGURE 2–5.

"We can cancel out the $(x^2 + 5)$ and the $(-x^2 - 5)$," Recordis noted.

$$f'(x) = \lim_{\Delta x \to 0} \frac{2x\,\Delta x + \Delta x^2}{\Delta x}$$

$$= \lim_{\Delta x \to 0} 2x + \Delta x$$

$$= 2x$$

"That's the same answer we got when Pal started pushing from point zero," Recordis said.

"It seems as though it should be the same answer," the king remarked.

"Pal pushes the same amount each time, so it seems as though the speed of the train at a given instant should not depend on where he started pushing."

"I remember one time that was very complicated," Recordis said. "The train started 5 units away from the ocean, and Trigonometeris started running inside the train. We made a table of his position." (Table 2–5.)

"I was very careful to make sure that I ran at a constant speed with respect to the train," Trigonometeris stated proudly.

"I know what position function we need to use," the king said. "It will be exactly the same as the time before, except this time we must add the distance that Trig has walked from the back of the train. Let's try the following function."

$$(\text{Trig's position at time } x) = f(x) = x^2 + (0.01)x + 5$$

We put that function into the formula for the derivative.

$$f(x) = x^2 + (0.01)x + 5$$

$$f'(x) = \lim_{\Delta x \to 0} \frac{[(x+\Delta x)^2 + (0.01)(x+\Delta x) + \mathbf{5}] - [x^2 + (0.01)x + \mathbf{5}]}{\Delta x}$$

$$= \lim_{\Delta x \to 0} \frac{x^2 + 2x\,\Delta x + \Delta x^2 + \mathbf{(0.01)}x + (0.01)\Delta x - x^2 - \mathbf{(0.01)}x}{\Delta x}$$

$$= \lim_{\Delta x \to 0} \frac{2x\,\Delta x + \Delta x^2 + (0.01)\,\Delta x}{\Delta x}$$

$$= \lim_{\Delta x \to 0} 2x + \Delta x + 0.01$$

$$f'(x) = 2x + 0.01$$

While we were admiring this answer, the king said, "I just noticed something: $2x$ is the speed of the train, and 0.01 is the speed of Trig as he walks along inside the train. It looks as though you just add them together."

"Fascinating," the professor stated. "The original function was $f(x) = x^2 + (0.01)x + 5$. The first term represents the position of the train, the second term represents the position of Trig with respect to the train, and the third term is a constant which, of course, has a derivative of zero. Maybe, if you have a sum of functions, you can take the derivative of each term and add them together to get the derivative of the whole function."

"Let's see whether we can prove that in general," the king said. "Suppose we have any two functions, say $f(x)$ and $g(x)$, and we make a new function—call it $q(x)$—which equals $f(x) + g(x)$. Let's plug that into the formula and see whether we can find the derivative of $q(x)$."

$$q(x) = f(x) + g(x)$$

$$q'(x) = \frac{dq}{dx} = \lim_{\Delta x \to 0} \frac{f(x + \Delta x) + g(x + \Delta x) - f(x) - g(x)}{\Delta x}$$

"Now we're stuck," Recordis mourned.

"I think we can rearrange these terms," the king said.

$$q'(x) = \lim_{\Delta x \to 0} \frac{f(x + \Delta x) - f(x) + g(x + \Delta x) - g(x)}{\Delta x}$$

$$= \lim_{\Delta x \to 0} \frac{f(x + \Delta x) - f(x)}{\Delta x} + \lim_{\Delta x \to 0} \frac{g(x + \Delta x) - g(x)}{\Delta x}$$

"I recognize those two expressions!" Recordis exclaimed in delight. "Those are two derivatives." The final answer became:

$$q'(x) = f'(x) + g'(x)$$

"It does work!" The professor said in amazement. "This rule will make life much simpler. This means that, whenever we have a function made up of a whole glob of little functions added together, we can take the derivative of each little function and add all the derivatives together to get the derivative of the whole glob."

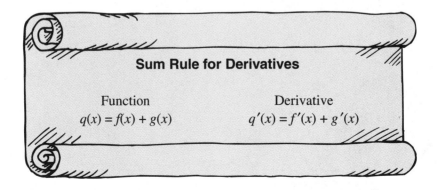

Sum Rule for Derivatives

Function	Derivative
$q(x) = f(x) + g(x)$	$q'(x) = f'(x) + g'(x)$

"We could write the same rule if we had three functions added together," Recordis added.

$$q(x) = f(x) + g(x) + h(x)$$

$$q'(x) = f'(x) + g'(x) + h'(x)$$

"Or even four functions added together," Recordis continued, getting carried away.

$$q(x) = f(x) + g(x) + h(x) + i(x)$$

$$q'(x) = f'(x) + g'(x) + h'(x) + i'(x)$$

"Yes, we know what you mean," the professor said quickly, before Recordis had a chance to say that they could write the same rule for five functions added together. "I think we should go on to something else."

"I remember a long time ago when we made a list of crazy functions," Recordis remarked. "I wonder if this calculus jazz will help us with any of these. Here's a good function: $f(x) = xxx$, or $f(x) = x^3$." (Figure 2–6.)

FIGURE 2–6.

"We can try to find the derivative," the professor said.

$$f(x) = x^3$$

$$f'(x) = \lim_{\Delta x \to 0} \frac{(x + \Delta x)^3 - x^3}{\Delta x}$$

$$= \lim_{\Delta x \to 0} \frac{(x^2 + 2x\,\Delta x + \Delta x^2)(x + \Delta x) - x^3}{\Delta x}$$

$$= \lim_{\Delta x \to 0} \frac{x^3 + 2x^2\,\Delta x + x\,\Delta x^2 + x^2\,\Delta x + 2x\,\Delta x^2 + \Delta x^3 - x^3}{\Delta x}$$

The x^3's canceled each other out; and then we combined the like terms:

$$f'(x) = \lim_{\Delta x \to 0} \frac{3x^2\,\Delta x + 3x\,\Delta x^2 + \Delta x^3}{\Delta x}$$

We factored out the Δx, which Recordis then gleefully canceled with the Δx in the denominator. (Recordis likes to cancel things.)

$$f'(x) = \lim_{\Delta x \to 0} 3x^2 + 3x\,\Delta x + \Delta x^2$$

The last two terms went to zero when we took the limit:

$$f'(x) = 3x^2$$

"Fascinating," the king said.

"It makes sense when you look at the graph," the professor told him. (Figure 2–6.)

"When $x = 0$, we're saying that the slope of the curve is zero, which is the way it looks in the picture. As x gets bigger, the slope becomes steeper. And even if x is negative, the slope is still positive because $3x^2$ is positive."

"How about x^4?" Recordis asked, wondering how complicated the world could get.

Igor slowly went through the algebra, and our eyes got tired as more and more symbols kept floating across his picture screen. "There's got to be a simpler way!" the professor said. When all of the algebra was finished, the final answer looked like this:

Function	Derivative
$f(x) = x^4$	$f'(x) = 4x^3$

"I'm afraid to suggest we try it with x^5," Recordis commented.

"Still, we must find some way to come up with the answer," the professor said, thinking wistfully about how she could impress people by telling them that she could find the derivative of a fifth-degree polynomial.

"Let's make a list of our results and see if we see a pattern."

Function	Derivative
$f(x) = x^2$	$f'(x) = 2x$
$f(x) = x^3$	$f'(x) = 3x^2$
$f(x) = x^4$	$f'(x) = 4x^3$

We stared at that table a long time. "Suppose we had to guess the result for $f(x) = x^5$," the professor wondered. "What do you think it would be?"

"I see a pattern," the king announced slowly. "It looks as though $f(x) = x^5$ should have the derivative $f'(x) = 5x^4$."

"In general, it looks as though $f(x) = x^n$ should have the derivative $f'(x) = nx^{n-1}$," Recordis said.

"We must find a way of proving that it always works," the professor said. "I will never be able to sleep at night if we try to use that formula without proving it is true."

"Do we have any way of testing that formula?" Trigonometeris asked.

"We could try," the professor said.

$$f(x) = cx^n$$

$$f'(x) = \lim_{\Delta x \to 0} \frac{c(x + \Delta x)^n - cx^n}{\Delta x}$$

That expression looked pretty hopeless, until I began to remember something. "Did you ever develop a formula for figuring out an expression like $(a + b)^n$?" I asked.

"I remember something like that," the king answered. "We derived it a long time ago when we were working out algebra. Look it up, Recordis."

Recordis fumbled through his giant book. "This might be useful," he said. "It's something called a *binomial formula*."

$$(a + b)^n = a^n + na^{n-1}b + c_2 a^{n-2}b^2 + c_3 a^{n-3}b^3 + c_4 a^{n-4}b^4 + \cdots + nab^{n-1} + b^n$$

"The c's are constants that come from a complicated formula that I'll have to look up on another page," Recordis said sheepishly. "I don't remember why this works."

"That's not important now," the king said. "We did prove it in the past so we know it works." We used the formula for $(x + \Delta x)^n$:

$$(x + \Delta x)^n = x^n + nx^{n-1}\,\Delta x + c_2 x^{n-2}\,\Delta x^2 + c_3 x^{n-3}\,\Delta x^3 + c_4 x^{n-4}\,\Delta x^4 + \cdots + nx\,\Delta x^{n-1} + \Delta x^n$$

We put this formula into the formula for the derivative for $y = x^n$:

$$f'(x) = \lim_{\Delta x \to 0} \frac{(x + \Delta x)^n - x^n}{\Delta x}$$

$$= \lim_{\Delta x \to 0} \frac{\boldsymbol{x^n} + nx^{n-1}\Delta x + c_2 x^{n-2}\Delta x^2 + c_3 x^{n-3}\Delta x^3 + c_4 x^{n-4}\Delta x^4 + \cdots + nx\Delta x^{n-1} + \Delta x^n - \boldsymbol{x^n}}{\Delta x}$$

We noticed the x^n and $-x^n$ canceled out:

$$f'(x) = \lim_{\Delta x \to 0} \frac{nx^{n-1}\Delta x + c_2 x^{n-2}\Delta x^2 + c_3 x^{n-3}\Delta x^3 + c_4 x^{n-4}\Delta x^4 + \cdots + nx\Delta x^{n-1} + \Delta x^n}{\Delta x}$$

We noticed that now every term in the top of the fraction contained Δx, so we factored that out:

$$f'(x) = \lim_{\Delta x \to 0} \frac{\boldsymbol{\Delta x}(nx^{n-1} + c_2 x^{n-2}\Delta x + c_3 x^{n-3}\Delta x^2 + c_4 x^{n-4}\Delta x^3 + \cdots + nx\Delta x^{n-2} + \Delta x^{n-1})}{\boldsymbol{\Delta x}}$$

We canceled the Δx from the top and bottom of the fraction:

$$f'(x) = \lim_{\Delta x \to 0} (nx^{n-1} + c_2 x^{n-2}\,\Delta x + c_3 x^{n-3}\,\Delta x^2 + c_4 x^{n-4}\,\Delta x^3 + \cdots + nx\,\Delta x^{n-2} + \Delta x^{n-1})$$

"You'd better look up that formula for the c's," the professor told Recordis, but the king suddenly noticed:

"We don't need that formula! When we take the limit, all the terms with Δx will disappear!" There was only one term left:

$$f'(x) = nx^{n-1}$$

"So when you take the derivative of a variable raised to a power, the exponent n comes down in front and multiplies, and the new exponent is one less than the old exponent," Recordis noted.

We were amazed that such an elegant formula could result from such a complicated process. Now that we knew the derivative of $y = x^n$, we needed to find the derivative of $y = cx^n$, where c was a constant. The professor thought we could derive a general rule for multiplication by a constant. If $y = cf(x)$, then we found $y' = cf'(x)$:

$$y = cf(x)$$

$$y' = \lim_{\Delta x \to 0} \frac{cf(x + \Delta x) - cf(x)}{\Delta x}$$

We factored out the constant c:

$$y' = (c) \lim_{\Delta x \to 0} \frac{f(x + \Delta x) - f(x)}{\Delta x}$$

We then recognized the formula for $f'(x)$:

$$y' = cf'(x)$$

We added these two rules

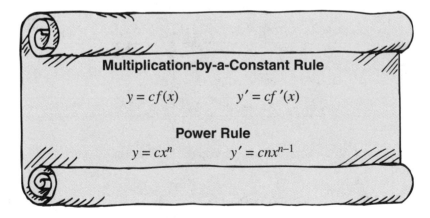

Multiplication-by-a-Constant Rule

$$y = cf(x) \qquad y' = cf'(x)$$

Power Rule

$$y = cx^n \qquad y' = cnx^{n-1}$$

"Amazing!" Trigonometeris said. "We already established that it works if $n = 2$ or $n = 3$ or $n = 4$."

"It also works if $n = 1$," the professor stated. "Then we would have $f(x) = cx$, and we already know that in that case the derivative is cx^0, which is just equal to c."

Igor displayed the results of our work: (In these rules, c and n represent any constant number.)

Function	Derivative
$y = c$	$y' = dy/dx = 0$
$y = cx$	$y' = dy/dx = c$
$y = cf(x)$	$y' = dy/dx = cf'(x)$
$y = f(x) + g(x)$	$y' = dy/dx = f'(x) + g'(x)$
$y = cx^n$	$y' = dy/dx = cnx^{n-1}$

WORKSHEET

Determine the derivative for each of these functions:

	$f(x)$	$f'(x)$
1.	6	
2.	32	
3.	68	
4.	$3x + 125$	
5.	$-8x - 643$	
6.	$-15x - 549$	
7.	$8x^2 + 4x + 419$	
8.	$7x^2 - 7x - 7652$	
9.	$3x^2 + 6x + 16$	
10.	$10x^3 + 2x^2 - 18x - 86$	
11.	$6x^3 - 6x^2 + 12x + 788$	
12.	$4x^3 + 9x^2 + 20x + 604$	
13.	$2x^4 + 3x^3 + 12x^2 + 16x - 13$	
14.	$4x^4 + 5x^3 - 8x^2 - 44x + 42$	
15.	$8x^4 + 7x^3 + 5x^2 + 25x + 121$	
16.	$h - \frac{1}{2}gx^2$	

(For question 16, treat h and g as constant numbers.)

"With these rules we can find the derivative of any polynomial," Recordis said. "Remember functions like $x^2 + 3x - 5$ or $2x^4 - 3x^3 + 2x^2 - 1$? I think that this just about wraps up the subject of calculus." (Recordis thinks that any problem that can't be expressed using polynomials is not worth bothering with.)

Just as he was saying these words, there was a loud thud out in the courtyard, and the next thing we knew an ominous figure had darted in through the window. Everyone in the room cowered in fear. Although I had never seen the strange apparition before, I could tell from the start that he meant trouble.

"So you think you can outwit me, do you?" he exclaimed, his voice ringing with wicked laughter.

"That's the gremlin," the professor whispered to me. "The Spirit of Hopelessness and Impossibility. He is our arch-enemy."

"You have no idea what you are getting yourselves into," the gremlin cried. "Just wait. First, with your last rule, you have never even thought about what happens if n is a fraction. Ah, but even that is too simple. I can show you curves that you have no hope of unraveling."

He held out his cape, and in it we could see misty pictures of strangely oscillating curves of every conceivable shape, which appeared to be floating in space. They seemed to be trying to reach out and strangle us. "What about any of these?" he cried, and a whole chain of algebraic symbols flew out in the air past us.

"This time I am sure to win!" he laughed, as he slowly folded his cape and flew out the window.

NOTE TO CHAPTER 2

It is important to note that the derivative can be defined for a particular function only if the limit

$$\lim_{\Delta x \to 0} \frac{f(x + \Delta x) - f(x)}{\Delta x}$$

has a definite value. Some functions, such as $y = |x|$ (the *absolute value*, defined by $y = x$ for $x \geq 0$ and $y = -x$ for $x < 0$), will not have derivatives defined at all points of the function. In this case, the function has no derivative at the point where $x = 0$ (Figure 2–7).

In general, any function with a cusp in it, like the ones in Figure 2–8, will not have a derivative defined at the point where the cusp is located.

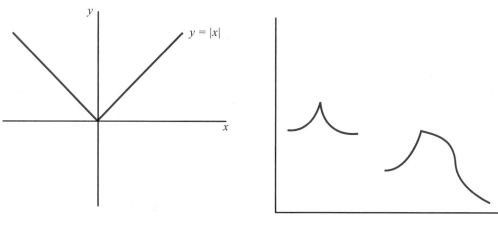

FIGURE 2–7.   FIGURE 2–8.

Here is the complete binomial formula:

$$(a + b)^n = \binom{n}{0}a^n + \binom{n}{1}a^{n-1}b + \binom{n}{2}a^{n-2}b^2 + \cdots + \binom{n}{n-1}ab^{n-1} + \binom{n}{n}b^n$$

The expression $\binom{n}{j}$ is called the binomial coefficient. It is defined to be

$$\binom{n}{j} = \frac{n!}{(n-j)!\,j!}$$

The exclamation mark indicates factorial. The factorial of a whole number is the product of all of the numbers from 1 up to that number. For example:

$$0! = 1$$

$$1! = 1$$

$$2! = 2 \times 1 = 2$$

$$3! = 3 \times 2 \times 1 = 6$$

$$4! = 4 \times 3 \times 2 \times 1 = 24$$

$$\binom{n}{0} = \frac{n!}{n!0} = 1$$

$$\binom{n}{1} = \frac{n!}{(n-1)!1!} = n$$

$$\binom{n}{2} = \frac{n!}{(n-2)!2} = \frac{n(n-1)}{2}$$

$$\binom{n}{n-1} = \frac{n!}{1!(n-1)!} = n$$

$$\binom{n}{n} = \frac{n!}{0!n!} = 1$$

Exercises

Find the derivatives of the following functions. Then evaluate the derivative for the given value of the independent variable.

1. $y = 3x^3 + 2x^2 + x + 5$; evaluate y' when $x = 3$

2. $y = 4x^5 + x^2$; evaluate y' when $x = 10$

3. $y = x^{35}$; evaluate y' when $x = 1$

4. $y = \frac{1}{3}x^3 + \frac{1}{2}x^2 + x + 1$; evaluate y' when $x = 6.42$

5. $y = (4.34)x^2 + (0.98)x$; evaluate y' when $x = -4$

6. $f(t) = (2t - 5)(3t + 4)$; evaluate $f'(t)$ when $t = \frac{1}{2}$

Find formulas for the derivatives with respect to x for each of these functions. (Treat a, b, and c as constants.)

7. $y = ax + 4$

8. $y = ax^2 - x$

9. $y = ax^3 + bx^2 + cx$

10. $y = ax^b + c$

11. $y = (a + x)(b + x)$

12. If $y = c \times u(x)$, where c is a constant and u is a function of x, then use the definition of the derivative to prove that $dy/dx = c \times du/dx$.

13. Write an expression for $f(x + \Delta x) - f(x)$ for the function $f(x) = ax^2 + bx + c$. Then use the definition of the derivative to find $f'(x)$.

14. Using the definition of the derivative, show that, if $y = f(x) + g(x) + h(x)$, then $dy/dx = f'(x) + g'(x) + h'(x)$.

15. When Pal throws his beach ball straight up in the air, its height h at time t is given by $h = -\frac{1}{2}gt^2 + v_0 t$. (a) Find the velocity of the ball at time t. (b) Find the velocity of the ball when $t = 0$. (c) Find out how long the ball takes to reach its highest point (i.e., at what value of t does $dh/dt = 0$?).

16. When Pal drops his ball off the Hasselbluff Mountain Viewpoint, its height above the ground at time t is given by $h = 64 - \frac{1}{2}gt^2$. (a) What is the velocity at time t? (Is the velocity positive or negative?) (b) How long will the ball take to hit the ground (i.e., at what value of t does $h = 0$)? (c) How fast is the ball going the instant before it hits the ground? (d) The quantity g is known as the *acceleration of gravity* and is measured in meters per second^2. Find a numerical value for g if the ball takes 3.61 seconds to fall to the ground.

17. Find the values of x where the slope of the curve $y = \frac{1}{3}x^3 - x^2 + 3x + 5$ is equal to 3.

18. For what values of b is the line $y = 8x + b$ tangent to the curve $y = x^3$?

19. The *mean value theorem* states that, if a function $y = f(x)$ has a derivative defined everywhere between $x = a$ and $x = b$, then there is some value of x (call it x_0) such that $a < x_0 < b$ and $f'(x_0)$ equals the slope of the secant line between the points $(a, f(a))$ and $(b, f(b))$. Consider the function $f(x) = -x^2 + 10x - 15$, and two points on the graph of that function: $(2, 1)$ and $(6, 9)$. Find the value of x_0 that is predicted by the mean value theorem (i.e., find x_0 such that

$$f'(x_0) = \frac{f(b) - f(a)}{b - a}$$

for $a = 2$ and $b = 6$).

20. *Newton's method* provides an iterative method for estimating the x intercept of complicated functions. The goal of the method is to find x_0 such that $f(x_0) = 0$. First, make a guess (x_1) that is reasonably close to the true value of x_0. Then calculate a better guess according to the formula $x_2 = x_1 - f(x_1)/f'(x_1)$. The method can be repeated to yield a still better guess, $x_3 = x_2 - f(x_2)/f'(x_2)$. Keep going until you are satisfied that the result is close enough to the true answer. Now, use Newton's method to estimate the cube root of 7. Start with $x_1 = 2$, and find the x intercept of the function $f(x) = x^3 - 7$. Perform a total of three iterations, and compare the results with the true value. (See Figure 2–9.)

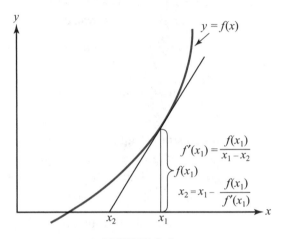

FIGURE 2–9.

Finding Maximum and Minimum Points

Panic-stricken, Recordis raced out of the room as soon as the gremlin disappeared. "Come back!" the professor cried, and chased after him down the hall. The rest of us followed with deep concern.

Recordis tore into his chamber and slammed the door of a closet, then lay panting on the floor, blocking the door.

"What is this about?" the king demanded.

"I left my problem closet open," Recordis gasped after regaining his breath.

We stared at him uncomprehendingly.

"I throw unsolvable problems into my problem closet," Recordis explained sheepishly. "I was terrified the gremlin would see the door open and taunt us with the problems we can't solve."

"Wouldn't it be better to solve the problems?" the professor asked.

"As if we could!" Recordis moaned.

"Show us an example," the king ordered.

Recordis knew he must obey, so he gently opened the door. One paper blew off the giant stack and fell to the floor. Recordis picked it up and explained, "Here's one problem you stuck me with. You might remember last month when the torrential downpour delayed our scheduled hike up Column mountain. While we were waiting for the rainstorm to end, the professor started worrying that it might rain on the day of an important ball game, so you said we should build a retractable roof for the stadium."

"We wanted to make sure that the roof would be high enough so that a high fly ball would not hit it," the king remembered.

"You stuck me with the problem of figuring out how high the ball might go," Recordis continued. "I watched the ballplayers practice hitting fly balls, and drew a sketch showing the arc traced out by the ball's course. I had a sneaking suspicion that we'd seen this curve before." He showed us the sketch that had fallen from the problem closet. (See Figure 3–1.)

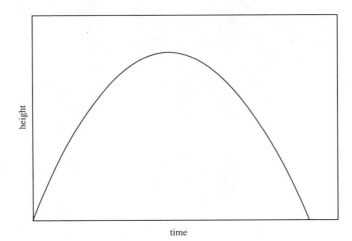

FIGURE 3–1.

"It looks like a parabola!" the professor said. "That means we could write its equation in this form:

$$h = at^2 + bt + c$$

where h = height, t = time, and a, b, and c are constants that depend on the particular fly ball being investigated."

"If only we knew how to find the value of t that leads to the maximum value of h," the king wondered.

"We'll think about the problem during our hike, which we may as well do tomorrow," the professor said.

Our hike up the side of Column mountain started out with a very steep climb. (See Figure 3–2.) As we continued the climb, the slope gradually lessened, but we were becoming tired.

"Are we at the top yet?" Recordis asked.

"We're not at the top because we're still climbing," the professor said.

This conversation was repeated a few times, until finally we reached the top. Then the professor said, "Now that we're no longer climbing, we are at the top." (See Figure 3–3.) From the top we had a tremendous view of the still-roofless stadium, which reminded us of the fly ball problem.

Suddenly, the king had an idea. "If you're still climbing, then the slope will be positive, and you're not at the top."

"Right," the professor agreed.

"After you've passed the top and are on the way down, the slope will be negative."

"Right again."

"At the very top, you're neither going up nor down, so the slope has to be zero."

The professor suddenly realized the implications. "That's the clue we need to locate the top of a curve! At the top of any curve the slope will be zero!"

"That would help if we knew the value of x at the top of the curve, because then we could verify that the slope was zero at that point. However, the problem is that we don't know the value of x," Recordis cautioned.

"But if we know the slope is zero, that should let us track down the value of x!" the professor said.

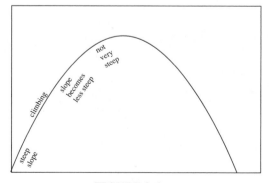

FIGURE 3–2.

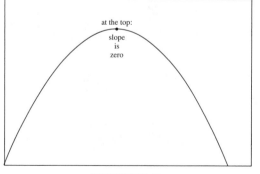

FIGURE 3–3.

"Does this work in practice?" Trigonometeris wondered. "I'll make up a formula for a parabola, and let's see if you can find the highest point on the curve."

$$y = -4x^2 + 24x + 42$$

Find the derivative of $-4x^2$, which is $-8x$
Find the derivative of $24x$, which is 24.
Find the derivative of 42, which is zero (since 42 is a constant)
Add these together:

$$\frac{dy}{dx} = -8x + 24$$

In order to find the top of the curve, we set the derivative equal to zero:

$$0 = -8x + 24$$

We solved the equation to find $x = 3$.

"So, if our method works, we should find that the curve will reach its highest point where $x = 3$."

"I could sleep much easier at night if we could make a graph of the curve so we could verify that," Recordis said.

We calculated some values:

With the table of values we could create a graph. (See Figure 3–4).

The professor described the three steps required to find the maximum point for a curve. We applied the method to the general form of a parabola:

$$y = ax^2 + bx + c$$

x	$y = -4x^2 + 24x + 42$
−5	−178
−4	−118
−3	−66
−2	−22
−1	14
0	42
1	62
2	74
3	78
4	74
5	62
6	42
7	14
8	−22
9	−66
10	−118
11	−178

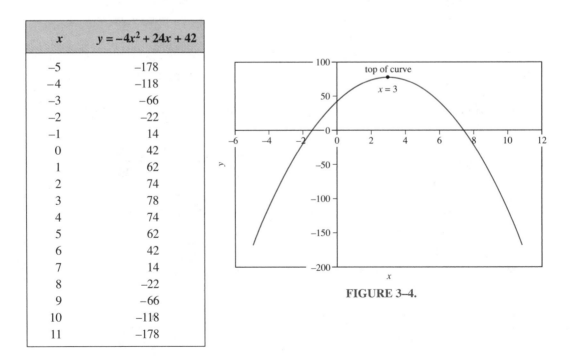

FIGURE 3–4.

Step 1: Find the derivative dy/dx (keeping in mind that a, b, and c are known constants):

$$\frac{dy}{dx} = 2ax + b$$

Step 2: To find the top of a curve, set the derivative equal to zero:

$$0 = 2ax + b$$

Step 3: Solve for the variable (in this case x):

$$-b = 2ax$$

$$\frac{-b}{2a} = x$$

While we were resting at the top of the mountain, another hiker joined us. "She's the publisher of *Carmorra* magazine!" the professor whispered to us excitedly. (*Carmorra* magazine had once done a feature on the professor, and the professor was hoping it would do another one someday.)

"We're planning an issue that features hikes along mountains in Carmorra," the publisher explained to us. "I hope we will be able to afford to include pictures for all of the hikes. At first we were short of revenue, so we decided to raise the subscription price. But then we found that the higher price discouraged people from subscribing, so our revenue went down."

"There must be some value that is a compromise between charging too low of a price or too high of a price," the king reasoned.

"I can estimate how the number of subscribers will change depending on the subscription price," the publisher said. She gave us this equation:

$$Q = 60,000 - 1,500P$$

where Q is the quantity of subscriptions that are sold, and P is the price of subscriptions, based on past observations. "For example, if we charge \$40, then nobody will subscribe ($Q = 60,000 - 1,500 \times 40 = 0$). If we give the magazine away for free, then the total number of subscriptions will be 60,000, but in either of those cases our revenue will be zero."

The revenue (R) is price times quantity, so we wrote the equation:

$$R = PQ$$

Substitute in the formula for Q:

$$R = P(60,000 - 1,500P)$$

Multiply the P across the parentheses, using the distributive property:

$$R = 60,000P - 1,500P^2$$

"Could we make a graph of this relation?" Recordis asked. "I always say, one picture is worth a thousand equations."

See Figure 3–5.

x	$R = 60,000P - 1,500P^2$
0	0
2	114,000
4	216,000
6	306,000
8	384,000
10	450,000
12	504,000
14	546,000
16	576,000
18	594,000
20	600,000
22	594,000
24	576,000
26	546,000
28	504,000
30	450,000
32	384,000
34	306,000
36	216,000
38	114,000
40	0

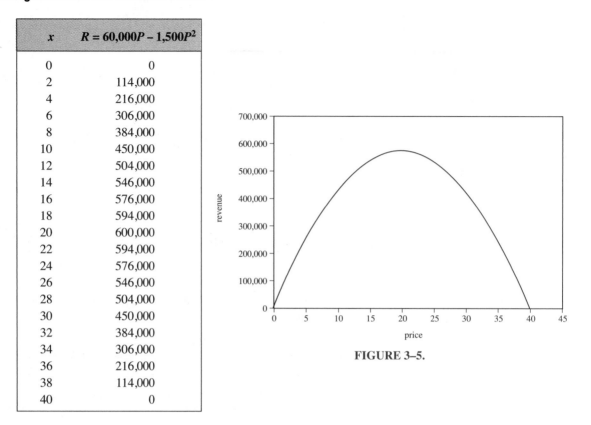

FIGURE 3–5.

Recognizing the curve was a parabola, the professor said, "Our tried-and-true method of finding the top of a parabola will work to find the top of the revenue curve."

Step 1: Find the derivative.

$$R = 60,000P - 1,500P^2$$

$$\frac{dR}{dP} = 60,000 - 3,000P$$

Step 2: Set the derivative equal to zero:

$$0 = 60,000 - 3,000P$$

Step 3: Solve the resulting equation for the variable (in this case P):

$$3,000P = 60,000$$

$$P = \frac{60,000}{3,000} = 20$$

From the graph (Figure 3–5), we could see that the high point of the curve does happen where $P = 20$.

"This method might be useful for other types of businesses if we could find a way to generalize it," the king said thoughtfully. Suppose the relation between the price and the quantity sold is a straight line that crosses the vertgical price axis at P_0 and the horizontal quantity axis at Q_0, which therefore has a slope of $-P_0/Q_0$. (See Figure 3–6.)

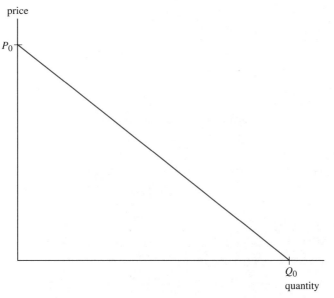

FIGURE 3–6.

$$P = P_0 - \frac{P_0}{Q_0} Q$$

(The two intercepts P_0 and Q_0 are constants.)

Since revenue equals price times quantity:

$$R = PQ = \left(P_0 - \frac{P_0}{Q_0} Q \right) Q = P_0 Q - \frac{P_0}{Q_0} Q^2$$

Step 1: Find the derivative $\dfrac{dR}{dQ}$

$$\frac{dR}{dQ} = P_0 - 2\frac{P_0}{Q_0} Q$$

Step 2: Set the derivative equal to zero:

$$0 = P_0 - 2\frac{P_0}{Q_0} Q$$

Step 3: Solve for the independent variable (in this case Q):

$$2 \frac{P_0}{Q_0} Q = P_0$$

$$2 P_0 Q = P_0 Q_0$$

$$2Q = Q_0$$

$$Q = \frac{Q_0}{2}$$

Given the value of Q, we could find the value of P by inserting $Q = \frac{Q_0}{2}$ into the formula for P:

$$P = P_0 - \frac{P_0}{Q_0} Q = P_0 - \frac{P_0}{Q_0} \left(\frac{Q_0}{2} \right) = P_0 - \frac{P_0}{2} = \frac{2P_0}{2} - \frac{P_0}{2} = \frac{P_0}{2}$$

See Figure 3–7.

"I'm not the least bit surprised. Clearly if you charge a price of $P = P_0$, then your quantity sold will be zero, and your revenue will be zero. Also, revenue will be zero if you charge a price $P = 0$. So, the revenue-maximizing price is halfway between the two extremes."

Before we began our descent from the mountain we took one last look at the view. We could see the Craggy Island lighthouse in the distance. "That reminds me of another problem I tossed in my closet last month," Recordis said. "The lighthouse keeper wants us to help test the new foghorn. We need to sail in a straight line from Westbay Point to Eastbay Point, and determine if we can hear the foghorn when we are at our maximum distance from the lighthouse."

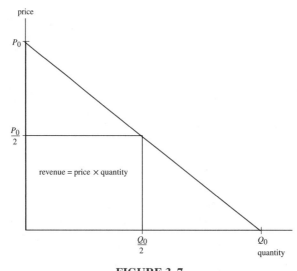

FIGURE 3–7.

"You should have told us about this problem before!" the king cried in frustration.

"I didn't know how to figure out which point on the course is the farthest from the lighthouse," Recordis said.

"Show us the course," the professor said. Recordis showed us a map which used an xy (Cartesian) coordinate system (Figure 3–8). He showed us the course that went from point ($x = 20, y = 70$) to point ($x = 80, y = 250$). The course line can be represented by this equation:

$$y = 10 + 3x$$

The coordinates of the lighthouse were (62, 156).

We knew the formula for the distance (D) from a point (x, y) to the point (62, 156), using the Pythagorean theorem:

$$D = \sqrt{(x-62)^2 + (y-156)^2}$$

Now substitute $10 + 3x$ in the place of y for the points on our course:

$$D = \sqrt{(x-62)^2 + [(10+3x)-156]^2}$$

$$D = \sqrt{(x-62)^2 + (3x-146)^2}$$

We didn't yet know how to find the derivative of an expression involving a square root sign. "I'd like to square both sides to get rid of the square root sign," Recordis said.

$$D^2 = (x-62)^2 + (3x-146)^2$$

He made the definition $z = D^2$:

$$z = (x-62)^2 + (3x-146)^2$$

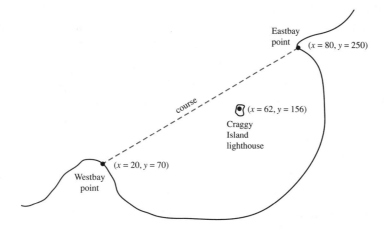

FIGURE 3–8.

Recordis insisted we try to solve this revised, easier problem. At first the professor grumbled that you shouldn't be able to make the problem easier before solving it, but then she realized that the same value of x that maximized the distance would also maximize the square of the distance.

We used the trusty old algebra rule:

$$(a - b)^2 = a^2 - 2ab + b^2$$

to rewrite the expression for z:

$$z = x^2 - 2 \times 62x + 62^2 + (3x)^2 - 2 \times 3 \times 146x + 146^2$$

$$z = x^2 - 124x + 62^2 + 9x^2 - 876x + 146^2$$

$$z = 10x^2 - 1{,}000x + 25{,}160$$

Find the derivative dz/dx:

$$\frac{dz}{dx} = 20x - 1{,}000$$

Set the derivative equal to zero:

$$0 = 20x - 1{,}000$$

Solve for x:

$$1{,}000 = 20x$$

$$x = \frac{1{,}000}{20} = 50$$

Solve for y:

$$y = 10 + 3 \times 50 = 160$$

As we hiked down the mountain, we were excitedly looking forward to being able to tell the lighthouse keeper that we knew how to find the maximum distance. The next day we traveled to Westbay Point and set sail, being careful to follow the planned course. We arranged to meet the lighthouse keeper when we reached the end of our voyage at Eastbay Point. We kept our eye on the distant lighthouse, expecting to see the distance increasing as we approached the point ($x = 50$, $y = 160$). However, something peculiar was happening. "Is it just me, or does it seem like the lighthouse is getting closer rather than farther?" Trigonometeris wondered.

We reached the designated point ($x = 50$, $y = 160$). "That lighthouse looks so close it seems that I could reach out and touch it," Trigonometeris said. As we watched with increasing perplexity, the lighthouse got farther and farther away as we passed the point that we had thought would have been the maximum distance from the lighthouse.

"I have never been so embarrassed in my life!" Recordis screamed. "The point we thought was the maximum distance point was really the minimum distance point!"

"How could our maximum-finding method go wrong like that?" the king asked in great distress.

As we got off the boat, we made up a quick excuse for the puzzled lighthouse keeper and then rushed back to the Main Conference Room to figure out what was going on.

"A graph would help," the professor suggested.

We created a table of values and looked at the graph:

With the table of values, we could create a graph. (See Figure 3–9.)

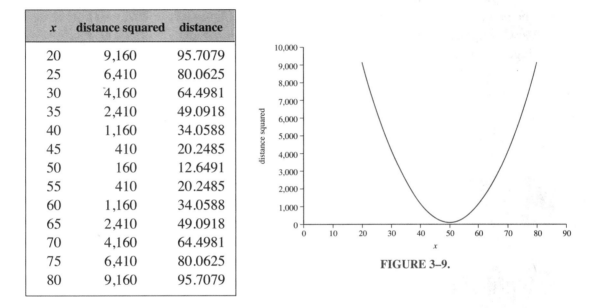

x	distance squared	distance
20	9,160	95.7079
25	6,410	80.0625
30	4,160	64.4981
35	2,410	49.0918
40	1,160	34.0588
45	410	20.2485
50	160	12.6491
55	410	20.2485
60	1,160	34.0588
65	2,410	49.0918
70	4,160	64.4981
75	6,410	80.0625
80	9,160	95.7079

FIGURE 3–9.

We stared at the graph, and the nature of our predicament slowly dawned on us.

"This graph reminds me of our hike across Deep Scoop Valley," the king reminisced. "As we climb down into the valley, our path has a negative slope. While we're climbing back out of the valley, our path has a positive slope. At the exact moment we're at the bottom of the valley, our path has a slope of zero." (See Figure 3–10.)

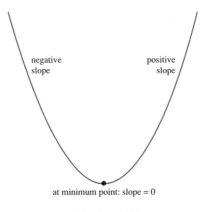

negative slope

positive slope

at minimum point: slope = 0

FIGURE 3–10.

"This is hopeless!" Recordis cried in despair. "The slope is zero at a minimum point, just the same as it is at a maximum point! I'm having nightmares where we're looking for maximum points but instead finding minimum points!"

"I always try to look at the bright side," the king said. "This means that we don't need to work on a new method to find minimum points. The method we've used to find maximum points will also work to find minimum points."

"If only we had a way to tell the difference between maximum points and minimum points," the professor said wistfully.

"How hard can that be?" Trigonometeris asked. "Think of the curve as a bowl. If the bowl is right-side-up, then it will hold water and it has a minimum point. If the bowl is upside down, then it will spill water, and it has a maximum point." (See Figure 3–11.)

concave-upward curve: holds water

concave-downward curve: does not hold water

FIGURE 3–11.

WORKSHEET

For each question, find the point on the given line that is closest to the given point. Let z be the square of the distance from a point on the line to the given point. Then find dz/dx, set it equal to zero, and solve for x.

1. line: $y = 60$; point $(82, 12)$

2. line: $y = 100 - x$; point $(15, 45)$

3. line: $y = 30 + 2x$; point $(18, 86)$

4. line: $y = 2 + 100x$; point $(402, 198)$

5. line: $y = 72 - 12x$; point $(27, 38)$

6. line: $y = 32 + 0.125x$; point $(21, 384)$

7. line: $y = b + mx$; point (h, k)

$(b, m, h,$ and k are constants)

"We can easily tell which way the curve faces when we're looking at the graph, but it would be good to know a formula that would tell us which way it faces," the professor said.

"Give me one good reason why we can't just look at the graph," Recordis demanded.

"Because creating the graph takes some work," the professor said.

"Oh. That's a good reason," Recordis agreed.

"We invented a term we can use to distinguish the two sides of a bowl," the king reminded us. "Remember the time Pal was running along Sinkhole Meadow and the ground caved in? It made a giant hole, so we used the term *concave* for something shaped like a hole."

"We could say that these curves (Figure 3–12) are oriented so their concave side is downward," the professor said.

"I know how you can draw curves that are concave upward," the king stated. Igor obligingly drew some curves that were concave upward (Figure 3–13).

FIGURE 3–12.   FIGURE 3–13.

"Because these are curves, the slopes change at different points along the curve. There must be a way to visualize how the slope is different when the curve is concave up versus concave down," the king said.

"In order to visualize the slopes, we have to visualize the tangent lines," the professor said. "I have an idea. Igor, provide us with a bug with an arrow strapped to its back."

The Visiomatic Picture Chalkboard Machine looked startled, but a few minutes later a little hatch opened in the side of the screen and out flew a purple bug with a red arrow strapped to its back.

"I want you to walk along this function," the professor said, pointing to the function we had looked at earlier:

$$y = -4x^2 + 24x + 42$$

She oriented the bug so that the arrow was tangent to the curve. (See Figure 3–14.)

"Amazing!" Trigonometeris gasped. "I have never figured out how bugs are able to walk on walls."

"Look at what's happening to the arrow," the professor said as the bug walked along the function.

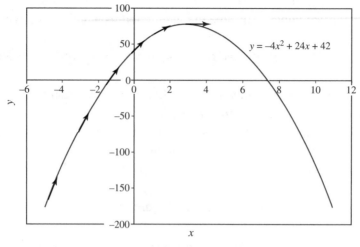

FIGURE 3–14.

"It's slowly getting flatter," the king observed. "So the slope is decreasing."
We worked out a table showing how the slope changed:

"For a concave down curve, we can see that the slope is *decreasing* as we move along the curve from left to right." Igor sketched a graph of the derivative curve $y' = -8x + 24$ (Figure 3–15).

x	value of original function $y = -4x^2 + 24x + 42$	value of derivative $y' = -8x + 24$
−5	−178	64
−4	−118	56
−3	−66	48
−2	−22	40
−1	14	32
0	42	24
1	62	16
2	74	8
3	78	0
4	74	−8
5	62	−16
6	42	−24
7	14	−32
8	−22	−40
9	−66	−48
10	−118	−56
11	−178	−64

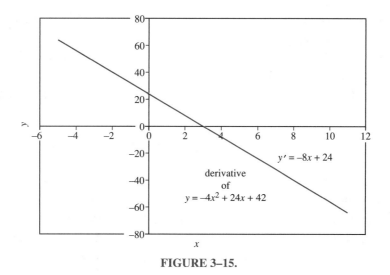

FIGURE 3–15.

"If only we had a way to determine the slope of the derivative curve $y' = -8x + 24$." He paused. "If only we could take the derivative of $y' = -8x + 24$." He paused again. "I suppose there is nobody that can stop me."

original function:	$y = -4x^2 + 24x + 42$
derivative of the original function (or, take the derivative of the original function once):	$y' = -8x + 24$
derivative of the derivative of the original function (or take the derivative of the original function twice):	$y'' = -8$

"This provides the clue we need!" the professor exclaimed. "We'll call the derivative of the original function the *first derivative*, since we get it by taking the derivative once. We'll symbolize it with one prime: y'. Then we'll call the derivative of the derivative the *second derivative*, since we get it by taking the derivative of the original function twice. We'll symbolize it with two primes: y''. If the original function is oriented concave downward, then its slope will be decreasing, and its second derivative will be negative."

We also worked out a notation for the second derivative that corresponds to the dy/dx notation for the first derivative. Since hitting y with d/dx turns it into the first derivative dy/dx, we figured that hitting dy/dx with d/dx would turn it into the second derivative:

$$\text{second derivative} = \frac{d}{dx}\frac{dy}{dx} = \frac{ddy}{dxdx} = \frac{d^2y}{dx^2}$$

We settled on the notation d^2y/dx^2 for the second derivative, which we would read as "d squared y over dx squared."

"If the second derivative is negative for a concave downward curve, I'll guess it will be positive for a concave upward curve."

Igor displayed the curve representing the distance to the lighthouse (see Figure 3–16).

$$z = 10x^2 - 1{,}000x + 25{,}160$$

The bug began walking along this curve. This time the tangent line started out with a steep negative slope, but it became less steep as the bug approached the minimum. At the exact minimum point, the slope was zero, and after that it turned positive. The positive slope continued to increase as the bug continued to walk along the curve.

In general, the slope is increasing for a concave upward curve, so the second derivative is positive.

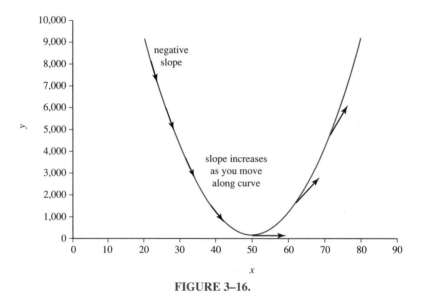

FIGURE 3–16.

WORKSHEET

Determine the values of x and y at the maximum or minimum points for the following functions. Use the second derivative to determine whether the curve has a maximum or minimum.

	Function	Derivative	x	y . . .	Second Derivative	Max or Min?
1.	$y = 2x^2 + 44x + 6$					
2.	$y = 6x^2 - 84x + 3$					
3.	$y = 8x^2 - 80x + 4$					
4.	$y = -12x^2 + 184x + 9$					
5.	$y = -10x^2 + 36x + 18$					
6.	$y = 20x^2 + 23x + 34$					
7.	$y = -15x^2 - 18x + 72$					
8.	$y = 30x^2 - 9x + 65$					
9.	$y = -42x^2 - 96x + 112$					
10.	$y = -22x^2 - 82x + 360$					
11.	$y = 16x^2 + 42x + 94$					
12.	$y = -11x^2 + 18x + 178$					
13.	$y = -4x^2 - 16x + 3$					
14.	$y = 9x^2 + 20x + 5$					

Another problem arrived with the mail that day. "Our baseball uniform company might go out of business!" the professor said in anguish. "This has been a good company because they have always included everything we need with the uniform."

The letter detailed the problem. "We cannot charge more than $126 because the market is very competitive," the professor read the letter. "We can choose the quantity of uniforms we produce per day. We find that if we produce a small quantity, the factory is not very efficient, but if we produce a large quantity, we bump up against capacity limits that increase costs."

The letter contained a graph of the cost function (Figure 3–17).

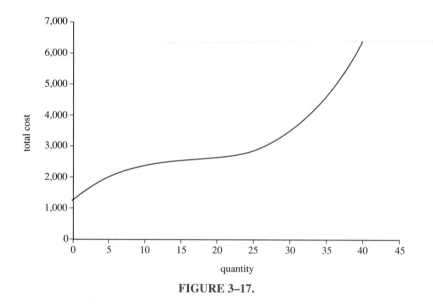

FIGURE 3–17.

"Do we know the formula for a curve that starts out steep, then becomes flatter, then becomes steeper again?" Recordis wondered.

The professor continued, "According to the letter, the cost function can be represented by this third-degree polynomial."

$$C = 0.25Q^3 - 12Q^2 + 210Q + 1,218$$

(C = total cost, Q = quantity of uniforms produced per day.)

"We can work out the quantity that will maximize the profit, and if we're lucky, that quantity will result in a positive profit," the king decided.

"After we've found the value of Q that makes the first derivative equal to zero, we have to check the second derivative so we can be sure we have a maximum," Recordis said. "I'm not going to fall for that trick again."

Since revenue (R) is price times quantity, we wrote:

$$R = 126Q$$

We let Z represent profit, equal to revenue minus cost:

$$Z = R - C$$

$$Z = 126Q - (0.25Q^3 - 12Q^2 + 210Q + 1,218)$$

$$Z = 126Q - 0.25Q^3 + 12Q^2 - 210Q - 1,218$$

$$Z = -0.25Q^3 + 12Q^2 - 84Q - 1,218$$

In order to find the value of Q that maximizes Z, we found the derivative:

$$\frac{dZ}{dQ} = -0.75Q^2 + 24Q - 84$$

and set the derivative equal to zero:

$$0 = -0.75Q^2 + 24Q - 84$$

Now, solve for Q. "Solving a quadratic equation involves a square root sign," Recordis moaned.

We wrote out the quadratic formula:

$$\text{if } ax^2 + bx + c = 0$$

$$\text{then } x = \frac{-b \pm \sqrt{b^2 - 4ac}}{2a}$$

For our problem, we identified the values $a = -0.75$, $b = 24$, and $c = -84$, and then we inserted these numbers into the formula:

$$x = \frac{-24 \pm \sqrt{24^2 - 4 \times (-0.75) \times (-84)}}{2 \times (-0.75)}$$

"The square root sign is the least of our worries," the professor said. "The quadratic formula usually results in two different answers; how are we going to tell which is the right one?"

We worked out the result:

$$x = \frac{-24 \pm \sqrt{576 - 3 \times 84}}{-1.5}$$

$$x = \frac{-24 \pm \sqrt{576 - 252}}{-1.5}$$

$$x = \frac{-24 \pm \sqrt{324}}{-1.5}$$

$$x = \frac{-24 \pm 18}{-1.5}$$

$$x = \frac{-24 + 18}{-1.5} \text{ or } \frac{-24 - 18}{-1.5}$$

$$x = \frac{-6}{-1.5} \text{ or } \frac{-42}{-1.5}$$

$$x = 4 \text{ or } 28$$

"Which of these two values of Q corresponds to the maximum profit?" Recordis wondered.

"We should check the second derivative for both of them and see if that gives us a clue," the king ordered.

We found the second derivative by taking the derivative of dZ/dQ:

$$\text{first derivative} = \frac{dZ}{dQ} = -0.75Q^2 + 24Q - 84$$

$$\text{second derivative} = \frac{d^2Z}{dQ^2} = -1.5Q + 24$$

Inserting $Q = 4$ into the second derivative formula:

$$\frac{d^2Z}{dQ^2} = -1.5 \times 4 + 24 = -6 + 24 = 18$$

"The second derivative is positive at that point, so the curve must be concave upward, and the point must be a minimum," the professor reasoned.

Inserting $Q = 28$ into the second derivative formula:

$$\frac{d^2Z}{dQ^2} = -1.5 \times 28 + 24 = -42 + 24 = -18$$

The professor continued, "The second derivative is negative at that point, so the curve would be concave downward, and the point must be a maximum."

"How can a curve have both a maximum point and a minimum point?" Recordis wondered.

We calculated a table of values, and Igor displayed the graph (Figure 3–18).

Q	Z = profit
0	−121
2	−243
4	−281
6	−247
8	−153
10	−11
12	167
14	369
16	583
18	797
20	999
22	1,177
24	1,319
26	1,413
28	1,447
30	1,409
32	1,287
34	1,069
36	743

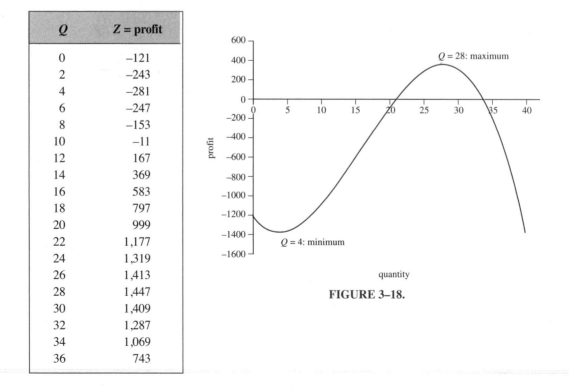

FIGURE 3–18.

From the graph, we could see that $Q = 28$ was indeed the maximum, corresponding to a profit of $1,447. Fortunately this number was positive, so we had hope that the uniform company could stay in business after we let them know this result.

"Wait a minute," Recordis cried. "You can't say that $Q = 28$ is the highest point that the curve ever gets, because it will reach higher values if you extend the curve to the left." He had Igor display a graph of the curve from $Q = -30$ to $Q = 50$. (See Figure 3–19.)

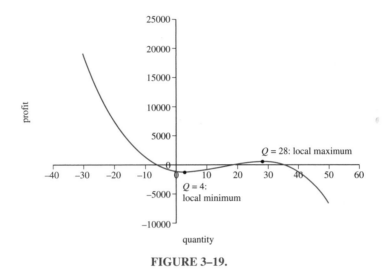

FIGURE 3–19.

"We'll call a point like $Q = 28$ a *local maximum*, because it is higher than all of the points around it," the king decreed. "A local maximum is not necessarily the highest point that a curve ever reaches. This curve goes off to infinity as x goes to negative infinity."

"We'll call a point like $Q = 4$ a *local minimum*, because it is lower than all of the points around it," the professor said. "A local minimum is not necessarily the lowest point that a curve reaches. This curve goes off to negative infinity as x goes to positive infinity."

"In this case we know the quantity of uniforms cannot be negative, so the fact that the formula for profit becomes large when Q is negative is not relevant," the king declared. "That means $Q = 28$ is the best point for the uniform factory."

We summarized:

- If a point is higher than the points around it, it is called a *local maximum*. If a point is the highest point that a curve ever gets, it is called an *absolute maximum*.
- If a point is lower than the points around it, it is called a *local minimum*. If a point is the lowest point that a curve ever gets, it is called an *absolute minimum*.
- If a curve has one change of direction (such as a parabola), then there will be one point where the first derivative is zero. This point will be a local and absolute

maximum if the second derivative is negative (so the curve is concave downward). Ths point will be a local and absolute minimum if the second derivative is positive (so the curve is concave upward).

- If a squiggly curve has two changes of direction (such as some third-degree polynomials), there will be two points where the first derivative is zero. One of these points will be a local maximum, and the other will be a local minimum.

Igor showed us a graph of the profit function, its first derivative, and its second derivative on the same screen, so we could see how these were related. (Figure 3–20.)

"I just thought of something else," the king stated. "What happens if the second derivative is zero?"

"Fiddlesticks!" the professor said. "I knew somebody would think of a complication."

We looked at the graph of our original curve again (Figure 3–18).

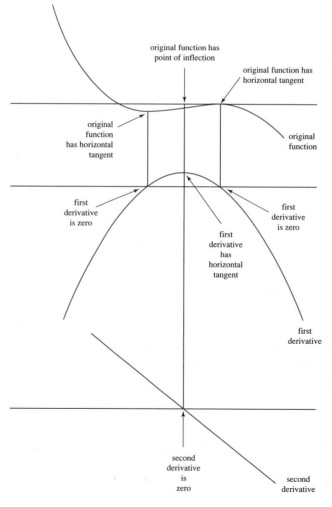

FIGURE 3–20.

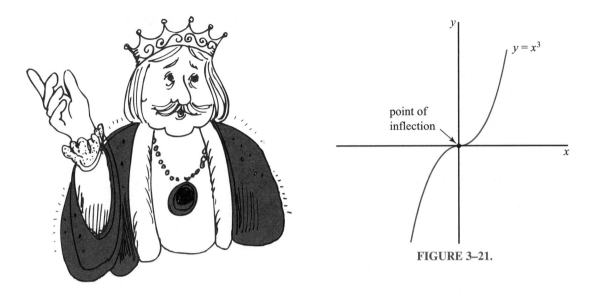

FIGURE 3–21.

"The value of the second derivative is zero when $x = 16$," the professor said.

"What does that mean?" the professor asked.

There was a long pause before Recordis began slowly. "I'll say what it looks like to me, but I don't think this will help much."

"Go ahead," the professor said.

"When x is less than 16, the second derivative is negative. That means that the curve is concave downward when x is less than 16."

"We know that," the professor said.

"And when x is greater than 16, the second derivative is positive, so the curve is concave upward."

"We know that," the king said.

"So the point where the second derivative is zero must be the *boundary* between the curve being concave upward the curve being concave downward."

"I think you have it!" the professor said.

"I do?" Recordis asked.

"We will have to think of a name for such a point," the professor stated. "But I think you just said the only thing that we need to know about it—that it is the boundary between the curve being concave upward and being concave downward."

Everyone looked at me after Recordis and the professor began arguing over what we should call such a point. "We could call it a point of *inflection*," I said, coming up with another name. "That has a nice ring to it."

"What if the first derivative is zero at the same point where the second derivative is zero?" Recordis wondered.

"That wouldn't ever happen, would it?" the professor asked.

"I just thought of a case where it does," the king said. "Suppose we have $f(x) = x^3$. Then $f'(x) = 3x^2$ and $f''(x) = 6x$. (See Figure 3–21.) Look at the point $(0, 0)$. Although

$f'(0)$ equals zero, we can't say that the point is either a local maximum or a local minimum."

"The point $(0, 0)$ is still a point of inflection, though," the professor said. "When x is less than zero the curve is concave downward, and when x is greater than zero the curve is concave upward."

"It still is a point with a horizontal tangent," Trigonometeris noted.

"We can summarize all the rules we've come up with for sketching curves," the professor said.

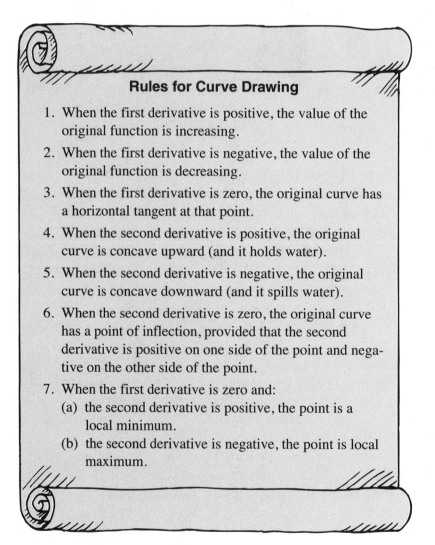

Rules for Curve Drawing

1. When the first derivative is positive, the value of the original function is increasing.

2. When the first derivative is negative, the value of the original function is decreasing.

3. When the first derivative is zero, the original curve has a horizontal tangent at that point.

4. When the second derivative is positive, the original curve is concave upward (and it holds water).

5. When the second derivative is negative, the original curve is concave downward (and it spills water).

6. When the second derivative is zero, the original curve has a point of inflection, provided that the second derivative is positive on one side of the point and negative on the other side of the point.

7. When the first derivative is zero and:
   (a) the second derivative is positive, the point is a local minimum.
   (b) the second derivative is negative, the point is local maximum.

WORKSHEET

For each function below, determine the derivative. Then find the values of x that make the value of the derivative zero. There will often be two of these values. You can find them by using the quadratic formula or by factoring. Then find the value of the second derivative at these two points. Finally, use your graphing calculator or spreadsheet to make a graph of the curve.

1. $y = 2x^3 - 33x^2 + 168x + 12$

   $y' = $ _____

Factor this expression:

   $y' = $ _____$(x - $ _____$)(x - $ _____$)$

List the two values of x that make the first derivative zero: $x = $ _____ and $x = $ _____ . Find the second derivative:

   $y'' = $ _____

Value of the second derivative at the first x value:

   $y'' = $ _____ $= $ _____

Value of the second derivative at the second x value:

   $y'' = $ _____ $= $ _____

2. $y = 2x^3 - 51x^2 + 360x - 164$

   $y' = $ _____

Factor this expression:

   $y' = $ _____$(x - $ _____$)(x - $ _____$)$

List the two values of x that make the first derivative zero: $x = $ _____ and $x = $ _____ . Find the second derivative:

   $y'' = $ _____

Value of the second derivative at the first x value:

   $y'' = $ _____ $= $ _____

Value of the second derivative at the second x value:

   $y'' = $ _____ $= $ _____

3. $y = 2x^3 - 72x^2 + 810x + 17$

   $y' = $ _____

Factor this expression:

$$y' = \underline{\hspace{1cm}}(x - \underline{\hspace{1cm}})(x - \underline{\hspace{1cm}})$$

List the two values of x that make the first derivative zero: $x = \underline{\hspace{1cm}}$ and $x = \underline{\hspace{1cm}}$. Find the second derivative:

$$y'' = \underline{\hspace{1cm}}$$

Value of the second derivative at the first x value:

$$y'' = \underline{\hspace{1cm}} = \underline{\hspace{1cm}}$$

Value of the second derivative at the second x value:

$$y'' = \underline{\hspace{1cm}} = \underline{\hspace{1cm}}$$

4. $y = 2x^3 + 48x^2 + 378x - 124$

$$y' = \underline{\hspace{1cm}}$$

Factor this expression:

$$y' = \underline{\hspace{1cm}}(x - \underline{\hspace{1cm}})(x - \underline{\hspace{1cm}})$$

List the two values of x that make the first derivative zero: $x = \underline{\hspace{1cm}}$ and $x = \underline{\hspace{1cm}}$. Find the second derivative:

$$y'' = \underline{\hspace{1cm}}$$

Value of the second derivative at the first x value:

$$y'' = \underline{\hspace{1cm}} = \underline{\hspace{1cm}}$$

Value of the second derivative at the second x value:

$$y'' = \underline{\hspace{1cm}} = \underline{\hspace{1cm}}$$

5. $y = 2x^3 + 6x^2 - 480x - 14$

$$y' = \underline{\hspace{1cm}}$$

Factor this expression:

$$y' = \underline{\hspace{1cm}}(x - \underline{\hspace{1cm}})(x - \underline{\hspace{1cm}})$$

List the two values of x that make the first derivative zero: $x = \underline{\hspace{1cm}}$ and $x = \underline{\hspace{1cm}}$. Find the second derivative:

$$y'' = \underline{\hspace{1cm}}$$

Value of the second derivative at the first x value:

$$y'' = \underline{\hspace{1cm}} = \underline{\hspace{1cm}}$$

Value of the second derivative at the second x value:

$$y'' = \underline{\hspace{2cm}} = \underline{\hspace{1.5cm}}$$

6. $y = 2x^3 - 51x^2 - 360x + 96$

$$y' = \underline{\hspace{1.5cm}}$$

Factor this expression:

$$y' = \underline{\hspace{1cm}}(x - \underline{\hspace{1cm}})(x - \underline{\hspace{1cm}})$$

List the two values of x that make the first derivative zero: $x = \underline{\hspace{1cm}}$ and
$x = \underline{\hspace{1cm}}$. Find the second derivative:

$$y'' = \underline{\hspace{1.5cm}}$$

Value of the second derivative at the first x value:

$$y'' = \underline{\hspace{2cm}} = \underline{\hspace{1.5cm}}$$

Value of the second derivative at the second x value:

$$y'' = \underline{\hspace{2cm}} = \underline{\hspace{1.5cm}}$$

7. $y = 2x^3 + 9x^2 - 60x + 84$

$$y' = \underline{\hspace{1.5cm}}$$

Factor this expression:

$$y' = \underline{\hspace{1cm}}(x - \underline{\hspace{1cm}})(x - \underline{\hspace{1cm}})$$

List the two values of x that make the first derivative zero: $x = \underline{\hspace{1cm}}$ and
$x = \underline{\hspace{1cm}}$. Find the second derivative:

$$y'' = \underline{\hspace{1.5cm}}$$

Value of the second derivative at the first x value:

$$y'' = \underline{\hspace{2cm}} = \underline{\hspace{1.5cm}}$$

Value of the second derivative at the second x value:

$$y'' = \underline{\hspace{2cm}} = \underline{\hspace{1.5cm}}$$

8. $y = 2x^3 + 60x^2 + 384x - 43$

$$y' = \underline{\hspace{1.5cm}}$$

Factor this expression:

$$y' = \underline{\hspace{1cm}}(x - \underline{\hspace{1cm}})(x - \underline{\hspace{1cm}})$$

List the two values of x that make the first derivative zero: $x =$ _____ and
$x =$ _____ . Find the second derivative:

$y'' =$ _____

Value of the second derivative at the first x value:

$y'' =$ _____ = _____

Value of the second derivative at the second x value:

$y'' =$ _____ = _____

9. $y = 2x^3 - 24x^2 + 96x + 44$

$y' =$ _____

Factor this expression:

$y' =$ _____$(x -$ _____$)(x -$ _____$)$

List the two values of x that make the first derivative zero: $x =$ _____ and
$x =$ _____ . Find the second derivative:

$y'' =$ _____

Value of the second derivative at the first x value:

$y'' =$ _____ = _____

Value of the second derivative at the second x value:

$y'' =$ _____ = _____

10. $y = 2x^3 - 9x^2 + 204x + 16$

$y' =$ _____

What happens with this function?

"The subject of calculus might be useful for more than just finding the speed of the train," the professor said thoughtfully. "If we could only solve the fly ball problem." (That problem continued to perplex us until Chapter 7.)

Exercises

For Exercises 1 to 5, find formulas for the derivative and the second derivative, and determine the value of x corresponding to maximum or minimum points (and say which it is).

1. $y = -3x^2 + 5x - 9$

2. $y = 13x^2 + 4x + 19$

3. $y = 17x^2 + 11x - 10$

4. $y = 19x^2 - 10x + 18$

5. $y = -9x^2 + 16x + 9$

6. The second derivative of a position function is known in physics as the *acceleration*. Find the acceleration of Pal's beach ball when he tosses it straight up into the air ($h = -\frac{1}{2}gt^2 + v_0 t$). Is the acceleration positive or negative?

7. Find the acceleration of the ball when Pal throws it off Hasselbluff Mountain ($h = 64 - \frac{1}{2}gt^2$).

8. Find the acceleration of Pal when he slides on Ice Skating Lake ($x = 3t$).

9. Consider the curve $y = x^2 - 3x$ on the interval $x = 0$ to $x = 5$. Find the absolute maximum value obtained by y on this interval. Find the absolute minimum value obtained by y on this interval.

10. Consider the curve $y = -x^4 + 8x^2$. What are the coordinates of the points where the curve has horizontal tangents? How many such points are there? Are these points local maximum points or local minimum points? In what interval is the curve concave downward? When is it concave upward?

11. Consider the curve $y = x^2 - \frac{1}{3}x^3$. Where is the curve rising? Where is it falling? Where is it concave upward? Where is it concave downward?

12. $y = x^3 + x^2 + x$; $d^4y/dx^4 = ?$

13. $y = ax^3 + bx^2 + cx$; $d^3y/dx^3 = ?$

14. $y = x^n$; $d^n y/dx^n = ?$

15. Create a spreadsheet that draws the graph of a polynomial and use it on the following problems.

16. $y = x^3 - 15x^2 + 48x + 12$

17. $y = x^3 - 36x^2 + 432x - 37$

18. $y = x^3 + 3x^2 + 243x + 600$

19. $y = x^4 - 11x^3 - 7x^2 + 155x + 150$

20. $y = x^4 - 19x^3 + 114x^2 - 256x + 160$

21. $y = x^4 - 4x^3 - 5x^2 - 2x + 10$

22. $y = x^4 - 6x^3 + 2x^2 + 16x + 32$

23. $y = x^4 - 11x^3 - 7x^2 + 155x + 800$

Derivatives of Complicated Functions

Recordis woke everyone early the next morning. "I have another problem!" he shouted. "I think we can solve it easily."

Everyone was still sleepy as we gathered in the Main Conference Room. Recordis had Igor draw a map. (See Figure 4–1.)

"You remember the field I have out in the country," he began. "There is a stream that runs in a perfect half-circle around one end, and there is a straight road that borders it on the other end. I've been wanting to build a house on the field for a long time, but I've never been able to figure out what its dimensions should be. I want the house to cover as much area as possible. It has to have square edges, of course, because the builders don't know how to make any other kind of house. It is obvious that you can't build a very wide house or a very thin house (Figure 4–2). The shape that has the largest area must be somewhere in the middle between these two extremes. We have to figure out exactly what the maximum area shape is."

"Why did you call us all together?" the professor asked. "What does this problem have to do with anything we have been doing?"

"Yesterday we worked on finding maximum points," Recordis said. "All we have to do is find the area of the house as a function of its width, and then we can find the derivative to solve for the particular width that is a maximum point for the area function."

"Can you do that?" the king asked timidly. "I thought functions had to be functions of x that were set equal to y."

"I don't see why that has to be," I replied. "We can make up some new letters to stand for the variables in the problem."

"Like what?" the professor said sleepily.

"The first thing we do know is the radius of the semicircle, right?" Recordis didn't say anything. "We do know the radius of the semicircle, don't we?"

"We could measure it very quickly," Recordis suggested.

"We don't need to know the exact length while we're doing the calculation," I said. "For now we will say that the length of the radius is equal to r."

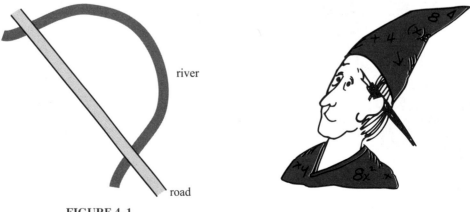

FIGURE 4–1.

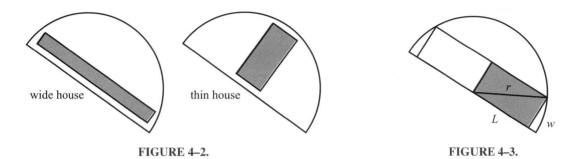

FIGURE 4–2.   FIGURE 4–3.

"It should be easy to remember," Recordis said, "that r stands for radius."

"The radius won't change," Trigonometeris offered helpfully.

"Right," I said, "r will be a constant. Unfortunately, everything else will be a variable."

"It would be easier if we tried to maximize the area of half of the house," the king pointed out. "We know that the total area is twice the shaded area (Figure 4–3), so if we maximize the shaded area it will be the same as maximizing the total area of the house."

"The shaded area will be where the living room will be," Recordis said.

"We'll call the length of the living room L and the width of the living room w," I stated. "Then we know that (area of living room) $= w \times L$."

"We can find a relationship between w and L," the professor said. "L and w will always form the sides of a right triangle, with the radius of the circle forming the hypotenuse. Then we can use the Pythagorean theorem to show that $L^2 + w^2 = r^2$."

"Now we can solve for L," Recordis said.

$$L^2 + w^2 = r^2$$

$$L^2 = r^2 - w^2$$

$$L = \sqrt{r^2 - w^2}$$

"And we can put the last expression in the equation for $A = wL$," the king added.

$$A = w \sqrt{r^2 - w^2}$$

"That's great!" Recordis exclaimed. "We now have A expressed as a function of only one variable: w. That means $A = f(w)$. All we need to do now is find dA/dw, and then set $dA/dw = 0$. Then we can solve for the value of w that maximizes A."

"Hold it!" the professor objected. "We can't find the derivative of that function! We don't know whether the $f(x) = x^n, f'(x) = nx^{n-1}$ rule works if n is a fraction, and in this case we have something raised to the ½ power. And we don't even have just *w* raised to a power; we have a weird expression like $(r^2 - w^2)$ raised to a power. Not only that, but we have a stray *w* in front of the complicated function that is multiplying everything else! There is no way we can do that problem!"

"You're right!" Recordis said, looking closely at the function. "That is hopeless. That is really impossible."

"Then why did you get us up so early?" the professor complained. Everyone began to get ready to go back to bed.

"There has to be something we can do!" the king said.

Everyone stopped. "It looks to me as though we have three problems," I said. "First, we have that loose *w* in front of the rest of the function. We know that it is pretty easy to find the derivative of a function if we have a constant number multiplied by a function. (See Chapter 2, Exercise 12.) We just need to find a new rule that tells us what to do if we have a variable times a function. Then we need to see what to do about an embedded function; for example, in this case instead of having $w^{1/2}$ we have $(r^2 - w^2)^{1/2}$. All this means is that we have one function of $w(f(w) = r^2 - w^2)$ embedded in another function $(g(q) = q^{1/2})$, giving us $g(f(w))$. We need to figure out what to do with a function like that. Finally, we need to see whether the rule for finding the derivative of a power works when we have fractional powers like square roots. If we can just solve these three problems, we can find the answer to the problem we started with."

"Right," the professor said. "And if I could just grow 20 feet and develop some huge muscles, I could beat Pal in a wrestling match."

"We might as well at least give it a try," the king stated.

"The first problem is what to do when we have two functions multiplied together. Suppose we have two functions, $u(x)$ and $v(x)$, and a function $q(x)$ that is set up so that $q(x) = [u(x)] \times [v(x)]$. We need to find out what dq/dx is."

"I know what would be really simple," Recordis said. "When we had $q(x) = f(x) + g(x)$, we just said that

$$\frac{dq}{dx} = \frac{df}{dx} + \frac{dg}{dx}.$$

Why don't we say that if $q(x) = [u(x)] \times [v(x)]$, then we have

$$\frac{dq}{dx} = \frac{du}{dx} \times \frac{dv}{dx}?"$$

"That would be too easy," the king objected.

"Besides, we can't just make something up like that," the professor said.

"I thought we were making most of this up anyway," Recordis protested.

"I don't think your suggestion will work," the king said. "For example, suppose we have the function $q(x) = 6x^2$. Let's make these definitions:

$$u(x) = 6 \qquad\qquad du/dx = 0$$
$$v(x) = x^2 \qquad\qquad dv/dx = 2x$$

"Then $q(x) = [u(x)] \times [v(x)]$. According to your proposed rule, we would have:

$$\frac{dq}{dx} = \frac{du}{dx} \times \frac{dv}{dx}$$

$$= 0 \times 2x$$

$$= 0$$

"However, in this case we know that dq/dx is equal to $12x$, not zero, so your proposed rule does not work. If we have a function that consists of two other functions multiplied together, we cannot simply multiply the two derivatives to get the derivative of the whole function."

"All right, we can look for a new rule," Recordis agreed. "I was just trying to make life simple for all of us."

I suggested, "First, let's see if we can find out what $q(x + \Delta x)$ is."

"That would be this expression," the professor said.

$$q(x + \Delta x) = [u(x + \Delta x)] \times [v(x + \Delta x)]$$

"We can rename $u(x + \Delta x)$," I said. "Show us a graph of some function, Igor." (Figure 4–4.)

"We could call the difference between $u(x)$ and $u(x + \Delta x)$ something like Δu," the king suggested.

$$u(x + \Delta x) - u(x) = \Delta u$$

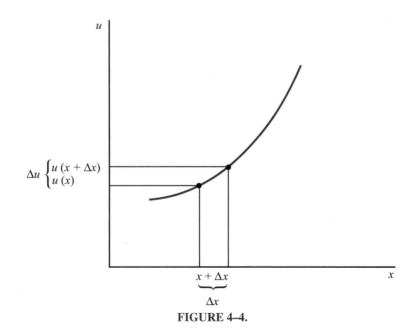

FIGURE 4–4.

"We can call the difference between $v(x + \Delta x)$ and $v(x)$ something like Δv," the professor said. "Then we can write two equations."

$$u(x + \Delta x) = u(x) + \Delta u$$

$$v(x + \Delta x) = v(x) + \Delta v$$

"We can put these expressions back in our equation for $q(x + \Delta x)$," the king noticed.

$$q(x + \Delta x) = [u(x + \Delta x)] \times [v(x + \Delta x)]$$

$$= [u(x) + \Delta u] \times [v(x) + \Delta v]$$

"We can multiply out the right side algebraically," Recordis offered.

$$q(x + \Delta x) = u(x)\, v(x) + u(x)\, \Delta v + v(x)\, \Delta u + \Delta u\, \Delta v$$

To save writing, Recordis decided to write u instead of $u(x)$ and v instead of $v(x)$. That works as long as we remember that both u and v are functions of x.

$$q(x + \Delta x) = uv + u\, \Delta v + v\, \Delta u + \Delta u\, \Delta v$$

"I know what uv is!" the king said. "That's equal to $q(x)$."

"We can subtract $q(x)$ from both sides," the professor said. She was beginning to get the gleam in her eye that she gets when she thinks she sees something that Recordis hasn't noticed yet.

$$q(x + \Delta x) - q(x) = u\, \Delta v + v\, \Delta u + \Delta u\, \Delta v$$

"That's beginning to look familiar," Recordis stated. "I just can't think of what it looks like."

"Let's divide both sides by Δx," the professor suggested.

$$\frac{q(x + \Delta x) - q(x)}{\Delta x} = u\frac{\Delta v}{\Delta x} + v\frac{\Delta u}{\Delta x} + \Delta u\frac{\Delta v}{\Delta x}$$

"Now I know what it is!" Recordis said. "We can take a limit on both sides, and then the left side will be the derivative of q with respect to x!"

$$\lim_{\Delta x \to 0} \frac{q(x + \Delta x) - q(x)}{\Delta x} = \lim_{\Delta x \to 0} u\frac{\Delta v}{\Delta x} + \lim_{\Delta x \to 0} v\frac{\Delta u}{\Delta x} + \lim_{\Delta x \to 0} \Delta u\frac{\Delta v}{\Delta x}$$

$$\frac{dq}{dx} = u \lim_{\Delta x \to 0} \frac{\Delta v}{\Delta x} + v \lim_{\Delta x \to 0} \frac{\Delta u}{\Delta x} + \lim_{\Delta x \to 0} \Delta u\frac{\Delta v}{\Delta x}$$

"I remember what Δv is," the king said. "We said that $\Delta v = v(x + \Delta x) - v(x)$. And Δu is the same—almost; $\Delta u = u(x + \Delta x) - u(x)$."

$$\frac{dq}{dx} = u \lim_{\Delta x \to 0} \frac{v(x+\Delta x)-v(x)}{\Delta x} + v \lim_{\Delta x \to 0} \frac{u(x+\Delta x)-u(x)}{\Delta x} + \lim_{\Delta x \to 0} \Delta u \frac{v(x+\Delta x)-v(x)}{\Delta x}$$

"Three of those things are just derivatives!" Recordis exclaimed, noticing gleefully that he could write the whole equation in a much shorter fashion.

$$\frac{dq}{dx} = u\frac{dv}{dx} + v\frac{du}{dx} + \frac{dv}{dx} \lim_{\Delta x \to 0} \Delta u$$

"I think we can get rid of that $\lim_{\Delta x \to 0} \Delta u$," the king said. "I have an idea. Remember that $\Delta u = u(x + \Delta x) - u(x)$. When we take the limit of Δx going to zero, we get $\Delta u = u(x) - u(x)$."

"That's equal to zero," the professor pointed out.

"That means we can erase it," Recordis said jubilantly. "If it is equal to zero, it doesn't make any difference whether it is multiplied by dv/dx or 752 or 3,569,204½."

Igor displayed the final rule:

$$q(x) = uv$$

$$\frac{dq}{dx} = u\frac{dv}{dx} + v\frac{du}{dx}$$

"That isn't nearly as bad as I thought it would be," Trigonometeris said.

"We still had better think of a name for it," Recordis told him. "It will be hard enough to remember as it is."

"The name will be easy," the professor said. "We can use this rule to find the derivative of a function if it is the product of two functions, so we'll call it the *product rule*. And I can think of a sentence that will make it a little bit easier to remember. To find the derivative of two functions multiplied together, take the first function, multiply it by the derivative of the second function, and add the result to the second function multiplied by the derivative of the first function."

"Let's make sure that this rule agrees with what we did before," Recordis suggested. "It would be a tremendous embarrassment if we derived a new rule like this and it turned out to give us a different result for some problem to which we already know the answer."

"We already found the result in a simple case where a constant function is multiplied by another function," the king said. "We had $y = cx$, where c was a constant. Let's try the product rule on that function."

"We can try," the professor said nervously. "First, we need the first function. That's c. Then we need the derivative of the second function. That's 1. That gives us

1 times c as the first term. Then we need to add the second function [the sweat was beginning to build on her brow], which is x, multiplied by the derivative of the first function, which is the derivative of c, which is . . . "

"Zero!" the king supplied. "We end up with this expression."

$$\frac{d(cx)}{dx} = (1)(c) + (0)(x) = c$$

"That is the same answer we got before," the professor said happily. "Our rule does work."

"You were lucky that time," Recordis objected. "I have another function. How about $f(x) = x^2$? We already know that $f'(x) = 2x$. In this case we know that $f(x) = x$ times x. Let's try your product rule on that function."

"All right," the professor said, slightly more confident. "We have the first function (which is x) times the derivative of the second function (which is the derivative of x, which is 1). Then we have the second function (which is also x) times the derivative of the first function, which is also 1. That means we end up as follows."

$$f(x) = x \cdot x$$

$$f'(x) = (1)(x) + (x)(1) = 2x$$

"That does equal $2x$!" the king said. "It is consistent with what we did before." Igor added the product rule to our list.

"One down and two to go," the professor said to me. "I think the other two problems are harder, though."

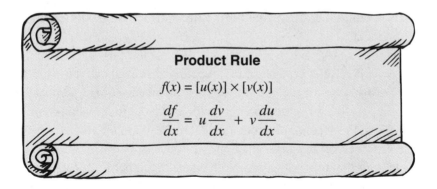

Product Rule

$$f(x) = [u(x)] \times [v(x)]$$

$$\frac{df}{dx} = u\frac{dv}{dx} + v\frac{du}{dx}$$

"All right," I said. "Now we have the problem of a function embedded in another function. (The mathematical term for a function of this form is *composite function*.) The problem is this: Find dy/dx when $y = f(g(x))$." (The expression $f(g(x))$ is read as "f of g of x.")

"What exactly does $f(g(x))$ mean?" Recordis asked.

"Both f and g are function machines," I said. "All the embedded function means is that we have Pal throw an x into the g machine. The g machine spits out some number $g(x)$, or we could call it u. Then we set up a machine so that the $g(x)$ number u gets thrown right into the f machine, which spits out y, which is the number Pal wanted to get out of the machine in the first place." (Figure 4–5.)

"It looks as though you would need a chain to hold the f machine and the g machine together," Recordis noted.

"Now suppose that Pal throws an $(x + \Delta x)$ into the machine," I said.

"Then the g machine will spit out the number $g(x + \Delta x)$," the professor offered.

"Let's call that number $(u + \Delta u)$," I said.

$$u + \Delta u = g(x + \Delta x)$$

"Then the number $(u + \Delta u)$ will fall into the f machine," the king said.

"That machine will then throw out the number $f(u + \Delta u)$," I added.

"I bet a hundred dollars you're going to call that number $y + \Delta y$," Recordis said.

$$y + \Delta y = f(u + \Delta u)$$

"We're looking for dy/dx," the professor said. "That means that we have to find the limit of $\Delta y/\Delta x$ as $\Delta x \to 0$. I don't see how you're going to do that."

"We can say the following," I suggested.

$$\frac{\Delta y}{\Delta x} = \frac{\Delta y}{\Delta u}\frac{\Delta u}{\Delta x}$$

"You can't just say something like that!" Recordis protested.

"Sure we can," I said. "The little numbers with the deltas are all real numbers, and they can be very, very small but we'll never let them be equal to zero. Then the equation we wrote is valid no matter how we define Δy, Δx, or Δu."

"We could try taking the limits of both sides now," the professor suggested.

$$\lim_{\Delta x \to 0}\frac{\Delta y}{\Delta x} = \lim_{\Delta x \to 0}\frac{\Delta y}{\Delta u}\frac{\Delta u}{\Delta x}$$

"The left-hand side is equal to dy/dx," Recordis noted.

$$\frac{dy}{dx} = \lim_{\Delta x \to 0}\frac{\Delta y}{\Delta u}\lim_{\Delta x \to 0}\frac{\Delta u}{\Delta x}$$

"The last part is equal to du/dx," the king said.

$$\frac{dy}{dx} = \lim_{\Delta x \to 0}\frac{\Delta y}{\Delta u}\frac{du}{dx}$$

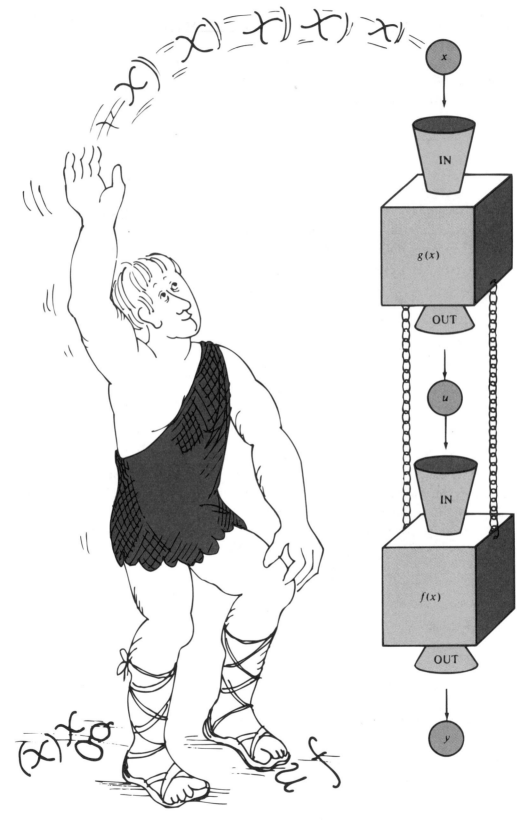

FIGURE 4–5.

"We found out a few minutes ago that, when Δx goes to zero, Δu goes to zero," Recordis reminded us.

"We could rewrite our expression like this," the professor said.

$$\frac{dy}{dx} = \lim_{\Delta u \to 0} \frac{\Delta y}{\Delta u}\, \frac{du}{dx}$$

"I know what that is equal to!" shouted the king.

$$\frac{dy}{dx} = \frac{dy}{du}\, \frac{du}{dx}$$

"Is that legal?" Recordis asked. "I thought we had to have a dx on the bottom all the time."

"I don't see why it has to be that way all the time," the professor answered. "In this case we have $y = f(u)$, so we should be able to say that the derivative is dy/du, just the same as we can say that if $y = f(x)$ the derivative is dy/dx."

"We must have a good name for this one," Recordis said.

"I know what we can call it," the king said. "In the original design for the composite function machine, we said that we needed a chain to hold the two functions together. We may as well call this rule the *chain rule*."

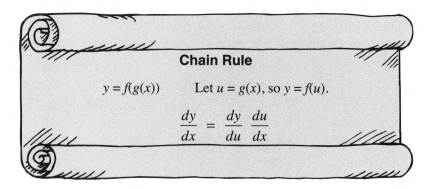

Chain Rule

$$y = f(g(x)) \qquad \text{Let } u = g(x), \text{ so } y = f(u).$$

$$\frac{dy}{dx} = \frac{dy}{du}\, \frac{du}{dx}$$

"I know a good way to remember the rule," the professor offered. "It looks as though derivatives behave as if they are two fractions being multiplied together, and the du's just cancel each other out."

"Let's try the rule on a simple example to make sure it works," the king cautioned. The simplest functions we could think of were straight-line functions, so we let $y = au + b$, and $u = cx + d$, where a, b, c, and d are constants. Then we could see that $dy/du = a$ and $du/dx = c$. The chain rule claimed:

$$\frac{dy}{dx} = \frac{dy}{du}\, \frac{du}{dx} = ac$$

By substitution we could see that $y = acx + ad + b$, and we could see that $dy/dx = ac$, just as the chain rule predicted.

"That's two problems out of three," the king said. "Now we come to the hard one. What do we do if we have a fractional power?"

"Let's start with $y = x^n$, only this time n isn't an integer," the professor suggested.

"That doesn't help much," the king said.

"How else can we write it?" the professor asked.

"If it is a fraction, we can write it as a ratio of two unknown integers," I suggested. "We could say that $y = x^{p/q}$."

"Will that work?" the professor demanded. "How do you know that p and q are integers?"

"Any rational number can be expressed as a ratio of two integers," Recordis proudly quoted from a page in his book. "That's the definition of a rational number."

"We're still stuck," the professor said. "Remember that an integer in the denominator of an exponent means to take that root of the number. For example, $x^{1/2}$ means the square root of x, and $x^{1/3}$ means the cube root of x. Now suppose we use the definition of the derivative. Then we get the following:

$$y' = \lim_{\Delta x \to 0} \frac{(x + \Delta x)^{p/q} - x^{p/q}}{\Delta x}$$

"The problem is going to arise when we have the quantity with the addition sign $(x + \Delta x)$ raised to a fractional power. That's the same thing as $\sqrt[q]{(x + \Delta x)^p}$. We don't have any way of evaluating a radical sign if we are adding together two numbers under it! We can't do anything with that kind of expression even if we know what the fractional power is, for example, $\sqrt{a + b}$ or $\sqrt[3]{a + b}$. We can't do anything with either of those expressions! It will be even harder when we don't know what root we're taking—in this case we know only that we're taking the qth root. We absolutely cannot do it this way."

"Then we must think of another way," the king stated simply.

$$y = x^{p/q}$$

"I have an idea," I said. "We know that our formula for the derivative of a power works if we have integer powers. That means we must set up the equation so that we are dealing only with integer powers."

"Just try," the professor told me.

"Suppose we take both sides of the equation and raise them to the power q," the king suggested.

Because of the properties of exponents, we could cancel the q's on the right-hand side:

$$y^q = (x^{p/q})^q$$

$$y^q = x^p$$

"Sure, that gets rid of the fraction," the professor said. "But look at how many more problems it causes! Now we no longer have a function defined explicitly!"

"Do we have to?" I asked.

"Don't we?" Recordis said. "Everything we have done before has had $y = f(x)$. We've never tried to apply a function to the y before."

"But we've run across this kind of thing in algebra," the king noted. "We derived an equation for a circle that was $x^2 + y^2 = r^2$. We called it an *implicit function*. We never said explicitly what y was as a function of x, but we could tell from the equation what values of y went with what values of x."

"All we need to do, then, is figure out how to take the derivative of an implicit function," I said. Slowly I began to see a vision of a chain surrounding the implicit function $y^q = x^p$. "We can use the chain rule," I suggested. "Suppose we have $y = (g(x))^2$ and we want to find dy/dx."

"We can use the chain rule," the king said. "We have $f(u) = u^2$, and $u = g(x)$. It looks like:"

$$\frac{dy}{dx} = \frac{dy}{du}\frac{du}{dx} = 2u\frac{du}{dx}$$

"How does that help?" the professor demanded. "Our problem is $y^q = x^p$."

"We can use what we just did to find the derivative of both sides of the equation," I said.

"Both sides?" Recordis asked. "How can you do that?"

"All the equation says is that we have two functions—one of them is y^q and the other one is x^p—and that they're equal for all x. That means that they're really two different ways of expressing the same function. And that means that their derivatives with respect to x should be equal. Therefore we can apply the operator d/dx to both sides."

$$\frac{d}{dx}y^q = \frac{d}{dx}x^p$$

"The part on the right is easy," Recordis said, "since we know that p is an integer."

$$\frac{d}{dx}x^p = px^{p-1}$$

We rewrote our equation:

$$\frac{d}{dx}y^q = px^{p-1}$$

"How can you take the derivative of a function if it has a y in it?" Recordis demanded.

"We can use what we just did," the king said. "We said that if (some function) = y^q, then (d/dx) (some function) = $qy^{q-1}(dy/dx)$."

"We don't know what dy/dx is," the professor objected. "That is what we are trying to solve for."

"But we can solve for it now," the king said. "Our original equation becomes as follows."

$$qy^{q-1}\frac{dy}{dx} = px^{p-1}$$

Recordis looked at the equation suspiciously, as if he wasn't sure it was a legitimate one. "You could try dividing both sides by qy^{q-1}," he suggested.

$$\frac{dy}{dx} = \frac{px^{p-1}}{qy^{q-1}}$$

"We can substitute $y = x^{p/q}$," the king noted.

$$\frac{dy}{dx} = \frac{p}{q}\frac{x^{p-1}}{(x^{p/q})^{q-1}}$$

"We can simplify that exponent in the denominator," the king noticed. (Remember that $(x^a)^b = x^{ab}$.)

$$\frac{dy}{dx} = \frac{p}{q}\frac{x^{p-1}}{x^{p-p/q}}$$

"We can use another exponent law to simplify the fraction with the x in it," Recordis noted. (Remember that $x^a/x^b = x^{a-b}$.)

$$\frac{dy}{dx} = \frac{p}{q}x^{(p-1)-(p-p/q)}$$

"That can be simplified," the professor noted.

$$\frac{dy}{dx} = \frac{p}{q}x^{p-1-p+p/q}$$

$$= \frac{p}{q}x^{-1+p/q}$$

$$= \frac{p}{q}x^{p/q-1}$$

"Now substitute n in place of p/q."

$$\frac{dy}{dx} = nx^{n-1}$$

"That's what we wanted to show!" Recordis cried happily. "That means that, if $y = x^n, y' = nx^{n-1}$, whether or not n is an integer."

"We should make up a name for this rule, since it seems to be true in general," the king said.

"That's easy," the professor announced. "We'll call it the *power rule*."

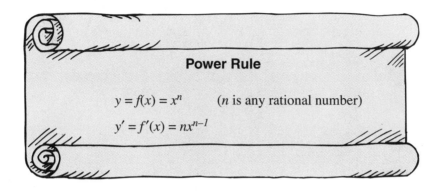

Power Rule

$$y = f(x) = x^n \qquad (n \text{ is any rational number})$$

$$y' = f'(x) = nx^{n-1}$$

(We later established that the power rule is valid for any real number exponent, including irrational numbers, such as π or $\sqrt{2}$. See Chapter 10, Exercise 12.)

WORKSHEET

Use the product rule to find the derivatives of the following functions of x (treating all other letters as constants). Take the first factor (call it u), multiply it by the derivative of the second factor (call it v), then add the second factor multiplied by the derivative of the first factor. (In some cases, there are also other ways to find the derivative, but you should get the same answer as you do by using the product rule.)

	$f(x)$	u	dv/dx	v	du/dx	$f'(x)$
1.	$(ax)(bx)$					
2.	$(a + x)(b + x)$					
3.	$(a + bx)(h + kx)$					
4.	$(ax^m)(bx^n)$					
5.	$(a + x)^m(b + x)^n$					
6.	$x\sqrt{1 - x^2}$					

"We should save the method we thought of to prove the power rule," Recordis said. "The implicit function method."

Igor wasn't sure how to generalize this, so I thought of a way to start.

"The problem will come when we have an expression like this: $f(y) = g(x)$. If we take the derivative with respect to x of both sides,

$$\frac{d}{dx} f(y) = \frac{d}{dx} g(x)$$

we know that the chain rule tells us that

$$\frac{df}{dy} \frac{dy}{dx} = \frac{dg}{dx}$$

Now we can solve for dy/dx:

$$\frac{dy}{dx} = \frac{\dfrac{dg}{dx}}{\dfrac{df}{dy}}$$

Igor displayed the function for which we could now allegedly find the derivative":

$$A = w(r^2 - w^2)^{1/2}$$

"Remember r is a constant, w is a variable, and we want to find dA/dw," the king said.

"First we should use the product rule," the professor said. "We would have dA/dw equals the first function (w) times the derivative of the messy function ($(r^2 - w^2)^{1/2}$) plus the derivative of w (which is 1) times the messy function."

$$\frac{dA}{dw} = (w) \frac{d}{dw} (r^2 - w^2)^{1/2} + (1)(r^2 - w^2)^{1/2}$$

"Now all we have to do is find the derivative of the messy function," the king noted. "I think we'll need both the chain rule and the power rule."

We made these definitions:

$$y = (r^2 - w^2)^{1/2}$$
$$u = r^2 - w^2$$
$$y = u^{1/2}$$

According to the chain rule:

$$\frac{dy}{dw} = \frac{dy}{du} \frac{du}{dw}$$

"We can find dy/du using the power rule with $n = \text{1/2}$," the professor said.

$$\frac{dy}{du} = \frac{1}{2}u^{1/2-1} = \frac{1}{2}u^{-1/2}$$

"I know how to find du/dw," Recordis said. "We have $u = r^2 - w^2$. The r^2 is constant, so its derivative is zero. The derivative of $-w^2$ is $-2w$. Therefore:"

$$\frac{d}{dw}(r^2 - w^2)^{1/2} = \frac{1}{2}(r^2 - w^2)^{-1/2}(-2w)$$

$$= -w(r^2 - w^2)^{-1/2}$$

"We can put that expression back in our equation for dA/dw," the king said.

$$\frac{dA}{dw} = -w^2(r^2 - w^2)^{-1/2} + (r^2 - w^2)^{1/2}$$

"Now we have to do what we did yesterday," the professor pointed out. "We set dA/dw equal to zero and then solve for the optimal value of w."

$$0 = -w^2(r^2 - w^2)^{-1/2} + (r^2 - w^2)^{1/2}$$

"Now we can do it," Recordis said. "That's just algebra."

$$w^2(r^2 - w^2)^{-1/2} = (r^2 - w^2)^{1/2}$$

Multiply both sides by $(r^2 - w^2)^{1/2}$:

$$w^2(r^2 - w^2)^{-1/2}(r^2 - w^2)^{1/2} = (r^2 - w^2)^{1/2}(r^2 - w^2)^{1/2}$$

$$w^2 = (r^2 - w^2)$$

$$2w^2 = r^2$$

$$w^2 = \frac{r^2}{2}$$

$$w = \frac{r}{\sqrt{2}}$$

"That's the answer!" Recordis shouted. "I need to design my house so that the width of the living room is equal to the radius of the circle divided by the square root of 2!"

"I remember what the radius of your yard is," the professor said. "It's 10 units. So that means that the width should be as follows." (Recordis spent a long time looking up the square root of 2 in a table and performing the division.)

$$w = (10)(2)^{-1/2} = \frac{10}{1.414} = 7.07$$

"Shouldn't we check the second derivative?" the king asked. "It would be embarrassing if this turned out to be a minimum point and Recordis ended up building the smallest possible house."

"I think we can just tell if we look at some points on the original function," the professor said. "That would save us from having to check the second derivative." (See Exercise 25.)

Igor made a table of values of the function $A(w) = w(100 - w^2)^{1/2}$ (Table 4–1 and Figure 4–6.)

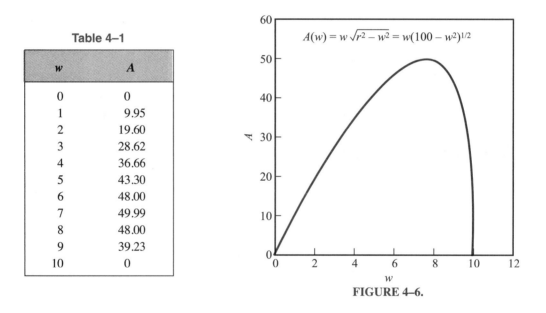

Table 4–1

w	A
0	0
1	9.95
2	19.60
3	28.62
4	36.66
5	43.30
6	48.00
7	49.99
8	48.00
9	39.23
10	0

$$A(w) = w\sqrt{r^2 - w^2} = w(100 - w^2)^{1/2}$$

FIGURE 4–6.

"Our answer is right!" Recordis said happily.

The professor was thinking about generalizing our results. "Will the power rule even work for a negative exponent?" she wondered. We tried an example:

$$y = x^{-n} \qquad (n > 0)$$

$$y = \frac{1}{x^n}$$

$$x^n y = 1$$

Using the product rule:

$$\frac{d}{dx}(x^n y) = \frac{d}{dx} 1$$

$$x^n \times \frac{dy}{dx} + y \times (nx^{n-1}) = 0$$

$$x^n \times \frac{dy}{dx} = -y(nx^{n-1})$$

Since $y = x^{-n}$:

$$x^n \times \frac{dy}{dx} = -x^{-n}(nx^{n-1})$$

Multiply both sides by x^{-n}, and then collect the exponents for x:

$$\frac{dy}{dx} = -nx^{-n-n+n-1} = -nx^{-n-1}$$

This result is what the power rule predicted, so the rule works for negative exponents as well as fractional or whole number exponents.

"Now that we know a rule for finding the derivative of two functions multiplied together, it seems as if we should find a rule for when two functions are divided," the professor continued.

$$y = \frac{u}{v}$$

We tried to find a way to use any of the derivative rules we already knew. "We can make it into a product if we do this," Recordis suggested.

$$y = u \times \frac{1}{v}$$

Then we figured out the next steps. We wrote the fraction with a negative exponent:

$$y = u \times v^{-1}$$

Use the product rule:

$$\frac{dy}{dx} = u\left(-v^{-2}\frac{dv}{dx}\right) + v^{-1}\frac{du}{dx}$$

$$\frac{dy}{dx} = v^{-1}\frac{du}{dx} - u\left(v^{-2}\frac{dv}{dx}\right)$$

Writing the negative exponents as fractions:

$$\frac{dy}{dx} = \frac{\frac{du}{dx}}{v} - \frac{u\frac{dv}{dx}}{v^2}$$

We multiplied the first fraction by v/v so we could have a common denominator, and we came up with our result:

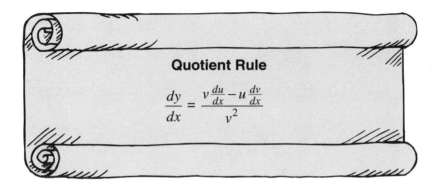

Quotient Rule

$$\frac{dy}{dx} = \frac{v\frac{du}{dx} - u\frac{dv}{dx}}{v^2}$$

The professor called this the quotient rule, but a heated argument broke out because Recordis insisted we didn't need to remember another rule because we could always use the product rule whenever we needed to find the derivative of u divided by v, since we could write u/v as $u \times v^{-1}$. We never did reach agreement on this issue.

WORKSHEET

Find the derivative:

1. $y = \dfrac{a+x}{b+x}$

2. $y = \dfrac{a+bx}{h+kx}$

3. $y = \dfrac{(a+bx)^m}{(h+kx)^n}$

4. $y = \dfrac{x}{\sqrt{1-x^2}}$

"It is amazing the uses we are finding for this subject of calculus. And I bet with these three new rules we can find the derivative of almost anything." Recordis was still talking cheerfully about plans for his house when we all went out to breakfast.

NOTE TO CHAPTER 4

The assumption that has been made in the demonstration of the chain rule in this chapter is that Δu does not equal zero. Since $\Delta u = g(x + \Delta x) - g(x)$, this condition means that $g(x + \Delta x)$ must not equal $g(x)$ for small values of Δx. This condition holds true for any function without any flat spots. For example, this demonstration of the chain rule will not work for the function in Figure 4–7. We later were able to establish, however, that the chain rule is indeed valid for all functions, whether or not they have any flat spots.

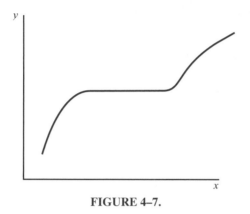

FIGURE 4–7.

Exercises

1. Use the product rule to find the derivative of the function $f(t) = (2t - 5)(3t + 4)$. (Does the answer agree with the result for Exercises 2–6?)

Find dy/dx for:

2. $y = (2x + 5)/(3x - 1)$

3. $y = 10x/(4x - 3)$

4. $y = (ax + b)/(cx + d)$

5. Use the product rule to find the derivative of $y = (x^2)(x^3)$.

6. Derive the triple product rule (i.e., find y' for $y = u(x)\,v(x)\,w(x)$).

Use the triple product rule to find the derivative of $y = x^3$.

7. Derive the quotient rule. (Find y' for $y = u(x)/v(x)$.)

8. Find dy/du for $y = u^{3/2}$. Find du/dx for $u = x^2 + 3$. Then find dy/dx for $y = (x^2 + 3)^{3/2}$.

9. Consider $y = \sqrt{1+x^2}$. Let $u = 1 + x^2$. Find dy/du. Find du/dx. Find dy/dx.

10. Consider $y = (3 + 4x)^{3/2}$. Let $u = 3 + 4x$. What is dy/du? What is du/dx? What is dy/dx?

11. Consider $y = \sqrt{ax^2 + bx + c}$. Let $u = ax^2 + bx + c$. What is dy/du? What is du/dx? What is dy/dx?

12. Consider $y = \sqrt{1+(x^2+4)^{-1}}$. Let $u = x^2 + 4$. Let $v = 1 + (x^2 + 4)^{-1}$. Find dy/dv. Find dv/du. Find du/dx. Find dy/dx.

13. Verify the power rule for $n = \frac{1}{2}$. Use the definition of the derivative.

14. Use the definition of the derivative to find y' for $y = 1/x$.

15. Derive a generalized power rule: for $u = u(x)$, find dy/dx for $y = u^n$.

Find dy/dx for:

16. $y = \sqrt{4+2x}$

17. $y = (4 + 3x)^{3.3}$

18. $y = \sqrt{3x^3 + 2x^2 + x}$

19. $y = 3/\sqrt{x^2 - 1}$

20. $y = \sqrt[3]{x}$

21. $y = \sqrt[3]{x^2 + 4}$

22. $y = \sqrt{1 + 1/x}$

23. $y = \sqrt{x + 1/x}$

24. $y = (x^2 + 4)^{-1}$

25. Find the second derivative of the function $A(w) = w\sqrt{r^2 - w^2}$ at the point where $w = r/\sqrt{2}$. Is the curve concave up or concave down at that point? Is Recordis building a house that will have maximum area or a house that will have minimum area?

26. (a) Use the implicit derivative method to find dy/dx for the circle $x^2 + y^2 = r^2$.
    (b) Solve for y as an explicit function of x for the semicircle where y is positive.
    (c) Using the chain rule and the power rule, find dy/dx for the function in (b). Does the result agree with that obtained in (a)?

27. Find dy/dx for the hyperbola $(y - 1)^2/9 - (x - 3)^2/4 = 1$.

28. Create a spreadsheet that draws the graph of a function together with its derivative. Have the computer approximate the derivative by finding the slopes of pairs of points that are close together.

Derivatives of
Trigonometric Functions

In the next few days I saw more of the pretty countryside of Carmorra. Recordis had announced he was starting construction of his house, and he promised to invite us all over for a visit when it was finished.

While I was walking along the balcony of the castle the next afternoon, I noticed Trigonometeris leaning against the railing. He apparently hadn't heard me the first time I called to him, so I became concerned and ran up to him. "Are you all right?" I asked.

He jumped and looked around at me. "Oh," he said. "Sure." His voice lacked its usual cheerful quality.

"Something's the matter," I insisted. "Are you sick?"

"Nothing really," he said. "Just a feeling of uselessness."

"Uselessness?"

"I suppose I can't stand in the way of progress," he said. "I guess this is what happens when new things are invented—the old inventors get left behind. I always wondered what happened to the inventors of drag-sleds after the wheel was invented, and now I know."

"What are you talking about?" I asked.

"My work was crucial when we were working on trigonometry," he said. "When we were discovering the new trigonometric identities, I was called upon to reach the limits of my skill. Trigonometry was the in-thing then. Now we're inventing all this new material, and trig just doesn't seem useful any more."

"Don't say that!" I exclaimed. "There's no telling what we'll run into if we keep going like this. Trigonometry may turn out to be one of the most important fields to investigate with the help of calculus."

Trigonometeris still didn't look very cheery, so I asked him to show me the Royal Triangles. He cheered up a bit and took me to the treasure room of the palace where the official triangles were kept. The triangles were intricate machines for accurately calculating the sine, cosine, and tangent of any angle. I was amazed at the accuracy they were capable of, but before I could look at the mechanics very closely we heard a desperate cry that echoed throughout the palace.

"Help! Everybody! We're in real trouble now!" Recordis was running around, trying to get the attention of everybody in the palace. We had no time to ask him any questions before we had followed him to the outskirts of town to a large brick building surrounded by a yard labeled "McCockle Chicken Farm."

"There!" Recordis cried, pointing to a horrible machine in the middle of the farmyard. A giant scaffolding had been erected, and hanging from it was a heavy spring attached to a large weight. The weight was jumping up and down because of the pull of the spring, and every time it came down all the chickens cackled in terror and ran about the yard.

"The gremlin put it there!" Recordis said, slowly getting his breath back. "He's trying to sabotage the chicken farm!"

By this time everybody from the palace had gathered around the yard (except Pal, who was terrified by the spring and was hiding behind a nearby mountain). "The gremlin put locks all around the spring so we can't take it apart," Recordis added.

"We've got to find some way to stop it," the king said. "Before long the chickens will have been scattered all over the place."

Builder, whom I had not seen since the day I arrived, looked thoughtfully at the block. "I could build a gremlin-block-stopping machine," he said. "All I need to know is how fast the block is moving at a given time."

"We can take a derivative!" the professor cried. "That will tell us how fast it is going."

"Take a derivative of what?" Recordis asked. "We can't tell how fast it is going before we know what a graph of its motion looks like."

"It's very simple," the king said. "We need to find some way to make a graph of its motion, then we find a function of time that matches the graph, and then we find the derivative of that function to find the speed of the block."

"I don't know any function that goes up and down like that!" Recordis protested. "And, besides, if you think I'm going to climb up on that block with my watch to time its position . . ."

"There must be another way to graph its motion," the professor said.

"We could use a camera," I suggested. I explained my idea, and we all worked quickly to set the camera up. Mr. McCockle was shouting at us to hurry because the chickens were quite hysterical by now.

"When the pictures are developed we will make life-sized prints of them, and then we can measure the position of the block," the professor said. For some reason there is very little friction in Carmorra, and the spring was still bouncing up and down at exactly the same rate while we took the pictures, had them developed, and laid them out on the floor of the Main Conference Room. (See Figure 5–1.)

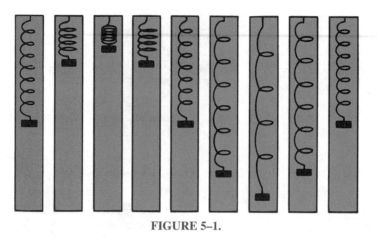

FIGURE 5–1.

"Make sure you put them in the right order," the professor advised. "It would surely confuse matters if you didn't."

"That looks familiar," Trigonometeris murmured, but nobody paid him any attention.

"Let's have Igor draw the graph of a function that fits these points," the professor said. (Figure 5–2.)

"I can't think of any algebraic function that looks like that," the king stated.

"I know what it is," Trigonometeris said softly. Finally someone paid attention to him. "We can use a sine function."

"You can't use a sine function!" Recordis exclaimed. "We don't have any angles! You can use a sine function only if you have an angle."

"But we made a graph of what a sine wave looks like," Trigonometeris said. "Igor, draw the graph of a sine wave for me." (Figure 5–3.)

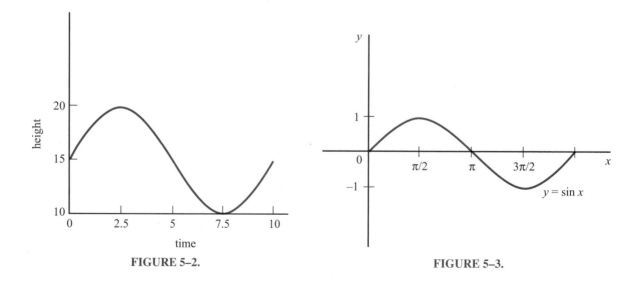

FIGURE 5–2.

FIGURE 5–3.

"That has exactly the same shape as the motion of the block," Trigonometeris continued. "We can use this function."

$$(\text{height}) = y = 15 + 5\sin\left(\frac{2\pi t}{10}\right)$$

"It seems to match exactly," the professor agreed. "Now all we have to do is find dy/dt if $y = 15 + 5\sin(2\pi t/10)$."

"Let's first find the derivative of $y = \sin x$," I suggested. "Then we can easily use the chain rule, the product rule, and the sum rule to find the derivative of the actual function that we have."

"We might as well start at the beginning," the professor said. "Igor, show us the definition of the derivative again."

$$\frac{dy}{dx} = \lim_{\Delta x \to 0} \frac{f(x+\Delta x) - f(x)}{\Delta x}$$

If $f(x) = \sin x$ then

$$\frac{dy}{dx} = \lim_{\Delta x \to 0} \frac{\sin(x+\Delta x) - \sin x}{\Delta x}$$

"Now we're stuck," Recordis said.

"I have a formula for the sine of the sum of two angles," Trigonometeris informed us triumphantly. He went to his book and read off the following formula:

$$\sin(A + B) = \sin A \cos B + \cos A \sin B$$

"We can put that formula in the equation for the derivative," Trig suggested.

$$\frac{dy}{dx} = \lim_{\Delta x \to 0} \frac{\sin x \cos \Delta x + \cos x \sin \Delta x - \sin x}{\Delta x}$$

"And we can separate the terms in the numerator," he added.

$$\frac{dy}{dx} = \lim_{\Delta x \to 0} \sin x \left(\frac{\cos \Delta x - 1}{\Delta x}\right) + \lim_{\Delta x \to 0} \cos x \frac{\sin \Delta x}{\Delta x}$$

$$\frac{dy}{dx} = \sin x \lim_{\Delta x \to 0} \frac{\cos \Delta x - 1}{\Delta x} + \cos x \lim_{\Delta x \to 0} \frac{\sin \Delta x}{\Delta x}$$

"Now we have to figure out what these limits are," Trigonometeris said. "Let's try this one: $\lim_{\theta \to 0} (\sin \theta)/\theta$." ($\theta$ is the Greek letter theta, which is often used in trigonometry to represent angles.)

"I think the limit will be zero," the professor said; "sin θ will approach zero, and that is on the top of the fraction. Whenever the top of the fraction goes to zero, the value of the fraction is zero."

"I think it will be infinity," Recordis objected. "Remember that we have a θ on the bottom. Whenever the bottom of the fraction goes to zero, the value of the fraction goes to infinity."

"Are we measuring θ in radian or degree measure?" the king asked.

"We had better use radian measure," Trigonometeris answered. "If I remember correctly, we found that degree measure was more convenient if we were actually building something or measuring a real angle, but for mathematical purposes we always used radian measure."

"Then I think the value of the limit will be 1," the king said. "Igor, draw a picture of a circle with radius 1." (Figure 5–4.)

"Remember that this dotted line is equal to sin θ, and this deep black curve is equal to θ. It looks as though they come closer and closer together as θ approaches zero."

"Fascinating," the professor said. "I bet you can't prove it, though."

I ventured a suggestion. "Suppose we draw another line." (Figure 5–5.)

"It looks as though the shaded area is smaller than the striped area, which is smaller than the area of the whole big triangle."

"We can write that as an inequality," the professor said.

$$(\text{area shaded}) < (\text{area striped}) < (\text{total area})$$

"Two of those areas are triangles, and one area is a sector of a circle," the professor went on. "We should be able to figure out what all those areas are. First, we'll label all the points. Then the area of the shaded triangle is $\frac{1}{2}(O - C)(C - D)$, and the area of the whole triangle is $\frac{1}{2}(O - A)(A - B)$." [$(O - C)$ is the distance from point O to point C.]

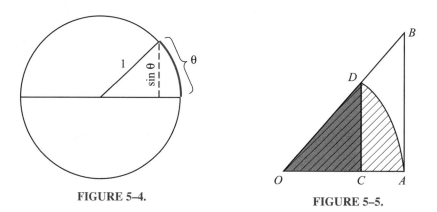

FIGURE 5–4.   **FIGURE 5–5.**

"We derived a formula for the area of a sector of a circle," Recordis said.

$$(\text{area of sector of circle}) = \tfrac{1}{2}\theta\, r^2$$

where θ is the angle enclosed by the sector and r is the radius of the circle.

"So that would make the striped area equal to $\tfrac{1}{2}\theta(1)^2 = \tfrac{1}{2}\theta$."

We then rewrote the inequality:

$$\tfrac{1}{2}(O - C)(C - D) < \tfrac{1}{2}\theta < \tfrac{1}{2}(O - A)(A - B)$$

"Now we can substitute trigonometric values in the expression," Trigonometeris said.

$$(O - C) = \cos\theta$$
$$(C - D) = \sin\theta$$
$$(O - A) = 1$$
$$(A - B) = \tan\theta$$

Igor displayed the rewritten inequality:

$$\tfrac{1}{2}\cos\theta\,\sin\theta < \tfrac{1}{2}\theta < \tfrac{1}{2}\tan\theta$$

"We can multiply everything by 2," the king noted. "That's legal with an inequality, as long as we multiply by a positive number."

$$\cos\theta\,\sin\theta < \theta < \tan\theta$$

"Remember that $\tan\theta = \sin\theta/\cos\theta$," Trigonometeris said.

$$\cos\theta\,\sin\theta < \theta < \frac{\sin\theta}{\cos\theta}$$

"We can divide everything by $\sin\theta$," the king suggested.

$$\cos\theta < \frac{\theta}{\sin\theta} < \frac{1}{\cos\theta}$$

Now we can find the limit!" Trigonometeris said. "As θ goes to 0, $\cos\theta$ goes to 1. That means $1/\cos\theta$ goes to 1, also. And that means that $\lim_{\theta\to 0}(\theta/\sin\theta)$ gets squeezed in between two numbers that are both going to 1!"

"That must be a tight fit," Recordis remarked.

"And if $\theta/\sin\theta$ approaches 1, then $\sin\theta/\theta$ must approach 1, too. That means the king was right."

$$\lim_{\theta\to0}\frac{\sin\theta}{\theta} = 1$$

Igor displayed the formula we had for the derivative of the sine function:

$$\frac{dy}{dx} = \sin x \lim_{\Delta x\to0}\frac{\cos\Delta x - 1}{\Delta x} + \cos x(1)$$

"Now we're stuck," Recordis mourned.

"No, all we have to do is evaluate that other limit," the king said.

That turned out to be tedious but not especially difficult, so I just wrote down the steps we took.

$$\lim_{\theta\to0}\frac{1-\cos\theta}{\theta} = \lim_{\theta\to0}\frac{1-\cos\theta}{\theta}\frac{1+\cos\theta}{1+\cos\theta}$$

$$= \lim_{\theta\to0}\frac{1-\cos^2\theta}{\theta(1+\cos\theta)}$$

$$= \lim_{\theta\to0}\frac{\sin^2\theta}{\theta(1+\cos\theta)}$$

$$= \lim_{\theta\to0}\frac{\sin\theta}{\theta}\lim_{\theta\to0}\frac{\sin\theta}{1+\cos\theta}$$

$$\lim_{\theta\to0}\frac{1-\cos\theta}{\theta} = \lim_{\theta\to0}\frac{\sin\theta}{1+\cos\theta}$$

As θ goes to 0, $\cos\theta$ goes to 1:

$$\lim_{\theta\to0}\frac{1-\cos\theta}{\theta} = \lim_{\theta\to0}\frac{\sin\theta}{2}$$

$$= 0$$

"Now we can solve the problem!" Trigonometeris rejoiced.

$$y = \sin x$$

$$\frac{dy}{dx} = \sin x \lim_{\Delta x\to0}\frac{\cos\Delta x - 1}{\Delta x} + \cos x \lim_{\Delta x\to0}\frac{\sin\Delta x}{\Delta x}$$

$$= \cos x$$

"Amazing!" the king exclaimed. "The cosine function is the derivative of the sine function! Does it make sense?"

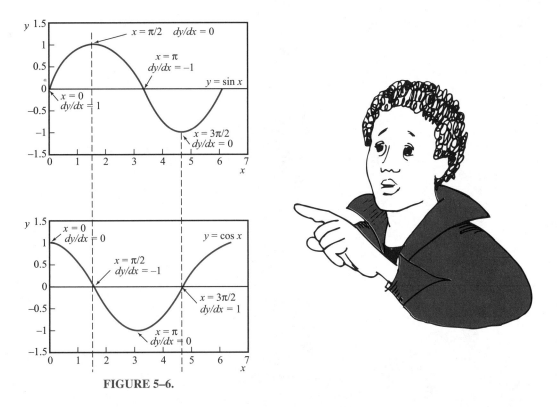

FIGURE 5–6.

"When $x = 0$ the sine curve is sloping gently upward," Trigonometeris said. (See Figure 5–6.) "Then $\cos x = 1$, so the slope of the curve should be 1. That looks reasonable."

"At $x = \pi/2$ the sine curve has a horizontal tangent," the professor said.

"And, sure enough, $\cos(\pi/2)$ is zero!" the king said.

We checked a few more points along the curve, and this surprisingly simple relationship seemed to check out all along the curve. We could see that the curve $y = \sin x$ is downward sloping when $\cos x$ is negative.

"I wonder what the derivative of $\cos x$ is," the king wondered.

"It's $\sin x$, I bet," Recordis said.

"Let's check that," the king said. (Figure 5–6.)

"The slope of the curve can't be $\sin x$," the professor objected. "You can tell that just by looking at these few points. When cosine has a negative slope, we can see that sine is positive, and when cosine has a positive slope, the sine is negative."

"It could be $-\sin x$," Recordis guessed.

"I know a quick way to check," Trigonometeris said. "Instead of trying to find the derivative of $y = \cos x$ directly, we could take advantage of the fact that $\cos x = \sin(\pi/2 - x)$. With the chain rule we should be able to find the derivative of that function easily."

$$y = \cos x = \sin\left(\frac{\pi}{2} - x\right)$$

Let $u = \pi/2 - x$. Then

$$y = \sin u$$

$$\frac{dy}{dx} = \frac{dy}{du}\frac{du}{dx}$$

"We know that $du/dx = -1$," Recordis said.
"And $dy/du = \cos u$, since that is what we just did," the king added.

$$\frac{dy}{dx} = (\cos u)(-1) = -\cos\left(\frac{\pi}{2} - x\right) = -\sin x$$

"Recordis, you were right!" the professor exclaimed.
"I was? I mean—of course, I was right." Recordis beamed happily.
We made a table of these rules:

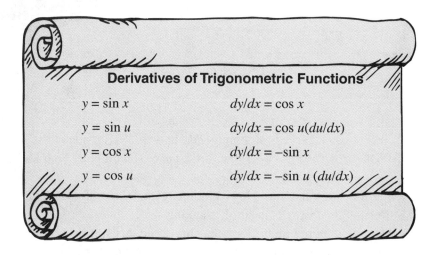

Derivatives of Trigonometric Functions

$y = \sin x$	$dy/dx = \cos x$
$y = \sin u$	$dy/dx = \cos u(du/dx)$
$y = \cos x$	$dy/dx = -\sin x$
$y = \cos u$	$dy/dx = -\sin u\,(du/dx)$

"Now we can solve the original problem and save those poor chickens," Trigonometeris said.

$$y = 15 + 5\sin\left(\frac{2\pi t}{10}\right)$$

$$= \frac{d}{dx}15 + \frac{d}{dx}5\sin\left(\frac{2\pi t}{10}\right) = 0 + 5 \times \frac{2\pi}{10}\cos\left(\frac{2\pi t}{10}\right)$$

$$\frac{dy}{dt} = \pi\cos\left(\frac{2\pi t}{10}\right)$$

Recordis quickly wrote the answer down and handed it to Pal, who had gotten over his fear of the spring. He ran to find Builder so Builder could begin work on the block-stopping machine.

"Let's find the derivative of $y = \tan x$," Trigonometeris said. "Then we can have a complete table."

"We can use the product rule," the professor pointed out.

$$y = \tan x$$

$$= (\sin x)(\cos x)^{-1}$$

$$\frac{dy}{dx} = \sin x \ \frac{d}{dx}(\cos x)^{-1} + (\cos x)^{-1} \ \frac{d}{dx}\sin x$$

$$= \sin x \ \frac{d}{dx}(\cos x)^{-1} + \frac{\cos x}{\cos x}$$

$$= (\sin x)(-1)(\cos x)^{-2} \frac{d}{dx}\cos x + 1$$

$$= (\sin x)(-1)(\cos x)^{-2}(-\sin x) + 1$$

$$\frac{dy}{dx} = \frac{\sin^2 x}{\cos^2 x} + 1$$

"That's not as complicated as I thought it would be," Recordis said.

"We can rewrite it, using $\tan x = (\sin x)/(\cos x)$," Trigonometeris told him.

$$\frac{dy}{dx} = \tan^2 x + 1$$

"I have another one," Trigonometeris said. "We derived this trigonometric identity."

$$\tan^2 x + 1 = \sec^2 x$$

We rewrote the derivative:

$$y = \tan x$$

$$\frac{dy}{dx} = \sec^2 x$$

We decided to find the derivatives for the rest of the trigonometric functions. For cotangent we merely inserted ctn $x = \tan (\pi/2 - x)$, and came up with $dy/dx = -\csc^2 x$. We puzzled over the derivative of the secant function for a minute before the professor realized we had already solved that problem since $\sec x = (\cos x)^{-1}$. We decided to write the answer as:

$$y = \sec x = (\cos x)^{-1}$$

$$y' = (-1)(\cos x)^{-2} (-\sin x)$$

$$y' = \frac{(\sin x)\times 1}{(\cos x)\times(\cos x)}$$

$$y' = \tan x \sec x$$

Igor displayed the results of our work for the day:

$y = \sin u$	$dy/dx = (\cos u)\, du/dx$
$y = \cos u$	$dy/dx = (-\sin u)\, du/dx$
$y = \tan u$	$dy/dx = (\sec^2 u)\, du/dx$
$y = \text{ctn } u$	$dy/dx = (-\csc^2 u)\, du/dx$
$y = \sec u$	$dy/dx = (\tan u \sec u)\, du/dx$
$y = \csc u$	$dy/dx = -(\text{ctn } u \csc u)\, du/dx$

"Note the negative sign in front of the derivatives of the 'co' functions (cosine, cotangent, and cosecant)," Trigonometeris commented, trying to make it easier for us to remember the results.

WORKSHEET

Use the product rule to find the derivatives of the following functions of x (treating all other letters as constants).

	$f(x)$	u	dv/dx	v	du/dx	$f'(x)$
1.	$\sin x \cos x$					
2.	$\sin (mx) \cos (nx)$					

Use the power rule to find the derivatives of the following functions.

	$f(x)$	$f'(x)$
3.	$\sin^2 x$	
4.	$\cos^2 x$	
5.	$\sin^2 x + \cos^2 x$	
6.	$\sin (x^n)$	
7.	$(\sin x)^n$	
8.	$\sec^2 x$	
9.	$\tan^2 x$	
10.	$\sec^2 x - \tan^2 x$	

We decided to return to the McCockle Chicken Farm and watch Builder destroy the gremlin's terrible machine. As we were walking out of the room, Trigonometeris started talking excitedly. "I never realized how fascinating this whole subject—this calculus—could be. We can make some use of these springs as toys for children, provided, of course, that we make them small enough so they won't scare any chickens. And now we can make sine-curve-shaped slides, and we will know what the slope is at any point! I'll have to start designing a new playground right away!"

By the time we reached the farm we found that Builder had indeed neutralized the deadly spring, and the chickens were all happily gathered about Mr. McCockle. And I was glad that this little adventure had provided new life for the career of someone as kind and polite as Alexanderman Trigonometeris.

Exercises

Find the derivatives of the following functions:

1. $y = \sin x^2$

2. $y = \sin^2 x$

3. $y = 1/(\sin x)$

4. $y = x \sin x$

5. $y = (\sin x)(\cos x)$

6. (a) Use the formula for sin $(A + B)$ to find y' for $y = \sin (x^2 + x)$. (b) Use the chain rule to find the derivative of the function in (a).

7. If A is an angle expressed in degree measure, find y' for $y = \sin A$.

8. Find the values of x where $y = \sin x$ has horizontal tangents. Which of these points are minima? Which are maxima? In what intervals is the curve concave upward? Concave downward?

9. Find the acceleration of the block in the gremlin's machine:
   $y = 15 + 5 \sin (2\pi t/10)$.

10. Trigonometeris knows that sin 30° = sin $(\pi/6) = \frac{1}{2}$. Find the equation of the tangent line to the curve $y = \sin x$ at the point $(\pi/6, \frac{1}{2})$. What is the radian equivalent for 32°? Use the tangent line derived above to find an approximate value for sin 32°.

11. Use the power rule and the chain rule to find the derivative $(dy/d\theta)$ of $y = (1 + \tan^2 \theta)^{1/2}$. What is a simpler way to express this function?

12. Consider the curve $y = \tan \theta$ in the interval $\theta = -\pi/2$ to $\theta = \pi/2$. Where does this curve have horizontal tangents? Where is it concave up? Where is it concave down?

13. Calculate the value of the function $f(x) = (\sin x)/x$ for each of the following values of x: $0.785, 0.5, 0.3, 0.1, 0.05$.

14. A block with mass $m = 2$ kg attached to a spring behaves according to the equation $-kx = m(d^2x/dt^2)$, where k is known as the *spring constant* for this particular spring (measured in kilogram-meters per second squared). The motion of the block is given by $x = 0.8 \sin 3t$. Find the value of k.

15. Imagine that the level of the tide in a bay follows a perfect sine curve when graphed as a function of time. When will the flow of the water into the bay be the fastest?

16. The voltage in an AC (alternating current) circuit is described by the following function of time: $V = A \sin(\omega t)$. (The symbol ω is the Greek lowercase letter omega. A and ω are constant.) (a) Calculate dV/dt. (b) What must the value of ω be if the frequency of the voltage is 60 hertz (60 cycles per second)?

17. A pendulum consists of a mass held up by a string. Let x represent the angle that the string makes with the vertical. At time t, the value of x is $A \sin(\omega t + h)$. (A, h, and ω are constants.) Find dA/dt. When is dA/dt the greatest?

18. Consider an object with mass m attached to a spring whose position at time t is given by this equation: $x = A \sin(\omega t)$. (A and ω are constant.) (a) The kinetic energy of a moving object is $\frac{1}{2}mv^2$, where v is the velocity ($v = dx/dt$). Calculate the kinetic energy as a function of time for this object. (b) Calculate the derivative of the kinetic energy with respect to time. (c) The potential energy for this object is given by the formula $\frac{1}{2}m\,\omega^2 x^2$. Write the potential energy as a function of time. (d) Calculate the derivative of the potential energy with respect to time. (e) The total energy is the sum of the kinetic energy and the potential energy. Write the total energy as a function of time. (f) Calculate the derivative of the total energy with respect to time. What does your result mean?

Optimum Values
and Related Rates

Carmorra Consolidated
Differentiating, Incorporated

The king had a new concern when we met in the Main Conference Room a few days later. "We need to start planning some way for us to make the results of everything we've done available to the people. It doesn't seem right that we should use calculus only to solve our own problems."

"I know what we can do!" the professor said. "We can make lots of money with these discoveries!"

"That's it!" Recordis cried. "We'll start a company that will solve calculus problems for people! I bet there are lots of calculus problems floating around that people haven't recognized before because they haven't known what calculus is. If we charge people to solve these problems, we should be able to make a nice profit," he said, with dollar signs flashing in his eyes.

"Let's think of a catchy name for our company," the professor said before the king had a chance to protest. "How about Carmorra Consolidated Derivative-Taking Business?"

"We need a shorter name," Recordis objected. "For instance, when we take the derivative of something, we have to say we're taking the derivative of it. That's three words! We need one word that stands for the process of taking a derivative."

Everyone turned to me. Another name popped into my head, so I suggested the word *differentiate*.

"That sounds good," the professor approved. "We'll say that when we differentiate a function we're taking the derivative of it."

"I like it because it's only one word," Recordis said.

"And if we're taking the derivative with respect to x, we can say that we're differentiating with respect to x," the professor added.

They went rushing on with their plans, although the king was uneasy about the whole idea.

"Now we need to have Igor draw an advertising brochure for us, which we'll have distributed all over the country," Recordis continued. A short while later they had helped Igor put together an artistic brochure telling about Carmorra Consolidated Differentiating, Incorporated.

Our brochures were printed and distributed all over the country. Builder constructed a small stand just outside the palace with a prominent sign bearing the company name and emblem. We worked out a schedule so we could trade off our various times for running the company, but at first we were so interested that we all crowded around awaiting our first customer.

"Remember the list of prices I made up," the professor whispered.

At first people weren't used to the business, and the first day there were no customers. We were all getting nervous during the second day, but in the afternoon our first customer finally showed up: our friend Farmer Floran. "Can you help me design a rectangular enclosure for my animals? I want an enclosure with the maximum possible area for a given amount of fencing."

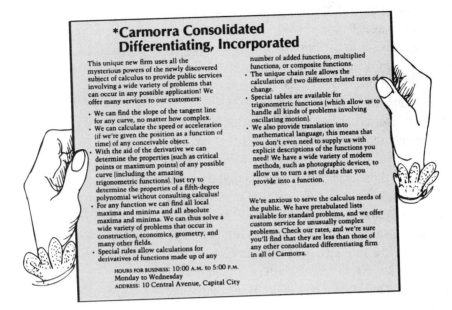

*Carmorra Consolidated Differentiating, Incorporated

This unique new firm uses all the mysterious powers of the newly discovered subject of calculus to provide public services involving a wide variety of problems that can occur in any possible application! We offer many services to our customers:

- We can find the slope of the tangent line for any curve, no matter how complex.
- We can calculate the speed or acceleration (if we're given the position as a function of time) of any conceivable object.
- With the aid of the derivative we can determine the properties (such as critical points or maximum points) of any possible curve (including the amazing trigonometric functions). Just try to determine the properties of a fifth-degree polynomial without consulting calculus!
- For any function we can find all local maxima and minima and all absolute maxima and minima. We can thus solve a wide variety of problems that occur in construction, economics, geometry, and many other fields.
- Special rules allow calculations for derivatives of functions made up of any

number of added functions, multiplied functions, or composite functions.
- The unique chain rule allows the calculation of two different related rates of change.
- Special tables are available for trigonometric functions (which allow us to handle all kinds of problems involving oscillating motion).
- We also provide translation into mathematical language; this means that you don't even need to supply us with explicit descriptions of the functions you need! We have a wide variety of modern methods, such as photographic devices, to allow us to turn a set of data that you provide into a function.

We're anxious to serve the calculus needs of the public. We have pretabulated lists available for standard problems, and we offer custom service for unusually complex problems. Check our rates, and we're sure you'll find that they are less than those of any other consolidated differentiating firm in all of Carmorra.

HOURS FOR BUSINESS: 10:00 A.M. to 5:00 P.M. Monday to Wednesday
ADDRESS: 10 Central Avenue, Capital City

We called the length x and the width y, so we would have the area = $a = xy$. See Figure 6-1. The total fence length equals the perimeter of the rectangle: $p = 2x + 2y$, where p must be a constant.

"Unfortunately x and y are both variables," Recordis whispered worriedly to us. "We can't find the maximum value for a since it depends on two variables!"

"But we can't choose whatever we want for x, and then choose whatever we want for y," the king noticed.

FIGURE 6–1.

"We're constrained by the equation $p = 2x + 2y$. This means that once we make our choice for x, we're forced to choose a particular value of y:

$$p = 2x + 2y$$
$$2y = p - 2x$$
$$y = 0.5p - x$$

"If we put that into the formula for area, then we'll have only one variable left!" the professor realized.

$$a = xy = x(0.5p - x) = 0.5px - x^2$$

$$\frac{da}{dx} = 0.5p - 2x$$

$$0 = 0.5p - 2x$$
$$2x = 0.5p$$
$$x = 0.25p$$

$$y = 0.5p - x = 0.5p - 0.25p = (0.5 - 0.25)p = 0.25p$$

"x and y end up with the same value. That means we have a square," Recordis said. "How simple."

"That makes sense," Floran said. "We wouldn't have much area if we made a long narrow enclosure (Figure 6–2). It makes sense that a square shape maximizes the area for a given perimeter."

The professor described the procedure. "We can apply this method to any problem where we are trying to maximize something that is a function of two variables, but there is another equation that links the two variables. In our case, our maximizing equation is $a = xy$, and our linking equation is $p = 2x + 2y$, with p being a constant. The two variables are not independent, because the linking equation forces the choice for the second variable after the first variable has been chosen. The general procedure is: use the linking equation to solve for one of the variables; then insert that formula into the maximizing formula. Now that you have a formula that depends on only one variable, use standard procedure to find the derivative, set it to zero, and solve." We called this type of problem a fence problem because we first encountered it in the case of determining the optimal-shaped fence enclosure. (In Chapter 16, we investigate what to do if we had a problem with two variables that really were independent.)

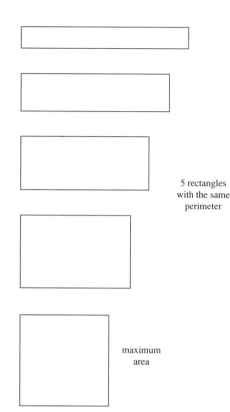

5 rectangles with the same perimeter

maximum area

FIGURE 6–2.

Floran was satisfied. We decided not to charge him in gratitude for helping me when I first washed up on the beach of Carmorra. The next morning another customer arrived.

"I hear you folks can solve optimum-value problems," the man drawled.

("I know him!" Recordis whispered to me. "That's Mr. Carrier. He's the man who runs the company where I buy my supply boxes for my pens and pencils.")

"Yes, we can," the professor said, glad that a customer had arrived while it was her turn to be in charge of the business.

"I need to make a box," the box maker said. "I have 2 square meters of wood to use. The box needs to have square ends and an open top. (See Figure 6–3.) Can you tell me the dimensions of the box that will allow me to get the maximum possible volume for my 2 square meters of material?"

"Sure," the professor answered. "First, we'll set up the variable names we need. We'll let x equal the length of one of the square edges, and y equal the length of one rectangular edge. Then the volume will be (volume) $= V = x^2y$. We can also figure out the sur-

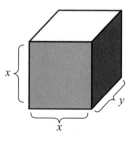

FIGURE 6–3.

face area, which will be the area of the two square ends plus the area of the three rectangular sides: (surface area) = $A = 2x^2 + 3xy$."

The professor was getting nervous, so the king helped her figure out what to do next. "We know that the surface area equals 2, since that is what you told us."

$$2 = 2x^2 + 3xy$$

"We can solve for y algebraically!" Recordis said.

$$2 - 2x^2 = 3xy$$

$$y = \frac{2}{3x} - \frac{2x}{3}$$

"We can put that expression for y back into the equation for the volume," the king pointed out.

$$V = x^2y = x^2\left(\frac{2}{3x} - \frac{2x}{3}\right) = \frac{2x}{3} - \frac{2x^3}{3}$$

"Now we have exactly what we want," the king continued. "The volume is expressed as a function of one variable (in this case x). All we need to do is find the derivative and set it equal to zero to solve for the optimum value of x."

$$V = \frac{2x}{3} - \frac{2x^3}{3}$$

$$\frac{dV}{dx} = \frac{2}{3} - 2x^2$$

$$0 = \frac{2}{3} - 2x^2$$

$$\frac{2}{3} = 2x^2$$

$$x^2 = \frac{1}{3}$$

$$x = \frac{1}{\sqrt{3}}$$

"Now we can easily figure out y," the professor said, regaining her composure. After some calculations, she told the customer, "You should make your box with edges 0.577 meter by 0.577 meter by 0.769 meter. That will give you a total volume of 0.256 cubic meters."

"Not bad," Mr. Carrier said. "A box that big will hold all I need."

Mr. Carrier was so satisfied with the solution to that problem that he asked for help with another problem: finding the shape of the optimal tin can. He asked us to design a cylindrical can of radius r and height h. The total volume (V) is:

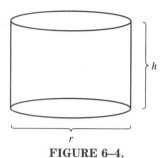

$$V = \pi r^2 h$$

See Figure 6–4.

FIGURE 6–4.

The amount of material (M) needed for the can depends on the area of the sides, top, and bottom:

$$M = 2\pi rh + 2\pi r^2$$

Our quest was to derive a formula for r that maximizes the volume V for a fixed amount of material M. (Note: M and π are constants.) The volume (V) depends on the two variables r and h, but they are linked by the equation for M, so we can use the fence problem procedure.

We used the second equation to solve for h (after deciding it would be much easier to solve for h instead of r):

$$M = 2\pi rh + 2\pi r^2$$
$$M - 2\pi r^2 = 2\pi rh$$
$$\frac{M - 2\pi r^2}{2\pi r} = h$$

Insert the formula for h into the formula for V:

$$V = \pi r^2 h$$
$$V = \pi r^2 \left(\frac{M - 2\pi r^2}{2\pi r} \right)$$
$$V = r \left(\frac{M - 2\pi r^2}{2} \right)$$
$$V = \left(\frac{Mr - 2\pi r^3}{2} \right)$$
$$V = 0.5Mr - \pi r^3$$

Find the derivative:

$$\frac{dV}{dr} = 0.5M - 3\pi r^2$$

Set the derivative to zero:

$$0 = 0.5M - 3\pi r^2$$

Solve for r:

$$\frac{M}{2} = 3\pi r^2$$

$$\frac{M}{6\pi} = r^2$$

$$r = \sqrt{\frac{M}{6\pi}}$$

We were so excited about solving these problems that we forget to hand Mr. Carrier a bill. So, the next day we went to the box shop, but we found Mr. Carrier very frazzled. "Sorry I don't have any boxes in stock for you. I'll have some for you tomorrow."

"Why don't you keep more inventory in stock?" the king asked.

"I tried that!" Mr. Carrier sputtered. "At first I made a large order of boxes at once, but I found it was expensive to store them all. Then I tried ordering a smaller number at a time, but it was expensive to keep ordering new batches in order to keep them in stock."

"Can't you compromise between the two extremes?" I asked.

We looked at the details of the problem. We let R represent the amount of boxes sold per year. Boxes were a very basic item that sold steadily. We let Q represent the quantity ordered at one time. If we timed it right, the next order would arrive just as soon as the inventory hit zero. After jumping up to Q, the inventory level would gradually decline until hitting zero just before it was time for the next order (see Figure 6–5). This meant the average inventory was $Q/2$, and the cost of holding inventory was $Q/2hc$, where hc stood for holding cost. Every time another order was made, there was a cost of A (standing for administrative costs). Since the number of orders per year would be R/Q, the total administrative cost during the year would be AR/Q. We put the two costs together into one equation:

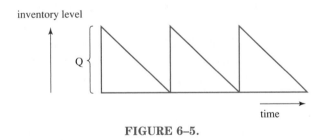

FIGURE 6–5.

$$TC = \frac{Q}{2} \times hc + \frac{AR}{Q}$$

R, c, h, and A represent known constants, so the total cost depends on only the one variable we get to choose: Q.

We took the derivative with respect to Q:

$$TC = \frac{1}{2}hcQ + ARQ^{-1}$$

$$\frac{dTC}{dQ} = \frac{hc}{2} - \frac{RA}{Q^2}$$

Set the derivative equal to zero:

$$0 = \frac{hc}{2} - \frac{RA}{Q^2}$$

Solve for Q:

$$\frac{RA}{Q^2} = \frac{hc}{2}$$

$$2RA = Q^2hc$$

$$Q^2 = \frac{2RA}{hc}$$

$$Q = \sqrt{\frac{2RA}{hc}}$$

(This formula is known as the *economic order quantity* or EOQ.) We calculated the result, using the numbers that Mr. Carrier gave us: $R = 3{,}000$ (amount sold per year); A = cost of order = \$20; $hc = 1.5$ = holding cost per unit of inventory.

From the formula:

$$Q = \sqrt{\frac{2 \times 3{,}000 \times 20}{1.5 \times 50}} = \sqrt{80{,}000} = 282.8 = \text{amount to order each time}$$

The professor started recalculating Mr. Carrier's bill (since we had now solved three problems for him), but then he presented us with another problem. He asked us to maximize the volume of an ice cream cone for a fixed amount of material making up the cone. Unfortunately, we immediately hit a brick wall: we did not know a formula for the volume or the surface area of a cone. We didn't want to charge Mr. Carrier until we had solved that problem, but this problem perplexed us all the way until Chapter 14 (see page 273).

"We need some paying customers now," the professor noted.

"I hope we get some other kinds of problems," Recordis said. "Not that I don't like maxima/minima problems, but I would prefer a little variety."

A while later a young mother came up to our desk, and it was the professor's turn to wait on her.

("I recognize her!" Recordis whispered. "She lives only a couple of houses away from me. She has such a nice family.")

"I understand you solve problems involving related rates of change," she said. "I'm planning a surprise birthday party for one of the children in the neighborhood. I worked out a very nice set of balloons that I would like to inflate suddenly at the moment of the surprise. I designed and built a special Variable Rate Air Pumper to inflate the balloons. I would like to have the balloons inflate so that their radii are increasing at constant rates. Can you tell me at what rate I should pump air into each balloon to make this happen?"

"Shouldn't you pump air into the balloon at a constant rate?" the professor asked. "If v represents the volume of the balloon and t represents time, then we would say in our notation that $dv/dt = c$, where c is some constant."

"That won't work because the balloon is getting bigger!" the customer said. "As the balloon gets bigger, we need to pump more air into it per given interval of time in order to keep the radius increasing at a constant rate."

"We need the chain rule, don't we?" the king suggested. "That rule tells us how to deal with related rates."

"If you say so," the professor said. "I always say that when in doubt you should write down what you know. We know that $dr/dt = 1$, where r is the radius. We also know from geometry that $v = (4/3)\pi r^3$. And we want dv/dt as a function of time. Now what?"

"We can find dv/dr," the king told her helpfully; "$dv/dr = 4\pi r^2$."

"I see how we can get dv/dt using the chain rule," the professor said suddenly.

$$\frac{dv}{dt} = \frac{dv}{dr} \frac{dr}{dt}$$
$$= (1)(4\pi r^2) = 4\pi r^2$$

"That's the answer!" Recordis said.

"Hold it!" the professor objected. "We need to find dv/dt as a function of time, but we have it as a function of the radius. Now we need to find out what r is as a function of time."

"We should be able to figure that out, since we know that the radius is increasing at the rate of 1 centimeter per second," the king said. "How big is the balloon when you start blowing air into it?" he asked the young mother.

"One centimeter in radius," she said.

"Then r must be equal to $t + 1$," the king said. We all agreed that this function had the desired properties: r was 1 when t was 0 and r increased at a constant rate of 1 centimeter per second.

"That gives us the answer for how much air to put in at a given time," the professor said.

$$\frac{dv}{dt} = 4\pi(1 + t)^2$$

"Thank you very much," the customer said. "I had guessed that the answer would involve the second power of time."

"Now about the charge . . ." the professor began.

"We can't charge someone who's planning a birthday party for some children!" Recordis cried. "How could you be so heartless! We'll let you have this answer free." The young mother went home grateful for her neighbor Recordis.

"We must have some paying customers now," Recordis said. "Otherwise we won't be able to stay in business."

A few minutes before closing time a young man came up to our stand outside the palace.

("I know him!" the professor whispered. "He's the lifeguard at National Park Beach!")

"I understand you solve problems related to the speed of things. I'd like to figure out how fast my shadow moves," he said. "At night I turn on a light at the top of my lifeguard station. When I start running away from the tower at a constant speed, my shadow runs ahead of me. Can you figure out how fast the shadow is moving?"

"Certainly," Recordis said. "First, we'll call the height of the light h and we'll call your height L (for lifeguard). We'll call the distance you've moved from the tower x.

"In the notation we use, the speed at which you're running is represented by dx/dt. We'll call that constant speed v."

$$\frac{dx}{dt} = v$$

"We may as well call the distance from the shadow to the tower s." (Figure 6–6.)

"Now all we have to do is find ds/dt," he continued, "which we do by . . ."

"Using the chain rule," the king interrupted. "We can use a similar triangle relationship to tell us that $h/s = L/(s - x)$."

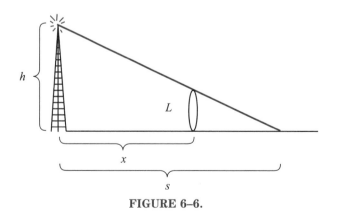

FIGURE 6–6.

"That's easy," the professor said. "We know that corresponding sides of similar triangles are in proportion."

We found dx/ds:

$$s - x = \frac{sL}{h}$$

$$x = s - \frac{sL}{h}$$

$$\frac{dx}{ds} = 1 - \frac{L}{h}$$

The king showed us how we could use the chain rule:

$$\frac{ds}{dt} = \frac{dx}{dt}\frac{ds}{dx} = \frac{\dfrac{dx}{dt}}{\dfrac{dx}{ds}}$$

We inserted the expressions we had found:

$$\frac{ds}{dt} = \frac{v}{1 - L/h}$$

"That's the answer!" Recordis said.

"But I want an answer that's a number," the lifeguard said.

"In that case, all you have to do is tell us what v, L, and h are."

"My lifeguard tower is 8 meters high, I'm 1.8 meters high, and I can run 6.7 meters per second."

Recordis did the arithmetic. The final answer turned out to be that the shadow moved with a speed of 8.6 meters per second. "Now about the charge. . ." Recordis began.

"We can't charge him!" the professor said. "Haven't you heard how many lives he's saved? One of my best friends was at the beach last summer and owes his life to this lifeguard. We'll let you have this answer free."

The lifeguard was very grateful, and he walked away just as it was time to close up our stand.

"How much money did we make this week?" Recordis asked.

The professor looked in the cash drawer. "We didn't make any," she said sadly.

"But we had lots of customers!" Recordis protested.

"Maybe it's better this way," the king said. "If we can help people solve calculus problems, I think that we should do it and not charge them for it."

"But what about our company?" the professor asked. "We spent so much time working on the brochures."

"We'll make it a nonprofit company," the king said. "That way we will still stay in business answering people's questions, but we won't need to worry about how much money we make."

We all decided to accept that plan, so before we returned to the palace we made a slight change in our sign so that it read:

Carmorra Consolidated Differentiating, Incorporated
A NONPROFIT AGENCY DESIGNED TO SERVE YOU!

Exercises

1. The conical Central Park reservoir is a units across and b units deep. Water is flowing into it at the rate of u cubic units per minute. How fast is the surface of the water rising when the water is h units deep? Suppose $a = 3$ meters, $b = 6$ meters, and $u = 0.2$ cubic meter per minute. What is dh/dt when $h = 3$ meters? (Volume of cone $= \pi r^2 h/3$, where $h =$ height and $r =$ base radius.)

2. One day Builder left a ladder of length L leaning against the side of the palace, and it started to slide. The bottom of the ladder slid along the ground with a constant speed of u meters per second. How fast was the top of the ladder sliding down the wall?

3. Find the point on the curve $y^2 = 8x$ that is closest to the point $(4, 2)$.

4. If a rectangle has a fixed perimeter equal to k, what is the shape of the rectangle that will have the maximum area?

5. If a rectangle has a fixed area of A, what shape will minimize the perimeter?

6. The profit function of a perfectly competitive firm is given by $Y = PQ - TC\,(Q)$. P, the price of output, is a fixed number beyond the control of the firm. $TC(Q)$ is the total cost function, which measures the cost of producing Q units of output. The marginal cost (MC) is the derivative of the cost function. Find a condition, stated in terms of MC and P, that the firm must meet if it is to maximize its profits.

For Problems 7 to 11, you are given the formula for the total cost (as a function of quantity) and the market price (P) for a perfectly competitive firm. Calculate the quantity of output that should be produced in order to earn the maximum possible profits.

7. $TC = 1.5Q^3 - 8Q^2 + 32Q + 680$; $P = \$1,858$.

8. $TC = 8Q^3 - 120Q^2 + 780Q + 2,384$; $P = \$2,124$

9. $TC = 2Q^3 - 30Q^2 + 210Q + 824$; $P = \$2,004$

10. $TC = 2Q^3 - 45Q^2 + 365Q + 8,498$; $P = \$3,929$

11. $TC = 3Q^3 - 60Q^2 + 800Q + 1,254$; $P = \$4,496$

12. The average cost of production is defined as the total cost divided by the quantity. Show that the average cost will increase when the quantity is increased if the marginal cost is greater than the average cost.

13. If the price of a good increases, then there will be a decline in the quantity of that good that people want to buy (called the quantity demanded of that good). Suppose the quantity demanded (Q) is given by the following function of price (P):

$$Q = a - bP$$

a and b are constants. Calculate a formula for the total revenue (expressed as a function of the quantity produced) and the marginal revenue (that is, the derivative of the total revenue with respect to quantity). What should the quantity be to earn the maximum possible revenue?

14. In economics, the *elasticity* of demand for a good is a way of measuring how responsive the demand is when the price changes. If the elasticity is zero, then the buyers always buy the same quantity, regardless of the price. In general, the elasticity of demand at a particular point is given by the formula:

$$\text{elasticity} = \left| \frac{dQ}{dP} \frac{P}{Q} \right|$$

The vertical lines stand for absolute value. Since the quantity demanded declines when the price increases, dQ/dP is negative. Because the absolute value is taken, the elasticity of demand will always be a positive number. If the demand curve is given by the formula $Q = a - bP$, determine a formula for the elasticity of demand at a particular quantity Q. What is the value of the elasticity at the revenue maximizing quantity you found in Exercise 13?

15. Derive a general formula for the marginal revenue, expressed in terms of the elasticity of demand and the price. Using this formula, what will be true about the elasticity at the quantity where the revenue is at a maximum? Show that, in general, an increase in quantity sold results in an increase in revenue if the elasticity is greater than one.

16. Consider a retail store that chooses its price by setting the marginal revenue equal to the marginal cost. The marginal cost for each item is constant (equal to the amount that must be paid to the wholesaler to obtain the item). The markup percentage is equal to the difference between the price and the marginal cost, divided by the marginal cost:

$$\text{markup percentage} = \frac{P - MC}{MC}$$

Derive a formula for the markup percentage in terms of the elasticity, and comment on the intuition of the result.

17. Derive a formula for the elasticity for this demand function: $Q = aP^{-e}$.

18. Recordis' special boxes have volume V and height h. The material to be used for the base and sides of the box costs $\$r$ per square unit, and the material to be used for the lid costs $\$2r$ per square unit. Find the dimensions of the box that minimize the total cost of the box.

19. Recordis has another kind of box with the top, three sides, and the base made of the material that costs $\$r$ per square unit, and the front side (dimensions x by h) made of the material that costs $\$2r$ per square unit. Find the shape of the box that minimizes the cost.

20. Find the shape of the right circular cylinder that has volume V and the minimum possible total surface area.

21. At what point is the slope of the curve $y = x^4 - 3x^3 + 5x - 10$ the greatest?

22. Recordis releases his toy boat from South Beach, and it travels due north at 5 cm/sec. At the same time, Pal releases his toy boat from East Beach, which is $20\sqrt{2}$ meters due northeast from South Beach. Pal's boat travels due west at 7 cm/sec. How far apart are the two boats at time t? At what time are they closest together? How close together are they at the moment that they are closest together?

23. Pal has a rope L units long that is strung over a pulley h units high. One end is attached to a large weight on the ground. Pal holds the other end and starts walking away from the pulley at a constant speed of u units per second. (When he starts walking the rope is taut.) How fast does the weight move up?

24. A car is to be driven on a trip D units long. The amount of fuel that the car uses per hour is given by $10v^2 - 100v + 290$, where v is the speed of the car in units per hour. The car will travel with a constant speed throughout the trip. What should v be so that the total fuel consumption for the trip will be minimized?

25. Derive a general formula for the point along the line $y = mx + b$ that minimizes the distance to the point (h, k). (This is a generalization of the Craggy Island lighthouse problem from Chapter 3.)

26. If you throw a ball with velocity v at an angle θ, its vertical velocity is

$$v_v = v \sin \theta$$

and its horizontal velocity is

$$v_h = v \cos \theta$$

Its height at time t is:

$$h = -0.5gt^2 + v_v t$$

The amount of time it stays in the air (t_{air}) is found by solving this equation:

$$0 = -0.5gt^2_{air} + v_v t_{air}$$

The distance it travels is $v_h t_{air}$. Determine the value of θ that maximizes the distance traveled (treating v as constant).

The Integral:
A Backward Derivative

The differentiation business went along smoothly. The professor told everyone that calculus was the most enjoyable invention since the Hasselbluff Mountain Sky Sled, and she was confident that we had discovered everything there was to know about mathematics.

Every day we held a meeting in the Main Conference Room. Slowly we began to notice that each day Recordis arrived at the meeting out of breath. A few days after the formation of the company he was panting so hard that the professor finally asked him what was happening.

"It's Rutherford, my dog," Recordis said. "Lately he's been very frisky. He dives off the diving board into the river. He keeps wanting me to raise the height of the diving board. I'm worried about raising it too high because I don't want him to get hurt. It would help if I knew how fast he was falling at the instant he hit the water."

"That's trivial," the professor sniffed. "Divide the distance he falls by the time he takes to fall and that will give you the speed."

"That calculation would tell us the average speed, but not the instantaneous speed at the moment he hits the water!" Recordis exclaimed. "He starts off falling at a slow speed, but as he falls further his speed increases."

"This does make it an interesting problem," the professor suddenly realized.

"In fact, if we can discover how Rutherford's speed changes as he falls, we might be able to apply the same method to the still-unsolved fly ball problem (see Chapter 3)," the king added.

We went to Recordis' house. Rutherford was so excited to see us that he jumped off the diving board again and again, so we were able to take lots of pictures and made careful measurements. (We used a strobe light so we could take pictures at very small intervals.)

One result gradually emerged from our measurements: Every time one second had elapsed, Rutherford's speed had increased by 9.8 meters per second, compared to what it has been one second earlier.

"I call the change in velocity divided by the change in time the *acceleration*," Builder said. So

Rutherford's acceleration while falling = 9.8

Since acceleration is change in velocity divided by change in time:

$$\frac{\text{change in velocity}}{\text{change in time}} = 9.8$$

Using the delta notation for "change in":

$$\frac{\Delta v}{\Delta t} = -g$$

where g is a constant equal to 9.8, and we use the negative sign because falling objects accelerate downward.

"This only works for very small intervals of time, the professor reminded us. She made us write this result as a limit:

$$\lim_{\Delta t \to 0} \frac{\Delta v}{\Delta t} = -g$$

When we took the limit, we recognized the left-hand side as the derivative:

$$\frac{dv}{dt} = -g$$

"This doesn't help!" Recordis screamed. "I don't need to know the derivative dv/dt. I need to know the velocity itself."

"But if we know the derivative of a function, we should be able to determine the function itself," the king reasoned. "We just have to work the derivative backwards."

"How could we do a derivative backwards!" Recordis exclaimed. "I have enough trouble doing a derivative forward."

"There should be some way to do it," the king said. "We need to think of a function of time ($v(t)$) whose derivative is equal to the constant $-g$."

"Try g itself," Trigonometeris guessed.

"No, g is a constant, so its derivative is zero," the professor replied.

"How about $1/g$?"

"No the derivative of $1/g$ is $1/g$."

"How about gt^2?"

"No," the exasperated professor said. "The derivative of gt^2 is $2gt$."

"Maybe $-gt$?"

"The derivative of $-gt$ is . . . It is $-g$! That's it!" the professor said excitedly.

"Maybe I can outwit that dog," Recordis said. "We must have a name for a backward derivative. I'm not going to write 'backward derivative' all the time."

"We could call it an *antiderivative*," the professor said. We agreed that that was one possible name.

$$\text{antiderivative} = v = -gt$$

$$\frac{dv}{dt} = -g$$

"That sounds too destructive," Trigonometeris said. "We should think of a constructive sounding name, too."

Everybody turned to me, and an interesting name suddenly popped into my mind. "We'll call it an integral," I said.

"That sounds impressive," Recordis agreed.

"We must think of a scientific way to calculate integrals, or antiderivatives," the king said. "Guessing won't work most of the time. How can we know ahead of time that the answer is $-gt$ and not $100t^3$ or $-gt + 99$ or something else?"

"We shall have to make a set of rules," Recordis answered. "We wrote a set of rules for calculating derivatives, so we should be able to develop a set of rules for calculating integrals."

"Hold everything!" the professor objected. "We have done something horribly wrong! How do we know that the answer is not $-gt + 99$?"

"We check the derivative," the king said.

$$x = -gt + 99$$

$$\frac{dx}{dt} = \frac{d}{dt}(-gt) + \frac{d}{dt}99 = -g + 0 = -g$$

The king stopped short.

"That function works, too!" Recordis cried. "They both work! However, one of them must be wrong, because I know that I have only one dog and he can be in only one place at one time. There is no way that two different functions can describe his velocity."

"You can't ever compute an integral," the professor said. "Suppose you know that $dx/dt = -g$. That will be the case if $x = -gt + 7$ or $x = -gt + 86.234567$ or, in fact, if $x = -gt + C$, where C is any constant number. That is hopelessly indefinite."

"We'll have to call it an *indefinite integral*," Recordis said.

"We'll have to forget about this and go back to our regular business," the professor stated firmly.

"I'm sorry I ever brought the subject up," Recordis said.

The professor was about to go on with the day's business when the king came over to me and whispered something in my ear. "I think you're right," I whispered back. "The king has an idea," I told everybody. "Igor, draw a graph of several curves that obey the relationships $dx/dt = -g$, $x = -gt + C$." (See Figure 7–1.)

"They all look like the same curve," the king said. "The only difference is that they have been moved up and down."

"That's why the idea of an integral should be useful," I explained. "I don't think it is hopelessly indefinite. For example, we know that a function with a different shape

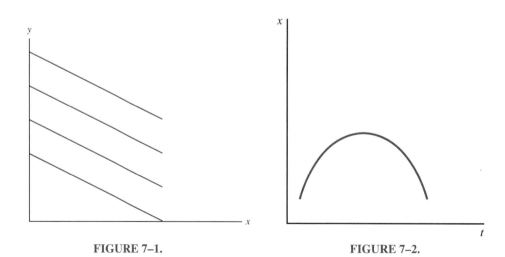

FIGURE 7–1.   FIGURE 7–2.

won't work (Figure 7–2) because its derivative isn't equal to −*g*. The only curves that will work are the curves on the first graph. It is useful to know that much."

"If we knew which one of these curves is the right one, then we would know exactly where Rutherford is all the time," the king said. "How are we going to tell which curve is the right one?"

We stared hard at the collection of curves on Igor's screen. We were looking for a clue that would make the single guilty curve stand out as the culprit responsible for Rutherford's devious motion.

"The difference between the curves seems to be how fast Rutherford is going at the start," the king said.

"We need to determine what Rutherford's velocity is a the moment he starts his dive," the professor said.

We observed Rutherford a few more times. Sometimes he would jump high in the air above the diving board before he slowed down and then began falling to the river. Other times he just slipped off the board without jumping.

We let v_0 represent Rutherford's initial velocity. Then we could track down the constant of integration C:

$$v = gt + C$$

initial condition: $v = v_0$ when $t = 0$

Therefore:

$$v_0 = g \times 0 + C$$
$$v_0 = C$$

"Aha!" the professor said. "So the arbitrary constant of integration C is equal to the initial velocity v_0."

The formula for v became:

$$v = -gt + v_0$$

where v = velocity (in meters per second), t = time (in seconds) $g = 9.8$ (the constant measuring the acceleration of gravity), and v_0 is the constant measuring the initial velocity (in meters per second), with upward motion having a positive velocity and downward motion having a negative velocity.

"We need a symbol for an integral," Recordis said.

"When we have a function $f(x)$ we call its derivative $f'(x)$," the professor offered. "Suppose we call its antiderivative $F(x)$."

"Using a capital letter!" Recordis said. "Ingenious! We need another symbol, too, to correspond to the dx/dt notation."

"We can figure something out," I said. "We start with $dv/dt = -g$, and we need some way to turn that dv/dt into just plain v."

"We can get rid of the dt by multiplying both sides by dt," Trigonometeris suggested.

"No, you can't," the professor objected. "That would work if dv/dt were a fraction, because then it would be dv divided by dt. But it's not a fraction. It's a derivative."

"Derivatives do seem to behave a little bit like fractions," the king said. "Remember the chain rule:

$$\frac{dy}{dx} = \frac{dy}{du}\frac{du}{dx}$$

"We can make a definition of what dx means," I said. "Igor, draw a picture of a curve with its tangent line (Figure 7–3). Let's let dx equal this deep black line, and dt equal this dotted line. Then dx divided by dt will equal the slope of the tangent line, which is what we know it must be."

We made the definition:

$$dx = \frac{dx}{dt}dt$$

"Still, dx is not a number that you can specify a value for," the professor pointed out. "We just said that it was some small number. We do need to think of a name for it. Since these numbers come from differentiating a function, we could call them *differentials*: dx means differential x, and dt means differential t." (We later established that the most important thing to remember about differentials is that a differential does not mean anything if it is all by itself. For example, the equation $dx = 4t$ does not mean anything. Differentials must always come in pairs, as in $dx = 4t\ dt$, or in a derivative, as in $dx/dt = 4t$, or with an integral sign, as in $x = \int f(t)\ dt$.)

We rewrote our equation for the derivative using differential notation:

$$\frac{dv}{dt} = -g$$

$$dv = -g\ dt$$

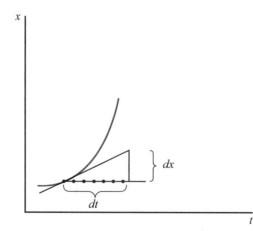

FIGURE 7–3.

"How do we get rid of the *d* in front of the *x*?" Recordis asked.

"We'll cross it out," I said cheerfully. However, Igor missed the *d* when he drew.

$$\int dv = -g\, dt$$

"Hold it!" the professor said. "That's an equation. If you do something to one side of the equation you must do the same thing to the other side!"

"All right," I said. "We'll do it like this."

$$\int dv = v = \int (-g)\, dt$$

"Pretty!" Pal muttered.

"If he likes it we had better keep it," Recordis said. "We'll call $\int$ the *integral sign*."

"It looks as though you need to have a *dt*, or a differential something, whenever you have an integral sign," the professor said.

$$x = \int (\text{function of } t \text{ that is to be integrated})\, dt$$

"We could call the function in the middle the *integrand*," the king remarked.

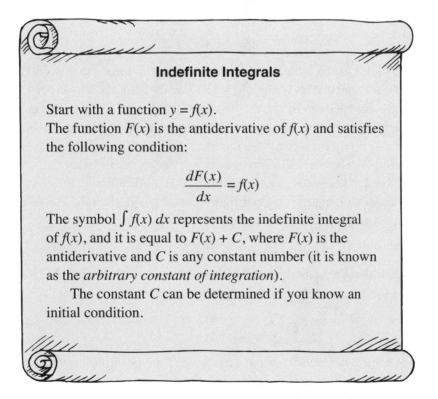

Indefinite Integrals

Start with a function $y = f(x)$.

The function $F(x)$ is the antiderivative of $f(x)$ and satisfies the following condition:

$$\frac{dF(x)}{dx} = f(x)$$

The symbol $\int f(x)\, dx$ represents the indefinite integral of $f(x)$, and it is equal to $F(x) + C$, where $F(x)$ is the antiderivative and C is any constant number (it is known as the *arbitrary constant of integration*).

The constant C can be determined if you know an initial condition.

(Note: In this box x is the independent variable, but at some other places in this chapter t is the independent variable and x is the dependent variable. At first this bothered Recordis, and he thought about asking the king to make a decree specifying that the independent variable would always be called x and never anything else. However, he soon realized that many times it is more convenient to use a different variable. For example, if the independent variable represents time, as it does in the problem of Rutherford's motion, it is more convenient to use t for the independent variable. In Chapter 4 the independent variable was w, which stood for width. The independent variable will usually be called x when a general rule is being stated, but in other specific situations Recordis reluctantly realized that it would be best to let other letters have a turn at that role.)

"The formula $v = -gt + v_0$ that works for Rutherford diving off the diving board will also work for the fly ball problem!" the professor said.

"No it won't!" Recordis objected. "Everybody knows that a heavy object like Rutherford falls faster than a light object like a baseball."

However, to our surprise, we found this wasn't the case. The velocity of any object moving straight up and down under the influence of the Earth's gravity is given by the formula:

$$v = -gt + v_0$$

where $g = 9.8$ and v_0 is the initial velocity of the object. It turns out that gravity pulls harder on heavier objects (this is obvious when you try lifting a heavy rock), but this effect is exactly canceled out by the fact that heavier objects are harder to move (they have more inertia). Therefore, all objects fall with the same acceleration when they are pulled on by gravity (except for the fact that air resistance sometimes has a big effect; for example, a feather falls very slowly because of air resistance).

"We don't need to know the velocity of the ball at a given time," Recordis said. "We need to know its height, so we can figure out how high the roof should be."

We looked at our formula for the velocity:

$$v = -gt + v_0$$

"We do know that the velocity is the derivative of the position. In this case, the position (x) is the height of the object at time t:

$$\text{velocity} = v = \frac{dx}{dt} = -gt + v_0$$

"Of course! Since the velocity is the derivative of the position function, we can determine the position function by taking the integral of the velocity!

First, multiply both sides of the equation by dt":

$$dx = (-gt + v_0)\, dt$$

"Now put the integral sign in front of both sides":

$$\int dx = \int (-gt + v_0) \, dt$$

$$x = \int (-gt + v_0) \, dt$$

"It would be useful if we could find a rule for the integral of the sum of two functions," the professor said. "Suppose we have the following."

$$Y = \int (f(x) + g(x)) \, dx$$

"I bet that's equal to this," Recordis said.

$$Y = \int f(x) \, dx + \int g(x) \, dx$$

"Why do you guess that?" the professor asked.

"That's the way it worked for derivatives," Recordis answered.

"That rule looks reasonable," the king agreed. "However, we need a way to determine if a rule for an integral is correct or not." (The king knew that Recordis sometimes suggested rules that looked nice but did not always work.)

The professor had a plan. "Suppose we guess that Y is the antiderivative for y. This means $dY/dx = y$. If we calculate the derivative of Y and find that it is not equal to y, then we know that Y is not really the antiderivative of y. In our case we have $y = f(x) + g(x)$ and $Y = \int f(x) \, dx + \int g(x) \, dx$. We want to see if dY/dx is equal to y."

$$Y = \int f(x) \, dx + \int g(x) \, dx$$

$$\frac{dY}{dx} = \frac{d}{dx} \int f(x) \, dx + \frac{d}{dx} \int g(x) \, dx$$

Since F and G are the antiderivative functions for f and g:

$$\frac{dY}{dx} = \frac{d}{dx} (F(x) + C) + \frac{d}{dx} (G(x) + C)$$

$$= \frac{d}{dx} F(x) + \frac{d}{dx} C + \frac{d}{dx} G(x) + \frac{d}{dx} C$$

$$= f(x) + 0 + g(x) + 0$$

$$\frac{dY}{dx} = f(x) + g(x)$$

"See! I was right again," Recordis said. "We'll call that the sum rule for integrals."

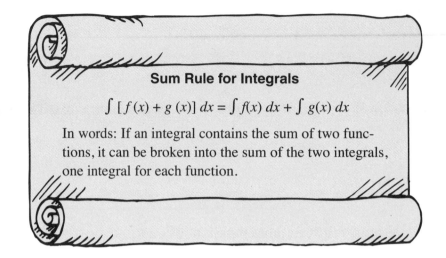

Sum Rule for Integrals

$$\int [f(x) + g(x)]\, dx = \int f(x)\, dx + \int g(x)\, dx$$

In words: If an integral contains the sum of two functions, it can be broken into the sum of the two integrals, one integral for each function.

Using the sum rule, we rewrote the integral we were trying to evaluate:

$$x = \int (-gt)\, dt + \int v_0\, dt$$

"It would be nice if we could figure out what to do with constants like v_0 when they occur inside an integral sign," Recordis said. "I wish we could pull that v_0 outside the integral sign, so we could write $v_0 \int dt$ instead of $\int v_0\, dt$. Integrals are nice, but I think the fewer symbols we have in the middle of them the better."

We tried Recordis' suggested rule. We wanted a function whose derivative was $y = nf(x)$, and Recordis' guess was that we should try $Y = n \int f(x)\, dx$.

$$\frac{dY}{dx} = \frac{d}{dx}\, n \int f(x)\, dx = n\frac{d}{dx} \int f(x)\, dx$$

(We know that constants move across derivatives. See page 28.)

$$\frac{dY}{dx} = n\, \frac{d}{dx}\, (F(x) + C)$$

$$= n\, (f(x) + 0)$$

$$= n\, f(x)$$

"It does work!" Recordis said. "We'll call that the *multiplication rule for integrals*."

"Let's extend that rule to the case where we have a variable number," Recordis went on eagerly. "I bet we can say that

$$\int q(x)f(x)\, dx = q(x) \int f(x)\, dx$$

where q is any variable number."

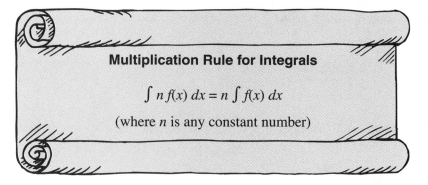

Multiplication Rule for Integrals

$$\int n\,f(x)\,dx = n\int f(x)\,dx$$

(where n is any constant number)

Igor took the derivative of $Y = q(x)\int f(x)\,dx$ to see whether it was equal to $q(x)f(x)$, as Recordis' theory said it would be.

$$Y = q(x)\int f(x)\,dx$$

$$\frac{dY}{dx} = \frac{d}{dx}\,q(x)\int f(x)\,dx$$

From the product rule:

$$\frac{dY}{dx} = q(x)\,\frac{d}{dx}\int f(x)\,dx + \frac{dq(x)}{dx}\int f(x)\,dx$$

$$= q(x)\,f(x) + \frac{dq(x)}{dx}\int f(x)\,dx \neq q(x)f(x)$$

"That doesn't work!" the king exclaimed.

"I was wrong," Recordis said apologetically.

The professor was irritated. "We'll have to add an amendment to the multiplication rule. It looks as though the integral sign acts as a filter. You can move *constants* across integral signs as much as you want, but you can never move *variables* across an integral sign."

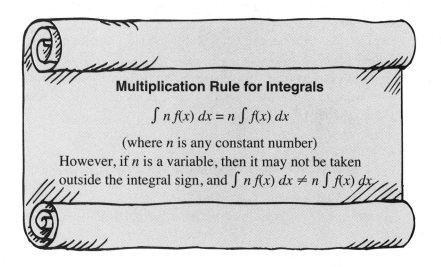

Multiplication Rule for Integrals

$$\int n\,f(x)\,dx = n\int f(x)\,dx$$

(where n is any constant number)
However, if n is a variable, then it may not be taken
outside the integral sign, and $\int n\,f(x)\,dx \neq n\int f(x)\,dx$

Using the multiplication rule, we rewrote the integral.

$$\frac{dx}{dt} = -gt + v_0$$

$$x = -g \int t \, dt + v_0 \int dt$$

(Since t is a variable, it had to stay inside the integral sign.)

"We know what $\int dt$ is," Recordis said. "That's just equal to $(t + C)$."

"How do you know that?" the professor asked.

"Because $\int dt$ is really the same thing as $\int (1) \, dt$, which is the same as $dx/dt = 1$. That means we need to find a function whose derivative is 1. The function that satisfies that condition is 1 times t, but of course we need to add that constant thingamajig to it."

We decided to make that a rule. Sometime later we agreed to call an integral of the form $\int dt$ a *perfect integral*, so we called the rule the *perfect integral rule*.

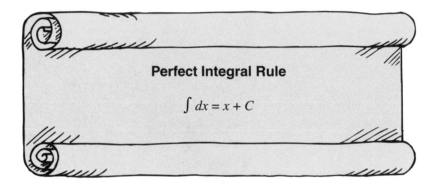

Perfect Integral Rule

$$\int dx = x + C$$

Igor rewrote our equation:

$$x = v_0 t + C - g \int t \, dt$$

"Now we have to take care of that t in the integral sign," the king said. We thought about this problem for a long time.

"We can write t as t^1 (t to the first power)" Recordis said, "although I don't see how that will help."

Finally the professor had an idea, "What we need to do is work the power rule backwards. The power rule says that, if $y = x^n$, then $dy/dx = nx^{n-1}$. Now suppose we have $dy/dx = x^n$, and we need to figure out what y is."

"I bet it will have an x^{n+1} in it," the king said. "If we take the derivative of x^{n+1}, we get $(n + 1)x^n$."

"But we don't want $(n + 1)x^n$," the professor objected. "We want to end up with just plain old x^n."

"It would work if we could find some way to get rid of the $(n + 1)$," the king said. "I know how we can get rid of it," Recordis contributed. "Try this."

$$y = \frac{1}{n+1} x^{n+1}$$

"Are you sure that will work?" the professor said, amazed that Recordis was coming up with so many answers in the same day.

$$y = \frac{1}{n+1} x^{n+1}$$

$$\frac{dy}{dx} = \frac{d}{dx} \frac{1}{n+1} x^{n+1}$$

$$= \frac{1}{n+1} \frac{d}{dx} x^{n+1}$$

$$= \frac{1}{n+1} (n+1)x^n$$

$$\frac{dy}{dx} = x^n$$

"It does work!" the king said. We called this rule the *power rule for integrals.*

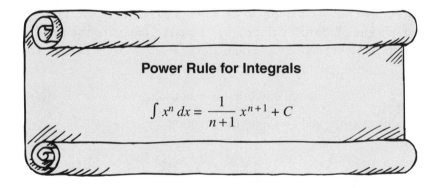

Power Rule for Integrals

$$\int x^n \, dx = \frac{1}{n+1} x^{n+1} + C$$

(However, this rule has one minor flaw, which we overlooked until we had a very unpleasant experience in Chapter 9.)

We reviewed tbe steps for finding the height of the fly ball, given, that we knew its velocity dx/dt:

$$\frac{dx}{dt} = -gt + v_0$$

(Note: g and v_0 are constants.)

Set up the integral:

$$x = \int (-gt + v_0)\, dt$$

Using the sum rule for integrals:

$$x = \int (-gt)\, dt + \int v_0\, dt$$

Since $-g$ and v_0 are constants, we moved them outside the integral signs:

$$x = -g \int t\, dt + v_0 \int dt$$

Write t as t^1:

$$x = -g \int t^1\, dt + v_0 \int dt$$

Using the power rule on the left and the perfect integral rule on the right:

$$x = (-g)\frac{1}{1+1}t^{1+1} + v_0 t + C$$

$$x = (-g)\frac{1}{2}t^2 + v_0 t + C$$

$$x = -\frac{1}{2}gt^2 + v_0 t + C$$

"Now we need an initial condition to track down the arbitrary constant C."

"The initial height of the ball will be about 1 meter," the professor said.

"That will be its height above the ground the moment it strikes the bat. To write the problem in general, call the initial height x_0."

initial condition: $x = x_0$ when $t = 0$

Therefore, insert $x = x_0$ and $t = 0$ into the equation for x:

$$x = -\frac{1}{2}gt^2 + v_0 t + C$$

$$x_0 = -\frac{1}{2}g \times 0^2 + v_0 \times 0 + C$$

$$x_0 = C$$

Since we could see that the arbitrary constant of integration would equal x_0, we wrote the equation for x:

$$x = -\frac{1}{2}gt^2 + v_0 t + x_0$$

"Now that we have the specific function giving the height of the fly ball, we can find the maximum height that it ever reaches." We found the derivative:

$$\frac{dx}{dt} = -\frac{2}{2}gt^1 + v_0$$

$$\frac{dx}{dt} = -gt + v_0$$

We realized this brings us back to the formula for the velocity. We set the derivative equal to zero, realizing that the highest point on the ball's trajectory would be the moment when its velocity was zero (it was about to stop its positive upward velocity and about to begin its downward negative velocity):

$$0 = -gt + v_0$$

Solve for the variable t:

$$gt = v_0$$

$$t = \frac{v_0}{g}$$

Now that we have the formula that tells us the time at which the ball reaches its greatest height, we can insert that time into the position function in order to find the height.

$$x = -\frac{1}{2}gt^2 + v_0t + x_0$$

$$x = -\frac{1}{2}g\left(\frac{v_0}{g}\right)^2 + v_0\left(\frac{v_0}{g}\right) + x_0$$

$$x = -\frac{1}{2}g\left(\frac{v_0^2}{g^2}\right) + \left(\frac{v_0^2}{g}\right) + x_0$$

$$x = -\left(\frac{v_0^2}{2g}\right) + \left(\frac{v_0^2}{g}\right) + x_0$$

$$x = \frac{v_0^2}{2g} + x_0$$

"In order to determine the height of the roof, we'll need to know the initial velocity v_0 of the fly ball," the king noted. "That could be a problem because there is a lot of variation in fly balls."

We realized this meant that we could not guarantee that no fly ball would ever hit the roof, because someday a player might hit a fly ball with an exceptionally large initial velocity. We hoped that nobody would hit a fly ball with initial velocity greater than $v_0 = 36$ meters per second. That would mean we could make the roof height equal to:

$$x = \frac{36^2}{2 \times 9.8} + 1$$

$$x = 67.1 \text{ meters}$$

Now to solve the problem with Rutherford.

$$x = -\frac{1}{2}gt^2 + v_0 t + x_0$$

x_0 is equal to the height of the diving board, since that is Rutherford's position at the moment he jumps. If he just slips off the diving board without jumping, then $v_0 = 0$, and the formula for x becomes:

$$x = -\frac{1}{2}gt^2 + x_0$$

At the moment he hits the water, x equals zero, so we can set up an equation to solve for the time when he hits the water:

$$0 = -\frac{1}{2}gt^2 + x_0$$

Now solve for t:

$$\frac{1}{2}gt^2 = x_0$$

$$t^2 = \frac{2x_0}{g}$$

$$t = \sqrt{\frac{2x_0}{g}}$$

Since $v = -gt + v_0 = -gt$, we could make a table showing the time it would take to hit the water, plus the velocity at the moment of impact, for some different values of the diving board height x_0:

diving board height, x_0 (meters)	time until impact (seconds)	velocity at impact (meters per second)
2	0.639	−6.261
4	0.904	−8.854
6	1.107	−10.844
8	1.278	−12.522
10	1.429	−14.000

We summarized:

To find the velocity of something if you know its acceleration (a, which is the derivative of the velocity function and the second derivative of the position function):

$$\frac{d^2x}{dt^2} = \frac{dv}{dt} = a(t)$$

Set up the integral:

$$v = \int a(t)\, dt = A(t) + C$$

where A is an antiderivative function such that $\frac{dA}{dt} = a(t)$ and C is the arbitrary constant of integration.

To find the value of C, you need an initial condition, which tells you what the velocity is at some specified time t.

To find the position of something if you know its velocity (which is the derivative of the position function):

$$\frac{dx}{dt} = v(t)$$

Se up the integral:

$$x = \int v(t)\, dt = V(t) + C$$

where V is an antiderivative function such that $\frac{dv}{dt} = v(t)$ and C is the arbitrary constant of integration.

To find the value of C, you need an initial condition which tells you what the position is at some specified time t. For example, for a falling object or a thrown object:

$$\text{acceleration} = \frac{d^2x}{dt^2} = \frac{dv}{dt} = -g = -9.8$$

$$\text{velocity} = \frac{dx}{dt} = v = -gt + v_0$$

$$\text{position} = x = -\frac{1}{2}gt^2 + v_0 t + x_0$$

"We'll have to evaluate a lot of integrals," Recordis said. "Sometimes Rutherford plays a game where he runs in a straight line in the yard. I know his velocity, but I can only catch him if I can figure out his position at a particular time so I can set the ladder in position and drop his collar on him. His velocity depends on the kind of dog food he eats that day, and he has many different kinds." We practiced with some more integrals (see Worksheet).

WORKSHEET

Determine the indefinite integral for each of these functions.

	$f(x)$	$\int f(x)\,dx$
1.	6	
2.	32	
3.	68	
4.	$4x + 12$	
5.	$-8x - 64$	
6.	$-16x - 54$	
7.	$6x^2 + 4x + 49$	
8.	$9x^2 - 8x - 762$	
9.	$12x^2 + 6x + 16$	
10.	$12x^3 - 21x^2 - 18x - 86$	
11.	$8x^3 - 30x^2 + 12x + 78$	
12.	$32x^3 + 18x^2 + 20x + 604$	
13.	$10x^4 + 16x^3 + 12x^2 + 8x - 13$	
14.	$20x^4 + 24x^3 + 24x^2 + 44x + 42$	
15.	$75x^4 + 20x^3 + 15x^2 + 24x + 121$	
16.	$x^4 - 3x^3 + 5x^2 - 7x + 6$	
17.	$x^3 - 7x^2 + 9x + 4$	
18.	$5x^3 + 4x^2 - 11x + 17$	

"I hate to cause more problems," Recordis said, "but I just thought of another kind of dog food that Rutherford has. What do we do if we have something other than x raised to a power? For example, what if we have $\int u^n\, dx$, where u is a function of x? Rutherford has one kind of dog food where his speed is given by $dx/dt = \sqrt{3t+5}$. In this case we have $\int u^{1/2}\, dt$, where $u = 3t + 5$."

We wrote the integral:

$$x = \int \sqrt{3t+5}\; dt$$

"We need to find some way to turn that dt into a du," the king told him. "Then we can use the power rule."

"I bet we can do that," the professor noted. "We have the derivative: $du/dt = 3$. We can write that in differential notation: $du = 3\, dt$, or $dt = \frac{1}{3}du$."

"That means we can write the integral!" the king said.

$$x = \int \sqrt{3t+5}\; dt = \int u^{1/2}\, \tfrac{1}{3}du$$

"We can pull that 1/3 outside the integral, since it is just a constant," Recordis said.

$$x = \tfrac{1}{3} \int u^{1/2}\, du$$

"Now we can use the power rule!" the professor exclaimed.

$$x = (\tfrac{1}{3})(\tfrac{2}{3})\, u^{3/2} + C$$

$$= \tfrac{2}{9}u^{3/2} + C$$

"But we want an answer in terms of t, not in terms of u," Recordis protested.

"We can always make the reverse substitution," the king said. "Since $u = 3t + 5$, we have $x = \tfrac{2}{9}(3t + 5)^{3/2} + C$."

"We had better check the derivative of that function to make sure that it works," the professor said.

$$x = \tfrac{2}{9}(3t + 5)^{3/2}$$

dx/dt had better equal $\sqrt{3t+5}$.

We realized that x was of the form cu^n, where $c = \tfrac{2}{9}$, $u = 3t + 5$, and $n = \tfrac{3}{2}$. Then the derivative comes from the formula:

$$\frac{dx}{dt} = cnu^{n-1}\,\frac{du}{dt} = \frac{2}{9} \times \frac{3}{2} \times (3t + 5)^{3/2-1} \times 3 = \frac{18}{18} \times (3t + 5)^{1/2} = \sqrt{3t+5}$$

This was exactly what we wanted: Since the derivative of $\frac{2}{9}(3t + 5)^{3/2}$ was equal to $\sqrt{3t + 5}$, it meant that the integral of $\sqrt{3t + 5}$ was equal to $\frac{2}{9}(3t + 5)^{3/2}$.

We called this method the *method of integration by substitution*.

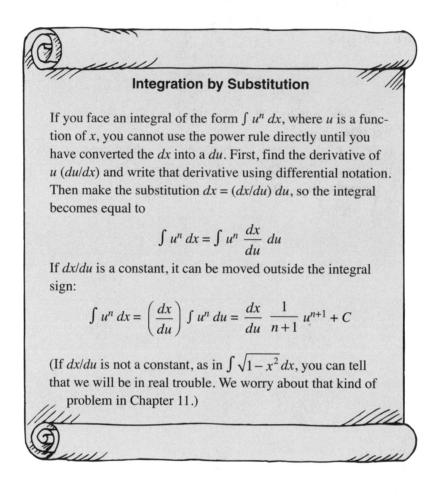

Integration by Substitution

If you face an integral of the form $\int u^n \, dx$, where u is a function of x, you cannot use the power rule directly until you have converted the dx into a du. First, find the derivative of u (du/dx) and write that derivative using differential notation. Then make the substitution $dx = (dx/du) \, du$, so the integral becomes equal to

$$\int u^n \, dx = \int u^n \, \frac{dx}{du} \, du$$

If dx/du is a constant, it can be moved outside the integral sign:

$$\int u^n \, dx = \left(\frac{dx}{du} \right) \int u^n \, du = \frac{dx}{du} \, \frac{1}{n+1} \, u^{n+1} + C$$

(If dx/du is not a constant, as in $\int \sqrt{1 - x^2} \, dx$, you can tell that we will be in real trouble. We worry about that kind of problem in Chapter 11.)

"Does Rutherford have any cans of dog food that make his speed a trigonometric function?" Trigonometeris asked, trying to be useful. "If he does, it should be no problem, since we can easily say that $\int \cos x \, dx = \sin x$, and $\int \sin x \, dx = -\cos x$."

"Now I better hurry home and catch him!" Recordis said, ignoring Trigonometeris.

WORKSHEET

Use a substitution to evaluate these intervals. (Treat all letters other than x, u, and y as constants.)

1. $y = \int \sin(mx)\, dx$ (let $u = mx$)

2. $y = \int \sqrt{a+x}\, dx$ (let $u = a + x$)

3. $y = \int \sqrt{ax+b}\, dx$

4. $y = \int (ax + b)^n\, dx$

5. $y = \int \cos x(\sin x)^3\, dx$ (let $u = \sin x$)

6. $y = \int \sin x(\cos x)^4\, dx$ (let $u = \cos x$)

7. $y = \int \cos x\, (\sin x)^n\, dx$ (let $u = \sin x$)

8. $y = \int x^{n-1} \sin(x^n)\, dx$ (let $u = x^n$)

9. $y = \int \dfrac{\sin(\sqrt{x})}{\sqrt{x}}\, dx$ (let $u = \sqrt{x}$)

10. $y = \int \cos x(a + \sin x)^n\, dx$ (let $u = a + \sin x$)

11. $y = \int \cos x \sqrt{a + \sin x}\, dx$ (let $u = a + \sin x$)

12. $y = \int (ax^2 + bx)^n(2ax + b)\, dx$ (let $u = ax^2 + bx$)

13. $y = \int 2ax(ax^2 + b)^n\, dx$ (let $u = ax^2 + b$)

14. $y = \int x(ax^2 + b)^2\, dx$ (let $u = ax^2 + b$)

15. $y = (a^2x^5 + 2ax^3b + b^2x)\, dx$ (Compare the answers for Exercises 14 and 15.)

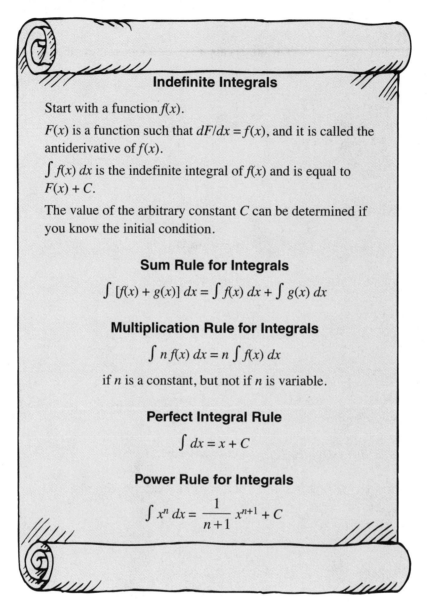

Indefinite Integrals

Start with a function $f(x)$.

$F(x)$ is a function such that $dF/dx = f(x)$, and it is called the antiderivative of $f(x)$.

$\int f(x)\,dx$ is the indefinite integral of $f(x)$ and is equal to $F(x) + C$.

The value of the arbitrary constant C can be determined if you know the initial condition.

Sum Rule for Integrals

$$\int [f(x) + g(x)]\,dx = \int f(x)\,dx + \int g(x)\,dx$$

Multiplication Rule for Integrals

$$\int n\,f(x)\,dx = n \int f(x)\,dx$$

if n is a constant, but not if n is variable.

Perfect Integral Rule

$$\int dx = x + C$$

Power Rule for Integrals

$$\int x^n\,dx = \frac{1}{n+1}\,x^{n+1} + C$$

NOTES TO CHAPTER 7

Velocity is similar to speed. Velocity also gives information about the direction of motion. A falling object is said to have a negative velocity, and an ascending object has a positive velocity. The velocity is actually a vector (see Chapter 16).

In the discussion of falling objects in this chapter, we assumed that the objects are moving straight up and down. If the object is not moving straight up and down, it will have a horizontal component to its velocity as well as a vertical component. The formulas in this chapter will work if the vertical component of velocity is used.

Exercises

Evaluate the following. (Remember that the answer to an indefinite integral is a function plus an arbitrary constant.)

1. $y = \int (x^2 + 3x + 5)\, dx$

2. $y = \int (ax^2 + bx + c)\, dx$

3. $y = \int (9x + 10)\, dx$

4. $y = \int (14)\, dx$

5. $y = \int (x^3 + 1)\, dx$

6. $y = \int (6x^5 + 10x^3 + 3x)\, dx$

7. $y = \int (4x^3 + 3x^2 + 2x + 1)\, dx$

8. $y = \int \left(\dfrac{x^3}{3} + \dfrac{x^2}{2} + x \right) dx$

9. $y = \int (x^m + x^n)\, dx$

10. $y = \int [(m + 1)x^m + (n + 1)x^n + (p + 1)x^p]\, dx$

11. $y = \int x^{100}\, dx$

12. $y = \int \sin \theta\, d\theta$

13. $y = \int \cos \theta\, d\theta$

14. $y = \int \sec^2 \theta\, d\theta$

15. $y = \int \left(\dfrac{1}{x^4} + \dfrac{1}{x^3} \right) dx$

16. $y = \int (x^2 + x^{-2})\, dx$

17. $y = - \int \csc^2 \theta\, d\theta$

18. $y = \int \sec \theta \tan \theta\, d\theta$

19. $y = - \int \csc \theta \operatorname{ctn} \theta\, d\theta$

20. $\int \tfrac{1}{2}(1 - \cos 2\theta)\, d\theta$

21. $\int \sin^2 \theta\, d\theta$

22. Evaluate $x \int x^2\, dx$ and $\int xx^2\, dx$.

Find $x(t)$ for each of the following situations. Think of an example of an object that might move according to each equation.

23. $dx/dt = 0$ $t = 0, x = 5$

24. $dx/dt = 4$ $t = 0, x = 2$

25. $dx/dt = 55$ $t = 8, x = 175$

26. $dx/dt = at$ $t = 0, x = 0$

27. $dx/dt = at$ $t = 0, x = x_0$

28. $dx/dt = \cos at$ $t = \pi/2a, x = \frac{1}{2}$

29. $dx/dt = t^4$ $t = 1, x = 1$

30. $d^2x/dt^2 = a$ $t = 0, x = x_0; t = 0, dx/dt = v_0$

31. $d^2x/dt^2 = 0$ $t = 0, x = x_0; t = 0, dx/dt = v_0$

Differentiate each of these functions and express the answer using differential (rather than derivative) notation:

32. $y = x^2$

33. $y = \sin x$

34. $y = cx$

35. $y = x^{1/2}$

36. $y = (1 + x^2)^{1/2}$

Solve the following integrals. Use the substitution indicated to convert the integral into a form where the power rule or some other simple rule can be used.

37. $y = \int x \sqrt{1-x^2} \, dx;$ let $u = 1 - x^2$

38. $y = \int x^2 \sqrt{a + bx^3} \, dx;$ let $u = a + bx^3$

39. $y = \int x \sin (x^2) \, dx;$ let $u = x^2$

40. $y = \int \sin^3 x \cos x \, dx;$ let $u = \sin x$

Evaluate:

41. $y = \int \dfrac{x}{(5 + 6x^2)^2} \, dx$

42. $y = \int x^{n-1} \sqrt{a + x^n} \, dx$

43. $y = \int \sec^2 x \tan^3 x \, dx$

44. $y = \int x^9 \sin (x^{10})\, dx$

45. $y = \int u^n\, (du/dx)\, dx$

46. $y = \int \sin \theta\, (d\theta/dx)\, dx$

47. When Pal throws his beach ball into the air, its acceleration is given by $dv/dt = -g$, with the initial condition $v = v_0$ when $t = 0$. (a) Find the velocity of the ball at time t. (b) Find the height of the ball at time t, using the initial condition $h = 0$ when $t = 0$.

48. When Pal drops his ball off Hasselbluff Mountain, its acceleration is given by $d^2h/dt^2 = -g$, with the initial conditions $h = 64$ and $dh/dt = 0$ when $t = 0$. Find the velocity of the ball at time t. Find the height of the ball at time t.

49. What is $\int (dy/dx)\, dx$? What is $d/dx \int y\, dx$?

Finding Areas with Integrals

Work on the stadium renovations continued, but the next day Builder told us of a problem he had with a support post that had a curved structure on top. "You'll have to tell me the area under the curve so I know how much paint to buy," Builder said.

"We're too busy working on calculus to find the formula for the area under a curve," the professor said dismissively.

"I wouldn't mind taking a break from calculus and working on another kind of problem," Recordis said. "What's the equation of the curve for the support post?"

Builder said the equation of the curve at the top was $y = x^2 + 5$, and we needed to find the area under the curve from $x = 3$ to $x = 6$. (See Figure 8–1.)

"How hard can that be?" Recordis asked.

"You're forgetting the difference between a curved figure and a rectangle!" the professor exclaimed. "We know the area of a rectangle is height times width, but that doesn't work for any shape with a curved boundary."

"We could try to fill the area of the curve with a lot of little rectangles," the king suggested. (Figure 8–2.)

"You will still have a lot of area left over that is not included in the rectangles!" the professor said.

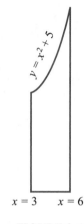

FIGURE 8–1.

"We'll at least be close to the area of the curve if we figure out the area of all the rectangles," the king pointed out. "Let's suppose that each rectangle has the same width."

"We can call that Δx," Recordis said.

"We need to figure out the boundaries of the region," the king noted. "Let's call the left-hand boundary $x = a$, the right-hand boundary $x = b$, and the lower boundary $y = 0$." (Figure 8–3.)

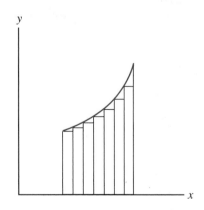

FIGURE 8–2.

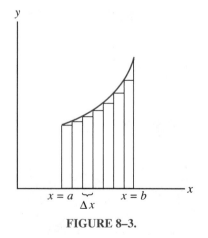

FIGURE 8–3.

"We can tell what the area of the first rectangle is," Recordis said.

$$(\text{area})_1 = (\text{height}) \times (\text{width}) = f(a)\,\Delta x$$

We also found the areas of the second and third rectangles:

$$(\text{area})_2 = f(a + \Delta x)\,\Delta x$$

$$(\text{area})_3 = f(a + 2\,\Delta x)\,\Delta x$$

"And the total area under the curve is approximately the sum of the first area plus the second area plus the third area plus . . . ," the king said.

"Hold it!" Recordis protested. "You must have a hundred rectangles up there, and if you want me to add up the areas of all of them you will have to pay me about a hundred times what you're paying me now!"

"Don't we have a shorter way for writing a sum like that?" Trigonometeris asked.

"Of course!" the king said. "Summation notation! Remember that we used a crooked letter s, Σ. (The symbol Σ is the Greek capital letter sigma.) We put where to start at the bottom: $\sum_{i=1}$, and where to stop at the top: $\overset{10}{\sum}$, and we put what we want to add up along the sides: $\sum_{i=1}^{10} i^2$. For example, we might have the following."

$$\sum_{i=1}^{10} i = 1 + 2 + 3 + 4 + 5 + 6 + 7 + 8 + 9 + 10 = 55$$

$$\sum_{i=1}^{10} 2i = 2 + 4 + 6 + 8 + 10 + 12 + 14 + 16 + 18 + 20 = 110$$

$$\sum_{i=1}^{10} i^2 = 1 + 4 + 9 + 16 + 25 + 36 + 49 + 64 + 81 + 100 = 385$$

"I remember how that worked," Recordis said "It certainly saved us a lot of writing."

"We can use summation notation for the area problem," the king told him. "Suppose we have n rectangles. Then we can say this."

$$(\text{area of all rectangles}) = \sum_{i=1}^{n} (\text{area of } i\text{th rectangle}) = \sum_{i=1}^{n} A_i$$

"We know what the area of the ith rectangle is," Recordis said. "It will be equal to $f(x_i)\,\Delta x$, if we define x_i right." (Figure 8–4.)

Igor wrote the equation for the area of all the rectangles:

$$(\text{area of rectangles}) = \sum_{i=1}^{n} A_i = \sum_{i=1}^{n} f(x_i)\,\Delta x$$

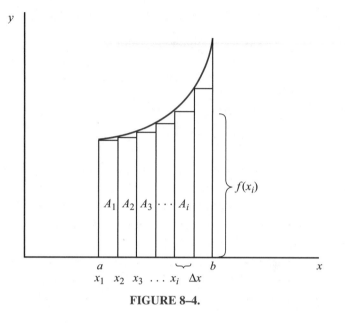

FIGURE 8–4.

"We still have a problem," Recordis said. "The summation notation makes it simple to write a sum involving hundreds of terms, but it doesn't make it any easier to do the actual hard work of adding them all together. And that's what I always get stuck doing."

"We have an even bigger problem," the professor pointed out. "This expression gives us the area of the rectangles, but it still doesn't give us the area of the curve."

"I know how we could get closer to the curve's area," the king said. "We could use twice as many rectangles. Then there would be much less difference between the area under the curve and the area of all the rectangles."

"You could get closer, but you still couldn't do it!" the professor objected. "There still is some area wasted between the curve and the rectangles. You'd have to use an almost infinite number of rectangles before you could get the right area."

"That's it!" the king shouted. "We'll use an almost infinite number of rectangles! We'll say that the area of the curve is the *limit* of the sum of the areas of all the rectangles as we let the number of rectangles go to infinity."

$$\text{(area under curve)} = \lim_{\Delta x \to 0, n \to \infty} \sum_{i=1}^{n} f(x_i)\, \Delta x$$

"You can't take a limit to infinity!" the professor said.

"I just did," the king responded.

"But look what will happen!" the professor went on. "Suppose $n = \infty$ and $\Delta x = 0$. Then $f(x_i)\, \Delta x$ would be zero all the time. You would be adding together an infinite number of zeros."

"But we can't ever let Δx actually equal zero," the king said. "We can let it become absolutely just-about-there close to zero. That's what 'limit' means."

"We used limits when we found derivatives," Recordis added. "We let Δx come very close to zero, but we never let it actually equal zero."

We were all very impressed until the professor suddenly realized that we had not come much closer to solving the problem. "This still does not tell us how to take a given curve, $y = f(x)$, and two given numbers, a and b, and come up with a number that is equal to the area under the curve." (See Exercise 24 for an example of using rectangles to calculate areas.)

We realized that our definition of the derivative in terms of limits was simple enough, because we could easily calculate the actual numerical value of the limit for a specific function. The area limit involved a sum with an infinite number of terms, so we all realized there was no way that we could directly figure out the area by adding all these terms together. "We couldn't solve this problem if we stayed here until Hotspot Caves freeze over," the king said sadly.

"We'll have to give up," the professor mourned. "That's very sad if we can never figure out the right amount of paint to buy."

We were about to return to the business of differentiation and integration when there came a sudden swoosh! through the window. There was an evil, cackling laugh, and the next thing we saw was . . . the gremlin!

"Surrender now to my supreme powers of evil. You have no hope of solving your new calculus problem."

"We're stuck on an area-under-a-curve problem, not a calculus problem," the professor said with a quivering voice.

"So you think this is not a calculus problem! Ha!" Suddenly the gremlin's face turned ashen. "Blast it!" he said under his breath. "If you really hadn't figured out this was a calculus problem, I shouldn't have given it away." In a moment his composure was back. "No matter," he cackled. "You'll never solve the area problem. Look out the window."

In the yard outside the Main Conference Room window we saw a big hole, with three straight sides and one curved side.

"You stole one of my support posts and pounded it into the ground to make that hole!" Builder accused him.

"So you recognize that shape," the gremlin mocked us. "Watch."

We saw flames suddenly appear and fill the hole. Fortunately the fire seemed unable to escape the hole.

The king was about to call the fire department, but the gremlin laughed at him. "These are magic flames. Only one thing can put them out . . . magic crystal water."

"We know where to get magic crystal water!" the king said defiantly.

"You must use exactly the right amount," the gremlin explained. "One drop too little—and the flames will escape and sweep over the entire kingdom of Carmorra! One drop too much—and a giant, sizzling, steaming flood will completely engulf Carmorra. Your only other choice is to give up now, and submit yourselves to my becoming king of Carmorra!"

"Never!" the king cried.

"We certainly cannot allow that!" Recordis said.

"Unless you pour in exactly the right amount of magic crystal water, the flames will escape at exactly sunset." He looked at his wrist hourglass. "You have exactly two hours." Before we knew what was happening, he had whipped his cape around himself and, with a tremendous blast of hot air, had disappeared out the window again.

Builder braved the flames to measure the depth of the hole while the king placed an emergency order for a large shipment of magic crystal water to be delivered to the palace.

"The volume of the pool will be depth times area, so all we need to do is figure out the area," the professor said.

"Which is what we can't do," Recordis moaned. "We could give the kingdom up to the gremlin."

"That would spoil everything," the king objected.

"But we can't let ourselves be burned to death," the professor said.

"Or flooded to death," Trigonometeris added.

Desperately we stared at the graph of the curve (Figure 8–5).

"Let's imagine there is a function that gives the area under the curve. Call it A, because A stands for area," Recordis said.

"It won't do any good to imagine such a function unless we can find out what the function actually is!" the professor exclaimed.

"If A is the function giving the area under the curve to the left of $x = a$, then clearly the area we are looking for is $A(b)$," the king contributed.

"We could also say that $A(a) = 0$; that is, the area under the curve to the right of $x = a$ and to the left of $x = a$ is always zero," Recordis noted.

"That's trivial. Not worth mentioning," the professor said scornfully.

After what seemed an eternity of fruitlessly staring at the problem, the king suddenly remembered something. "I recall the gremlin made one fatal mistake, and that may give us hope. He let slip that this is a calculus problem."

"Weird," Recordis said. "It doesn't look like a calculus problem. It's not like we're finding the derivative of the area function."

"That might be it!" the professor jumped in the air. "Maybe finding the derivative of the area function will be the key to solving the problem!"

"I hate to throw cold water on that idea, but remember that in previous instances where we found the derivative we *knew* what the function was! In this case we don't know the function $A(x)$. I remember the formula that defines the derivative requires us to find $A(x + \Delta x)$. How could we ever do that, if we can't even find $A(x)$?" Trigonometeris said.

Igor quickly sketched a graph (Figure 8–6).

"We can see that $A(x)$ is the light red area, and $A(x + \Delta x)$ is the sum of the light red area plus the dark red area," the professor said.

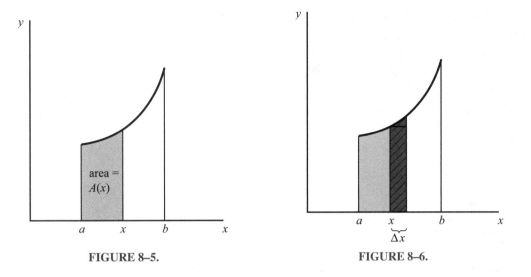

FIGURE 8–5.   FIGURE 8–6.

"The dark red area is equal to the difference," the king added hopefully:

$$A(x + \Delta x) - A(x) = \text{dark red area}$$

"If only the dark red area was a rectangle! Then its height would be $f(x)$, and its width would be Δx, so its area would be $f(x)\,\Delta x$," Recordis said wistfully.

"We could say this is true as an approximation," the king said.

$$A(x + \Delta x) - A(x) \approx f(x)\,\Delta x$$

"An approximation does us no good here!" the professor screamed. "You heard what the gremlin said."

"The approximation would be better if the rectangle is narrower—that is, if Δx is smaller," the king noted.

"Still not good enough. The only way to get it to work exactly would be to take the limit as Δx goes to zero!" the professor insisted.

"That's what we'll have to do," the king ordered. "Desperate times call for desperate measures!"

$$\lim_{\Delta x \to 0} A(x + \Delta x) - A(x) = f(x)\,\Delta x$$

"Wait . . . Wait . . . I'm just about ready to see something," Recordis said.
"Divide both sides by . . ."
"By delta x!" the professor finished his sentence.

$$\lim_{\Delta x \to 0} \frac{A(x + \Delta x) - A(x)}{\Delta x} = f(x)$$

Memories flooded back. "That's the derivative formula!" we cried in unison. We rewrote the left-hand side of the equation:

$$\frac{dA}{dx} = f(x)$$

"The derivative of the function giving the area under $f(x)$ is just $f(x)$ itself. Never in a million years would I have seen that one coming," Trigonometeris said in amazement.

"We know the derivative of A, but how do we find A itself?" Recordis cried. "Wait—we know how to do that! We discovered integrals just in the nick of time yesterday!"

Could we really find the area as an integral? We rewrote the equation as an integral:

$$\frac{dA}{dx} = f(x)$$

$$dA = f(x)\,dx$$

$$\int dA = \int f(x)\,dx$$

$$A = \int f(x)\,dx$$

Suppose $F(x)$ is the antiderivative function of f. Then

$$A = F(x)$$

"Wait! Remember that with integrals we have to add the arbitrary constant of integration!" the professor reminded us.

$$A = F(x) + C$$

"If we can't determine the value of the constant C, then we haven't solved the problem! Don't tell me that we might be defeated when we're this close to victory!" the king exclaimed. We could see the sun falling low in the west.

"We've tracked down arbitrary constants of integration before," the professor declared. "What we need is an initial condition."

"I know one," Recordis said, but then he became silent.

"I order you to talk," the king demanded.

Recordis gathered his courage. "Certain people think this is trivial," he gave a sidelong glance at the professor, "but we know $A(a) = 0$."

"That's the initial condition we need!" the professor said. "A million apologies for calling that trivial."

Inserting $A = 0$ when $x = a$ into the equation $A = F(x) + C$ gave us:

$$0 = F(a) + C$$

Solving for C:

$$-F(a) = C$$

Now putting this back into the equation $A = F(x) + C$:

$$A = F(x) - F(a)$$

"In order to find the whole area to the left of $x = b$, we need to evaluate the function where $x = b$:

$$A = (\text{total area under the curve}) = F(b) - F(a)$$

As fast as he could write, Recordis jotted down what he called the *fundamental theorem of integral calculus* (see below).

"Let's think of an easy way to write the area in terms of an integral," the professor said quickly. "Let's write the integral, and then write the boundary terms like this."

$$A = \int f(x)\, dx \,\Big|_a^b = F(b) - F(a)$$

(The boundary terms a and b are usually called the *limits of integration*.)

"Let's forget that little vertical line and write it like this," Recordis said.

$$A = \int_a^b f(x)\, dx = F(b) - F(a)$$

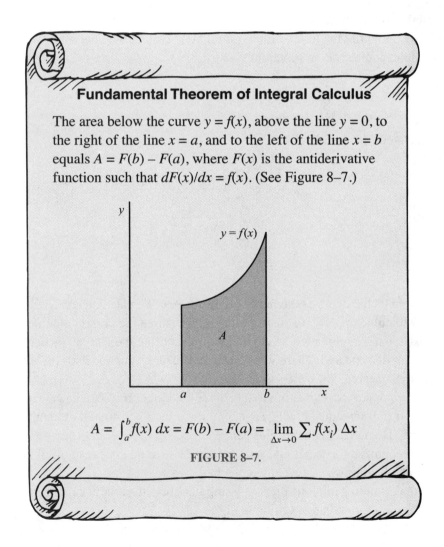

Fundamental Theorem of Integral Calculus

The area below the curve $y = f(x)$, above the line $y = 0$, to the right of the line $x = a$, and to the left of the line $x = b$ equals $A = F(b) - F(a)$, where $F(x)$ is the antiderivative function such that $dF(x)/dx = f(x)$. (See Figure 8–7.)

$$A = \int_a^b f(x)\, dx = F(b) - F(a) = \lim_{\Delta x \to 0} \sum f(x_i)\, \Delta x$$

FIGURE 8–7.

"That integral will tell us exactly what number is equal to the area."

"That is a definite integral if I ever saw one," Trigonometeris observed.

"All right, we'll call it a *definite integral* if it stands for an area," the professor said. "A definite integral will look almost the same as an indefinite integral, except that we have two integration limits written next to the integral sign.

We set up the definite integral, with the limits of integration written right next to the integral sign.

$$A = \int_3^6 (x^2 + 5)\, dx$$

"We know the antiderivative of x^2 is $\frac{1}{3}x^3$, and the antiderivative of 5 is $5x$," the professor said. "Therefore, the antiderivative of the whole thing is $F(x) = \frac{1}{3}x^3 + 5x$. According to the fundamental theorem, we need to evaluate the antiderivative at $x = 6$, and then subtract the antiderivative evaluated at $x = 3$."

We decided to write the limits of integration next to a vertical bar after we had determined the antiderivative function:

$$A = \left(\frac{1}{3}x^3 + 5x \right) \Bigg|_3^6$$

We calculated $F(6) - F(3)$:

$$= \left(\frac{1}{3} \right)6^3 + 5 \cdot 6 - \left[\left(\frac{1}{3} \right)3^3 + 5 \cdot 3 \right]$$

$$= \frac{216}{3} + 30 - \frac{27}{3} - 15$$

$$A = 78$$

Builder told us the depth of the pool was 10 meters. We calculated the volume as $10 \times 78 = 780$ cubic meters. We rushed outside, measured the correct amount of magic crystal water, and poured it into the pool just before the last ray of light from the setting sun disappeared. There was a steaming cataclysmic maelstrom, but then the flames disappeared. We were safe!

The gremlin appeared, expecting to take over the kingdom. Then he saw the doused flames and screamed "No!" Then a moment later he shouted defiantly: "It's not over yet!" He pulled out a piece of paper.

Builder recognized the paper and screamed, "He's stolen my list of all of the problems I've been having where I need to calculate area!"

One of the gremlin's minions threw a triangular-shaped stadium support post onto the ground, and it formed another flaming hole.

"We need to find the area under the function $y = hx/b$, from $x = 0$ to $x = b$. (See Figure 8–8.)

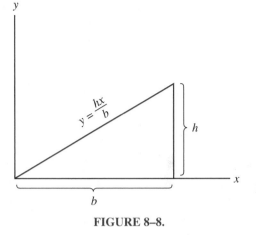

FIGURE 8–8.

"Set up the definite integral," the king ordered.

$$A = \int_{x=0}^{x=b} \left(\frac{hx}{b} \right) dx$$

h and b are constants (they don't depend on x), so they can be moved outside the integral sign:

$$A = \left(\frac{h}{b} \right) \int_{x=0}^{x=b} x^1 \, dx$$

Using the power rule for the integral of x^1:

$$A = \left(\frac{h}{b} \right) \frac{x^2}{2} \bigg|_{x=0}^{x=b}$$

Again we wrote the limits of integration to the right of a vertical line after we had evaluated the integral. Now to find $F(b) - F(0)$, where $F(x) = x^2/2$:

$$A = \left(\frac{h}{b} \right) \left(\frac{b^2}{2} - \frac{0^2}{2} \right)$$

$$A = \left(\frac{h}{b} \right) \left(\frac{b^2}{2} \right)$$

$$A = \frac{bh}{2}$$

"Of course! Just as we suspected from our knowledge of geometry—the area of a triangle is one half of the base multiplied by the height."

The gremlin threw another challenge at us: a trapezoidal-shaped support.

We needed to find the area under the curve $y = \frac{k-h}{b}x + h$ from $x = 0$ to $x = b$. (See Figure 8–9.)

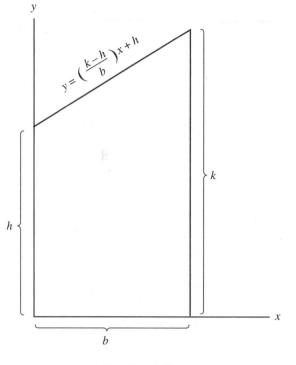

FIGURE 8–9.

We set up the definite integral:

$$A = \int_0^b \left(\frac{k-h}{b} x + h \right) dx$$

Using the sum rule for integrals:

$$A = \int_0^b \left(\frac{k-h}{b} x \right) dx + \int_0^b h \, dx$$

Moving the constants outside the integrals:

$$A = \frac{k-h}{b} \int_0^b x^1 \, dx + h \int_0^b dx$$

Evaluating the first integral using the power rule, and the second using the perfect integral rule:

$$A = \left(\frac{k-h}{b} \right) \frac{x^2}{2} \Big|_0^b + hx \Big|_0^b$$

Evaluating $F(b) - F(0)$ for each integral:

$$A = \frac{k-h}{b}\left(\frac{b^2}{2} - \frac{0^2}{2}\right) + hb - h \times 0$$

Simplifying:

$$A = \frac{k-h}{b}\left(\frac{b^2}{2}\right) + hb$$

$$A = \frac{(k-h)b}{2} + hb$$

$$A = \frac{kb - hb + 2hb}{2}$$

$$A = \frac{kb + hb}{2}$$

$$A = \frac{(k+h)b}{2}$$

"Both the triangle and the trapezoid are figures with only straight-line borders, so it's not the least bit surprising that our results for the area match what we found with geometry," the professor noted. "The problem will be if the gremlin starts having us find the areas of more curves, because then we will have to resort to our definite integral method."

"Very perceptive," the gremlin agreed.

"The next one is interesting," Recordis said. "The curve is $f(x) = -x^2 + 9$, with the boundary points -3 and 3."

$$A = \int_{-3}^{3}(-x^2 + 9)\, dx$$

$$= -\frac{1}{3}x^3\bigg|_{-3}^{3} + 9x\bigg|_{-3}^{3}$$

$$= -\frac{27}{3} - \frac{27}{3} + 27 + 27 = 54 - 18 = 36$$

WORKSHEET

Evaluate each definite integral:

1. $\int_{10}^{20} 17\, dx$

2. $\int_{3}^{8} (12x + 20)\, dx$

3. $\int_{2}^{10} (9x^2 + 8x + 60)\, dx$

4. $\int_{1}^{5} (24x^3 + 21x^2 + 6x + 30)\, dx$

5. $\int_{0}^{3} (25x^4 + 16x^3 + 6x^2 + 14x + 10)\, dx$

"The next pool is hard," Recordis told us. "Its area is not equal to the area under a curve; it is the area between two curves. The upper boundary is the curve $x^2/25 + 2$, and the lower boundary is the curve $4x^2/25 - 1$." (Figure 8–10.)

"We can find the area between two curves," the professor said quickly. "All we need to do is define a new function equal to the difference between the two curves," (See Exercise 29.)

$$y_1 = \frac{x^2}{25} + 2$$

$$y_2 = \frac{4x^2}{25} - 1$$

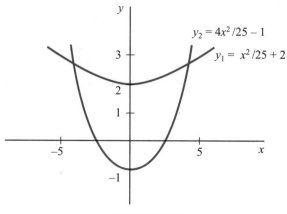

FIGURE 8–10.

"Let $y = y_1 - y_2 = x^2/25 + 2 - (4x^2/25 - 1)$. Now we can integrate that function from –5 to 5."

$$A = \int_{-5}^{5} (y_1 - y_2)\, dx = \int_{-5}^{5} \left(\frac{-3x^2}{25} + 3 \right) dx$$

$$= -\frac{1}{25} x^3 \Big|_{-5}^{5} + 3x \Big|_{-5}^{5}$$

$$= -\frac{1}{25}(5)^3 - \left[-\frac{1}{25}(-5)^3 \right] + 3(5) - 3(-5)$$

$$= -5 - 5 + 15 + 15 = 30 - 10 = 20$$

Our confidence was slowly growing as we met each definite integral challenge, and the gremlin was becoming more concerned as he desperately tried to come up with a problem that was too hard for us to solve.

"This one's for you," he told Trigonometeris. A new flaming pool appeared, shaped like one arch of a sine curve.

"That is the world's most beautiful shape," Trigonometeris said in a shaky voice. "I'm sure we can find the area."

We set up the definite integral for the area (Figure 8–11):

$$A = \int_0^{\pi} \sin x\, dx$$

We know the integral of sine is –cosine:

$$A = -\cos x \Big|_0^{\pi}$$

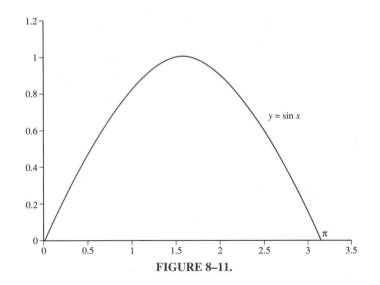

FIGURE 8–11.

Evaluating $F(\pi) - F(0)$:

$$A = -(\cos \pi - \cos 0)$$
$$A = -(-1 - 1)$$
$$A = -(-2)$$
$$A = 2$$

Trigonometeris' eyes glistened with tears of joy as he discovered the indescribably beautiful (to him) fact that the area under the sine curve was not an irrational number but actually a whole number.

Our success so far was causing the gremlin to panic. He tossed another paper our way and dared us to evaluate his definite integral:

$$A = \int_2^{10} (-x^3)\, dx$$

We pulled the −1 outside the integral:

$$A = -\int_2^{10} x^3\, dx$$

We used the power rule:

$$A = -\frac{x^4}{4}\bigg|_2^{10}$$

Evaluating $F(10) - F(2)$:

$$A = -\left(\frac{10^4}{4} - \frac{2^4}{4}\right)$$
$$A = -\left(\frac{10{,}000}{4} - \frac{16}{4}\right)$$
$$A = -(2{,}500 - 4)$$
$$A = -2{,}496$$

"Something went terribly wrong! You can't have an area that is a negative number!" We looked at the graph (Figure 8–12).

"I think I see the problem. The definite integral $\int_a^b f(x)\, dx$ gives the area under the curve $f(x)$ as long as $f(x)$ is positive everywhere between a and b. If $f(x)$ is negative everywhere in that interval, then the definite integral will give the *negative* of the area that is above the curve but below the x axis."

"Try this one!" the gremlin shouted desperately.

$$A = \int_0^{2\pi} \sin x\, dx$$

"This integral represents two arches of the sine curve," Trigonometeris said.

$$A = -\cos x\bigg|_0^{2\pi}$$

Evaluating $F(2\pi) - F(0)$:

$$A = -(\cos 2\pi - \cos 0)$$
$$A = -(1 - 1)$$
$$A = 0$$

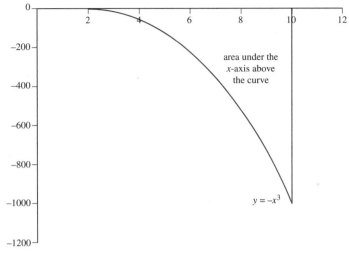

area under the
x-axis above
the curve

$y = -x^3$

FIGURE 8–12.

Now the gremlin chuckled while panic filled our eyes. "The area of two arches of the sine curve can't be zero!" Trigonomteris cried.

We looked at the graph (Figure 8–13).

"Of course," Trigonometeris explained. "Along the first arch of the curve ($x = 0$ to $x = \pi$), the function $\sin x$ is positive, so the definite integral comes up with a positive value for the area under the curve. However, along the second arch (from $x = \pi$ to $x = 2\pi$), the value of $\sin x$ is negative. As we have seen, when the function is negative, the definite integral will come up with a negative value. In this case we have a function that is positive sometimes and negative sometimes, so the definite integral has both a positive and negative part, and they happen to cancel each other out."

"This last one is sure to be your undoing," the gremlin cried desperately.

$$\int_6^3 (x^2 + 5)\, dx$$

"Haven't we already done this one?" Recordis asked.

We worked it out:

$$\left(\frac{x^3}{3} + 5x \right)\Bigg|_6^3$$

Evaluating $F(3) - F(6)$:

$$\left(\frac{3^3}{3} + 5 \times 3 \right) - \left(\frac{6^3}{3} + 5 \times 6 \right) =$$

$$\left(\frac{27}{3} + 15 \right) - \left(\frac{216}{3} + 30 \right) =$$

$$9 + 15 - (72 + 30) =$$

$$24 - 102 = -78$$

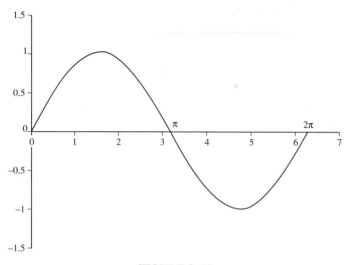

FIGURE 8–13.

"This one can't be negative! The function $x^2 + 5$ is positive everywhere between $x = 6$ and $x = 3$!" Recordis said.

The professor realized, "This is different than the one we did before."

$$\int_3^6 (x^2 + 5)\, dx = F(6) - F(3)$$

"Here is the new one."

$$\int_6^3 (x^2 + 5)\, dx = F(3) - F(6)$$

"It's obvious that the new one is the negative of the original one."

"Since the original integral represents the area under the curve, the new integral represents the negative of the area under the curve. What's the difference?" the king asked.

"The only difference is that one has the limits written as $\int_3^6$, and the other has $\int_6^3$," the professor described. We realized the definite integral $\int_a^b f(x)\, dx$ only represents the area under the curve $f(x)$ when b is greater than a. If $b < a$, then he definite integral will be the negative of the area.

The gremlin gave a loud wail of despair now that we had solved the last of his problems, and then he disappeared.

It was now long past sunset. We could finally relax now that we were safe. We talked about what we would do with all of the holes in the ground left by the gremlin and decided to turn them into decorative pools. We liked the pools so much we even added some new ones to represent some of the other curves we had investigated during our study of calculus.

"This is a very interesting unexpected turn our calculus investigation has taken," the king said. "I suspect we're only just beginning to see what we can accomplish with this study."

A Few Properties of Definite Integrals

1. The definite integral $\int_a^b f(x)\, dx$ represents the area under the curve $f(x)$ only if these two conditions are met:
   (1) $b > a$
   (2) $f(x) > 0$ for all x between a and b
2. If $b > a$ and $f(x)$ is negative everywhere between a and b, then the definite integral $\int_a^b f(x)\, dx$ represents the negative of the area above the curve and below the x axis. (See Figure 8–14.)

FIGURE 8–14.

3. If $f(x)$ is positive at some places and negative at other places between a and b, then the total area under the positive parts of the curve will contribute a positive amount to the definite integral, but the total area above the negative parts of the curve will contribute a negative amount. The final result may be positive, negative, or zero. (See Exercise 17.)
4. Reversing the two limits changes the sign of the integral:

$$\int_a^b f(x)\, dx = -\int_b^a f(x)\, dx \text{ since } F(b) - F(a) = -[F(a) - F(b)].$$

5. If $a < b < c$, then:

$$\int_a^b f(x)\, dx + \int_b^c f(x)\, dx = \int_a^c f(x)\, dx \text{ since } F(b) - F(a) + F(c) - F(b) = F(c) - F(a).$$

6. If you look carefully at a definite integral, you will see that the name of the variable of integration (the variable in the d-variable term) does not make any difference to the final value of the integral:

$$\int_a^b f(x)\, dx = F(b) - F(a)$$

$$\int_a^b f(q)\, dq = F(b) - F(a)$$

The value of the definite integral depends only on the function being integrated f (which determines the antiderivative function F) and the limits of integration a and b. Therefore, the variable of integration is sometimes called a *dummy variable*, not because it lacks intelligence but because it can be replaced with another variable without changing anything.

Notice that for an indefinite integral, however, the name of the variable of integration does matter:

$$\int x^2\, dx = \tfrac{1}{3}x^3 + C \neq \int y^2\, dy = \tfrac{1}{3}y^3 + C$$

NOTES TO CHAPTER 8

Sometimes the variable name is written in the limits. For example,

$$\int_{x=a}^{b} f(x)\, dx.$$

Exercises

---

1. Find the area under the curve $y = \cos x$ from $x = -\pi/2$ to $x = \pi/2$.

2. Find the area under the curve $y = \sin^2 x$ from $x = 0$ to $x = \pi$.

3. Find the area between the x axis and one arch of the curve $y = 3 \sin 5x$.

4. Find the area between the parabola $y = 2x^2 - 8$, the line $x = 2$, the line $x = -2$, and the x axis.

Evaluate the following.

5. $\int_0^1 (x^2 + 3x + 5)\, dx$

6. $\int_{-3}^5 (9x + 10)\, dx$

7. $\int_{d_1}^{d_2} (ax^2 + bx + c)\, dx$

8. $\int_{d_1}^{d_2} (at^2 + bt + c)\, dt$

9. $\int_0^a 14\, dx$

10. $\int_{-1}^1 (x^2 + 1)\, dx$

11. $\int_0^1 (5x^5 + 10x^3 + 3x)\, dx$

12. $\int_0^1 (4x^3 + 3x^2 + 2x + 1)\, dx$

13. $\int_0^a (x^3/3 + x^2/2 + x)\, dx$

14. $\int_{d_1}^{d_2} (x^m + x^n)\, dx$

15. $\int_0^1 [(m + 1)x^m + (n + 1)x^n + (p + 1)x^p]\, dx$

16. $\int_{-1}^1 x^{100}\, dx$

17. Compare the results for these definite integrals, and sketch the curve:

(a) $\int_0^{\pi/4} \sin\theta\, d\theta$

(b) $\int_0^{\pi/2} \sin\theta\, d\theta$

(c) $\int_0^{\pi} \sin\theta\, d\theta$

(d) $\int_0^{2\pi} \sin\theta\, d\theta$

(e) $\int_{-\pi}^{\pi} \sin\theta\, d\theta$

18. $\int_{-\pi/4}^{\pi/4} \cos\theta\, d\theta$

19. $\int_1^2 (x^{-4} + x^{-3})\, dx$

20. $\int_1^2 (x^2 + x^{-2})\, dx$

21. Use definite integrals to show that the area of a triangle is given by 1/2(base)(altitude).

22. Show that $\int_a^b f(x)\, dx = -\int_b^a f(x)\, dx$.

23. Evaluate the definite integral $y = \int_0^{\pi/2} \sin\theta\, d\theta$.

24. Use the table of values (Table 8–1) to estimate the area under the curve $y = \sin x$ from $x = 0$ to $x = \pi/2$, using (a) two inscribed rectangles and (b) eight inscribed rectangles.

Table 8–1

x	$\sin x$
0.1745	0.1736
0.3491	0.3421
0.5236	0.5000
0.6981	0.6428
0.8727	0.7661
1.0472	0.8660
1.2217	0.9397
1.3963	0.9848
1.5708	1.0000

25. The definite integral can be used to find the average value of a function over a given integral. The average value of the function $y = f(x)$ over the interval $x = a$ to $x = b$ is defined to be

$$y = [1/(b - a)] \int_a^b f(x)\, dx.$$

Find the average value of the function $y = x^2$ in the interval $x = 1$ to $x = 3$.

26. Find the average value of the function $y = x^{-2}$ over the interval $x = 1$ to $x = 4$.

27. The voltage in an alternating-current circuit is described by the function $V(t) = A \sin \omega t$. The peak-to-peak voltage (V_{pp}) is defined to be $V_{pp} = 2A$. The rms (for root-mean-square) voltage is a measure of the average value of the voltage in the circuit. It is defined to be $V_{rms} = \sqrt{\{|V(t)|^2\}_{average}}$. Find the average value of V^2 over the interval $t = 0$ to $t = \pi/\omega$.

Then take the square root of the result to find V_{rms}. (Remember that $\sin^2 \theta = \frac{1}{2}(1 - \cos 2\theta)$. What is V_{pp} when $V_{rms} = 1$?

28. The area under the parabola $y = x^2$ from $x = 0$ to $x = a$ is given by the expression:

$$A = \lim_{n \to \infty} \sum_{i=1}^{n} (x_i)^2\, \Delta x$$

You can evaluate this summation directly, if you know this tricky formula:

$$\sum_{i=1}^{n} i^2 = \frac{n}{6} (n + 1)(2n + 1)$$

Evaluate the summation, and compare the result with the definite integral $\int_0^a x^2\, dx$.

29. Show that the area between the curves $f(x)$ and $g(x)$ between $x = a$ and $x = b$ is given by the integral:

$$\int_a^b [f(x) - g(x)]\, dx$$

Assume that $f(x) > g(x) > 0$, for all $a \le x \le b$.

Natural Logarithms

The gremlin disappeared after his defeat, and we heard no more from him for some time. The whole kingdom was excited by our discovery of the method for finding areas.

"We should expand our business," the professor said. "We should call it the Differentiation and Integration Business."

"Now we can do just about anything!" Recordis boasted. "We can find areas or velocities-given-positions or positions-given-velocities or the slopes of tangents."

Business was great for the next few days. The first time we hit a snag occurred one morning when I was at the Differentiation and Integration Business office with Recordis. We were approached by a tall, elegantly dressed customer. "That's my neighbor, Count Q," Recordis whispered in my ear. "After we solve his problem, he will probably give us a gift large enough to support our business for a year."

"What can I do for you?" Recordis asked, making a special effort to impress the count.

"I have a horrible problem," the count said. "My children decided that they wanted to construct a life-size model of Hasselbluff Mountain, entirely out of beads. I was able to convince them to accept a one-tenth scale model, but that still required a huge amount of beads—4 million grams, as it turned out."

"Four million grams of beads!" Recordis breathed in astonishment.

"I had a large bead container built to hold them, but, as you might expect, my children knocked the container over this morning. The beads are all over the yard now. Fortunately, Rutherford, your dog, decided that it would be fun to pick them up. He started this morning. He does have a lot of energy, and he has been working fairly fast."

"What's the problem?" Recordis asked.

"I need to know how long it will take him to pick up the beads. At first the beads were easy to pick up, and he was able to scoop lots of them into the container every minute. As there are fewer and fewer beads, though, they're harder to pick up, so he can't pick them up as fast. I measured the rate at which he can pick up the beads, and I found out that each hour he can pick up one quarter of the beads that are in the yard at the start of the hour. Do you think you can find out how long it will take until he's finished?"

"Certainly," Recordis said proudly. "We invented an integral, symbolized by a squiggle: $\int$. Suppose you tell me what dx/dt is (that means the rate at which some variable x is changing with respect to some variable t). In your case x would be the number of beads left in the yard at a given time. If you tell us $dx/dt = f(t)$, where $f(\ )$ is some function of time, then we can easily calculate x as a function of t."

$$dx = f(t)\, dt$$

$$\int dx = \int f(t)\, dt$$

$$x = \int f(t)\, dt$$

Recordis was enjoying being able to show off his knowledge, but I could see that Count Q was becoming impatient.

"We can also do it this way, if you tell us $dx/dt = f(x)$."

$$\frac{dx}{dt} = f(x)$$

$$\frac{1}{f(x)}\, dx = dt$$

$$t = \int \frac{1}{f(x)}\, dx$$

"In your case we have $dx/dt = -1/4x$. So let's do this," Recordis said with a flourish.

$$-4\,\frac{dx}{dt} = x$$

$$-4\,\frac{1}{x}\, dx = dt$$

$$\int \left(-4\frac{1}{x}\right) dx = \int dt$$

"Then we use our Perfect Integral Rule," Recordis said, making sure that the count could hear the capital letters he had just given the name.

$$t = \int \left(-4\frac{1}{x}\right) dx$$

"Since -4 is a constant, we can use our Multiplication Rule," he continued.

$$t = -4 \int x^{-1}\, dx$$

"Now we have our Power Rule."

$$t = -4\,\frac{1}{(-1)\,+\,1}x^{-1+1} + C$$

As Recordis was about to put the grand finishing touch on the problem, he suddenly choked and a cold, desperate sweat broke out over his face. "Ah, yes, as I was saying, this is a very simple insert-the-numbers problem." He suddenly turned to me. "Why don't you tell the count why we'll have the answer for him tomorrow rather than today?"

I suddenly saw what Recordis had seen. "We have to go to lunch right now," I said.

"Lunch right now?" the count cried. "It's 11:23½ A.M.!"

"We always have lunch at 11:23½ A.M., on the dot," Recordis said. He quickly pulled the curtain in front of the Differentiation and Integration Business and ran into the palace.

"Help! Help! Meet right away!" Recordis shouted. A few minutes later we were all gathered in the Main Conference Room. Recordis told everyone our problem.

"The power rule doesn't work if $n = -1$!" he cried. "I have never been so embarrassed in my life."

"Calm down," the professor said. "I'm sure we can figure something out. Igor, write down the power rule."

$$\int x^n \, dx = \frac{1}{n + 1} x^{n+1} + C$$

"Just try putting $n = -1$ in there!" Recordis yelled.

"You get 1/0 times x^0, which is 1/0 times 1, or 1/0," the professor said.

"But that doesn't mean anything!" the king protested. "We found that out in algebra."

"Also we can see that it doesn't work," Trigonometeris said. "It is obvious that $(d/dx)(1/0)$ does not equal $1/x$."

"We better add that condition to the power rule," the king said.

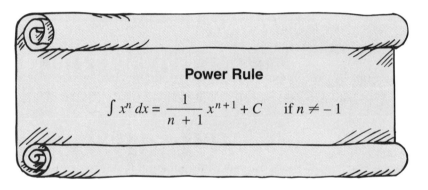

Power Rule

$$\int x^n \, dx = \frac{1}{n + 1} x^{n+1} + C \qquad \text{if } n \neq -1$$

"That must mean that there is no such thing as a function with derivative x^{-1}," the professor said. "If there is no such function, then the integral $\int x^{-1} \, dx$ does not exist. I can't think of any reason why we should worry about that integral."

"But there is such a function!" Recordis told him. He explained the problem with the beads. He also said that if they could find an answer by that night they would probably receive a generous gift from the count.

"Maybe we can use a definite integral to help us out," the king said. "We found that an integral represents the area under the curve, so let's draw a curve and find the area under it."

Igor drew a graph of the function $f(q) = 1/q$. (See Figure 9–1.)

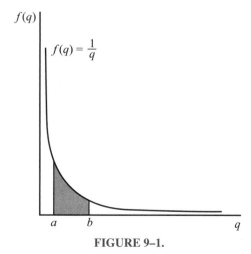

FIGURE 9–1.

"I know what this definite integral means," the professor said.

$$\int_a^b f(q)\, dq = \int_a^b \frac{1}{q}\, dq = \text{area above the } q \text{ axis, below the curve } f(q) = 1/q, \text{ to the}$$
$$\text{left of the line } q = b, \text{ and to the right of the line } q = a$$

"The integral function must exist!" Recordis said. "That area is clearly some real number that you could measure if you had to."

"All we have to do is set up a Mysterious Function that gives us this area and investigate what its properties are," the professor said. "If we find a function that gives us the area, then that means we have also found the function that gives us the antiderivative we need."

"That's from the fundamental theorem of integral calculus," Recordis added.

"We should fix the left-hand boundary of the area," the king said. "Then the area will be a function of just one variable: the right-hand boundary."

We decided to fix the left-hand boundary at $q = 1$, since it was clear that the Mysterious Function would do strange things if we fixed the left-hand boundary at the point where $q = 0$. 1 suggested that we use the letter L to stand for the Mysterious Function, so we made the following definition (Figure 9–2):

$$L(a) = \int_1^a \frac{1}{q}\, dq$$

"How do we find out what this function is?" Trigonometeris asked.

"I would suggest that the first thing we should do is look for properties," the professor said. "Does anybody see any obvious properties of the Mysterious Function?"

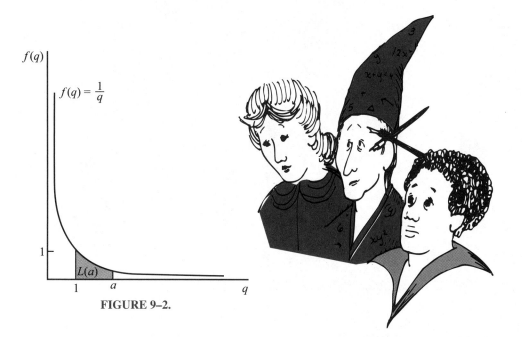

FIGURE 9–2.

"I see one obvious property," the king replied. "It looks as though $L(1) = 0$."

"That's a start," the professor said.

"We can also say that, when a is greater than 1, $L(a)$ is greater than 0," Recordis said.

"And from the way we set up our convention for the sign of an area, we can say that, when a is less than 1, $L(a)$ is less than 0," Trigonometeris added. (See the list at the end of Chapter 8 and Exercise 8–22.)

"What happens if a is less than or equal to 0?" Recordis asked.

"Then everything would blow up!" the professor said. "You couldn't define the area, so we may just as well say that $L(a)$ is undefined when a is less than 0."

We made a list of the properties we had found.

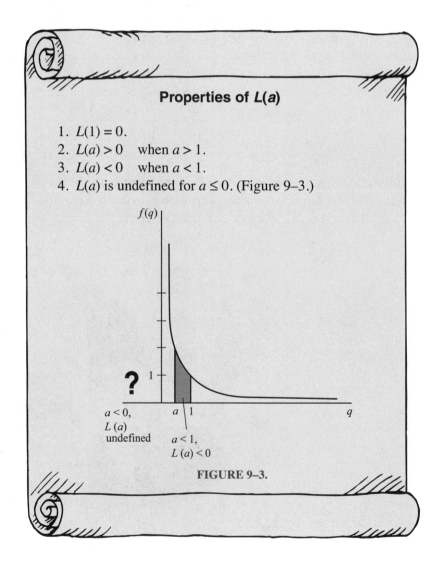

Properties of L(a)

1. $L(1) = 0$.
2. $L(a) > 0$ when $a > 1$.
3. $L(a) < 0$ when $a < 1$.
4. $L(a)$ is undefined for $a \leq 0$. (Figure 9–3.)

FIGURE 9–3.

"Does anybody recognize these properties?" the professor asked.

"I do!" the king said. "They do look familiar! The answer's right at the tip of my tongue, but I can't think of it."

Nobody else could think of any algebraic or trigonometric functions that fit these properties, but Trigonometeris came up with an idea that would allow us to calculate values for $L(a)$.

"Remember when we invented my functions—the trigonometric functions? We decided that they were more complicated than ordinary functions, so we constructed my triangles so we could measure the sine, cosine, or tangent of any angle. After we made the triangles, we drew up a table of values so we could just look in the table for the trigonometric functions of any given angle. All we have to do now is build something to calculate $L(a)$, and then make a table of values. Let's have Builder build a pool with a sliding wall (Figure 9–4). All he has to do is set the sliding wall at different values of a, and then measure the amount of water in the pool."

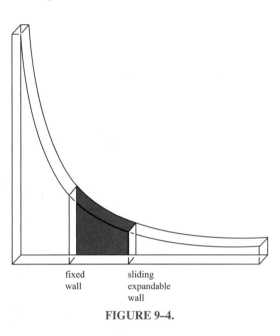

Builder was immediately summoned to the conference room. He was told the urgency of the project, and he promised to put his best craftspeople to work right away.

FIGURE 9–4.

"Make sure the pool is accurate!" the professor called after him as he went away.

We ate lunch while we waited for Builder to bring us a table of values. After a while we decided to look for some more properties. We tried to find an expression for $L(a + b)$, where a and b are any two numbers, but we soon gave up. We quickly established that $L(a + b)$ does not equal $L(a) + L(b)$ (let $a = b = 1$, and see what happens), but we could not think of anything else that worked. We decided to find an expression for $L(ab)$.

$$L(ab) = \int_1^{ab} \frac{1}{q}\, dq$$

"I know how we can simplify that expression," the king said.

$$L(ab) = \int_1^{ab} \frac{1}{q}\, dq = \int_1^a \frac{1}{q}\, dq + \int_a^{ab} \frac{1}{q}\, dq$$

$$= L(a) + \int_a^{ab} \frac{1}{q}\, dq$$

"How does that help?" the professor said. "We don't know what that second term is."

"If we could change the a in the lower bound into a 1, we would be in better shape," Recordis remarked.

"Maybe we can do that," the king said "We can make a substitution, so that instead of q as the integration variable we have some other variable."

Recordis suggested that we try $u = q/a$, $q = au$.

$$L(ab) = L(a) + \int_{q=a}^{q=ab} \frac{1}{au} \, dq$$

"Now we can't solve it because the integrand is in terms of u while the boundary limits and the d-variable terms are still in terms of q."

"Then we will have to change them so they are in terms of u," the king said. "Since $q = au$, we know that $dq = a \, du$."

"And I can't see any reason why we can't change the limits so that they are in terms of u," Recordis added. "If $u = q/a$, then $u = b$ when $q = ab$ and $u = 1$ when $q = a$."

We put these values in the integral:

$$L(ab) = L(a) + \int_{q=a}^{q=ab} \frac{1}{au} \, dq = L(a) + \int_{u=1}^{u=b} \frac{1}{au} \, a \, du$$

"Are you sure that is the same integral we started with?" the professor asked suspiciously.

"It should be," the king said. "All we've done is change the name of a variable."

We made a list of the procedure for the substitution method for definite integrals.

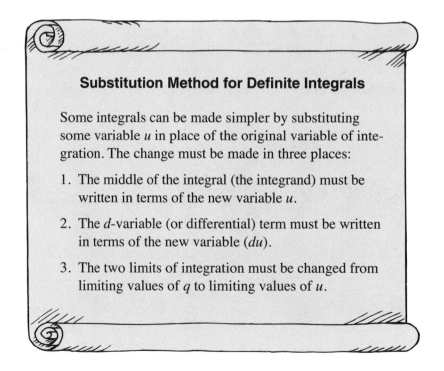

Substitution Method for Definite Integrals

Some integrals can be made simpler by substituting some variable u in place of the original variable of integration. The change must be made in three places:

1. The middle of the integral (the integrand) must be written in terms of the new variable u.

2. The d-variable (or differential) term must be written in terms of the new variable (du).

3. The two limits of integration must be changed from limiting values of q to limiting values of u.

"Now where were we?" Recordis said.

$$L(ab) = L(a) + \int_1^b \frac{a\,du}{au}$$

"We can cancel out the a's," the professor suggested.

$$L(ab) = L(a) + \int_1^b \frac{du}{u}$$

"We know what the second term is!" she said suddenly. "That's the definition of $L(b)$!"

$$L(ab) = L(a) + L(b)$$

"That's an amazingly simple property for this function," the professor added.

"Didn't we see this property before?" Recordis said, leafing through his old algebra book.

"I recognize it!" the king exclaimed triumphantly. "Remember the logarithm function!"

"The logarithm function!" the professor said in awe.

"The logarithm function!" Trigonometeris said in awe.

"The logarithm function!" Recordis said in awe. "What's the logarithm function?"

"Don't you remember?" the professor chided. "We said that, if $y^x = a$, then $x = \log_y a$ (this means that x is the logarithm to the base y of a)." The professor pointed to a page in the book that described the logarithm function.

"Oh," Recordis said. "Now I remember." (The people in Carmorra had developed a fairly extensive list of properties of the logarithm function. If you are not familiar with logarithms to the base 10, you can consult a book on algebra.)

"And the logarithm function has all the properties we found for our mysterious function," the king pointed out.

$$\log 1 = 0$$
$$\log a > 0 \qquad \text{when } a > 1$$
$$\log a < 0 \qquad \text{when } a < 1$$
$$\log a \text{ is not defined for } a \leq 0$$
$$\log (ab) = \log a + \log b$$

"But there is something else about the logarithm function," Recordis said. "We have to specify a *base*. We decided that, if we wrote $\log a$ without explicitly stating a base, we meant the logarithm to the base 10 ($\log a = \log_{10} a$), and we called that the *common logarithm*. But we could also take logarithms to any base: $\log_2 a$, $\log_{100} a$, or whatever."

"Except for the number 1," the professor added.

"That's right," Recordis said. "We found that there is no such thing as a logarithm to the base 1 ($\log_1 a$). But that doesn't help us know what base to use for the Mysterious Function, $L(a)$."

"We'll have to wait until Builder has brought us a table of values," the king stated. "Then we should be able to tell what base it is."

"I hope it's some easy number like 2 or 3," Recordis said. "We need to think of some letter to stand for the unknown base until we find out what it is."

They turned to me, and I came up with another suggestion: that we let the letter e stand for the unknown base.

"We still have to check this," the professor said. "You are saying that integrating a $1/x$ function gives you a logarithm function. If that's true, then *differentiating* a logarithm function will give you a $1/x$ function."

We set up the definition of the derivative:

$$y = \log_e x$$

$$y + \Delta y = \log_e (x + \Delta x)$$

$$\Delta y = \log_e (x + \Delta x) - \log_e x$$

"We should have done this a long time ago," Recordis remarked. "I thought that we had found the derivative of every single possible function already, but I completely forgot about the logarithm function."

"We can use a logarithm property here," the king said. "Remember that $\log a - \log b = \log(a/b)$."

$$\Delta y = \log_e \left(\frac{x + \Delta x}{x} \right) = \log_e (1 + \Delta x/x)$$

"We had better divide both sides by Δx," the professor advised. "That seems to be the standard way to proceed under these circumstances."

$$\frac{\Delta y}{\Delta x} = \frac{\log_e (1 + \Delta x / x)}{\Delta x}$$

"Now we're stuck," Recordis said.

We puzzled over this situation for several minutes, until I had a sudden inspiration. "Since we would like to end up with $1/x$, why don't we try multiplying the numerator and the denominator by x? That way we'll have at least one x in the denominator."

"It can't hurt too much," Recordis said.

$$\frac{\Delta y}{\Delta x} = \frac{x \, \log_e (1 + \Delta x / x)}{x \, \Delta x}$$

We rearranged the order of the factors:

$$\frac{\Delta y}{\Delta x} = \frac{1}{x}\,\frac{x}{\Delta x}\,\log_e\left(1 + \frac{\Delta x}{x}\right)$$

"Somewhere along in here you will need to put a limit as Δx goes to zero on both sides," Recordis pointed out.

$$\lim_{\Delta x \to 0}\frac{\Delta y}{\Delta x} = \frac{dy}{dx} = \lim_{\Delta x \to 0}\frac{1}{x}\,\frac{x}{\Delta x}\,\log_e\left(1 + \frac{\Delta x}{x}\right)$$

Recordis wanted a shorter way to write this expression, so we made the definition $w = \Delta x/x$. Then we could rewrite the derivative:

$$\frac{dy}{dx} = \frac{1}{x}\,\lim_{w \to 0}\frac{1}{w}\,\log_e(1 + w)$$

"I remember another property of logarithms," the professor said. "We discovered that $n \log x = \log(x^n)$."

$$\frac{dy}{dx} = \frac{1}{x}\,\lim_{w \to 0}\,\log_e(1 + w)^{1/w}$$

$$= \frac{1}{x}\,\log_e\,\lim_{w \to 0}\,(1 + w)^{1/w}$$

"I have an idea," Recordis stated. "If we could make this weird term, $\log_e \lim_{w \to 0}(1 + w)^{1/w}$, equal 1, then we would have the answer we want: $dy/dx = 1/x$."

"That's obvious," the professor said. "We can't just make it be equal to 1, though."

"He does have something!" the king exclaimed. "We might be able to use this expression to track down the mysterious number e! Remember that $\log_{(\text{some base})}$ (some base) = 1, always. Remember that $\log_2 2 = 1$, $\log_{10} 10 = 1$, and $\log_{17} 17 = 1$, because $2^1 = 2$, $10^1 = 10$, and $17^1 = 17$. So we should have $\log_e e = 1$! And we also want to have $\log_e \lim_{w \to 0}(1 + w)^{1/w} = 1$! That means $\lim_{w \to 0}(1 + w)^{1/w}$ equals e!"

"Of course!" Recordis said. "The old mysterious-number-e-caught-up-in-the-screwy-limit trick!"

"Can we calculate that?" the professor asked.

"It's just a matter of arithmetic," the king said. "We can't calculate e if $w = 0$, because then we would get $e = 1^\infty$, which doesn't help much. But we can calculate e for any value of w close to zero."

Igor and Recordis soon came up with a table of values (Table 9–1).

"Just exactly what I was afraid of!" Recordis moaned. "Look, e turns out to be some number between 2.7181 and 2.7196."

Table 9–1

w	$(1 + w)^{1/w}$
0.5	2.25
0.1	2.5937
0.01	2.7048
0.0001	2.7181
−0.001	2.7196
−0.01	2.7320
−0.1	2.8680

"But *why?*" the king asked. "Is there something special about a number that is approximately 2.718? I can't think of any reason why there should be."

"It must be an irrational number," Trigonometeris remarked. "I don't think you could ever find an exact decimal representation for it. Just like most trigonometric functions."

"Is there anything else like it?" the king asked in great distress. He believed very strongly that he should be an impartial ruler, and it bothered him that one number, more than any other, should be singled out as the base for the $L(x)$ function.

"I can think of only one other case," the professor said solemnly. "When we found the circumference of a circle, we decided that the circumference was $2\pi r$, where r is the radius. We found there was a special number, which we called pi, because it seemed to be so fundamental that it worked for any circle."

"I remember having to calculate a decimal approximation for π," Recordis added. "We came up with $\pi = 3.141592654$. The professor said that we could calculate π out to thousands of decimal places if we wanted to, but I made it clear that we were not going to calculate it out to thousands of places unless someone came up with a thousand very good reasons why we should."

"We decided that we had discovered a fundamental irrational number," the professor said. "I had thought π would be the only one. We'll have to add the number e to our list."

Recordis entered in his giant record book:

$$e = 2.718 \ldots$$

"Don't get carried away," Trigonometeris warned. "Remember that we still have to get the values of the $L(x)$ function from Builder. If you guys are right and $L(x) = \log_e x = \log_{2.718} x$, then $L(e) = L(2.718)$ must equal 1. If Builder comes back with a different value, we're sunk."

We waited nervously for Builder to come back with his table of values. Nobody wanted to rush him because we knew that he must do an accurate job. Finally there was a knock at the door, and Builder came in. "This table should have all the values you could possibly want," he said, exhausted.

"Read off all the values between 2.65 and 2.75," the professor requested. Builder looked puzzled, because these seemed like strange numbers, but he read them off (Table 9–2).

"It does work!" the professor exclaimed. "$L(x)$ does equal 1 somewhere between 2.71 and 2.72, just as it should. Then $L(x)$ must be the logarithm function."

Table 9–2

x	$L(x)$
2.65	0.975
2.66	0.978
2.67	0.982
2.68	0.986
2.69	0.990
2.70	0.993
2.71	0.997
2.72	1.001
2.73	1.004
2.74	1.008
2.75	1.012

"We had better write this as a rule," Recordis said. "First we need to think of a short way to write this function. We wrote log x to stand for $\log_{10} x$, so let's think of something to stand for $\log_e x$."

"We could write $\log_e x = \ln x$," the king offered. "I don't know why, but it looks pretty."

"That's good because it has fewer letters," Recordis said. "And we need to think of a name for a logarithm to the base e." I suggested the name *natural logarithm*, since it seemed as though this function had arisen in the natural course of our investigations of the properties of integrals.

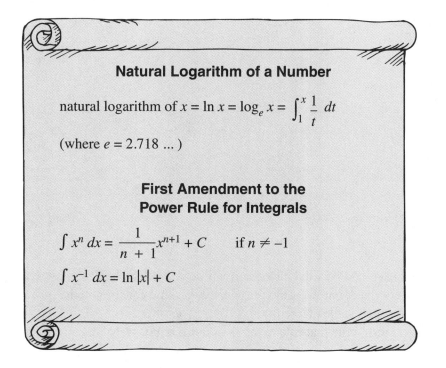

Natural Logarithm of a Number

natural logarithm of $x = \ln x = \log_e x = \int_1^x \frac{1}{t}\, dt$

(where $e = 2.718 \ldots$)

First Amendment to the Power Rule for Integrals

$\int x^n\, dx = \dfrac{1}{n + 1} x^{n+1} + C \qquad \text{if } n \neq -1$

$\int x^{-1}\, dx = \ln |x| + C$

WORKSHEET

Find the derivative for each of these:

1. $y = \ln x$

2. $y = \ln (a + x)$

3. $y = \ln (bx)$

4. $y = \ln (a + bx)$

5. $y = \ln (x^n)$

Evaluate each integral:

6. $\int \left(\dfrac{1}{x} \right) dx$

7. $\int \left(\dfrac{1}{a+x} \right) dx$

8. $\int \left(\dfrac{1}{bx} \right) dx$

9. $\int \left(\dfrac{2ax}{ax^2+b} \right) dx$ (Let $u = ax^2 + b$.)

10. $\int \left(\dfrac{x}{ax^2+b} \right) dx$

11. $\int_1^2 x^{-1} \, dx$

12. $\int_1^2 x^{-.9999} \, dx$

(We later investigated what happens if x is negative. The logarithm of a negative number is not defined, but we can still use the logarithm function in the power rule if we take the absolute value of x ($|x|$).)

"We could also write that rule in terms of definite integrals," the professor said.

$$\int_a^b x^{-1} \, dx = \ln b - \ln a$$

"No," Recordis said firmly. "We know that the connection between definite integrals and indefinite integrals always holds, so we definitely do not need to write each rule once in terms of indefinite integrals and once again in terms of definite integrals."

We went back to Count Q's problem:

$$t = -4 \int x^{-1} \, dx, \text{ with the initial condition } x = 4 \text{ million when } t = 0$$

"We can solve that right away," the professor said.

$$t = -4 \ln x + C$$

"Using the initial condition gives us the following expressions."

$$0 = -4 \ln (4 \text{ million}) + C$$

$$C = 4 \ln (4 \text{ million})$$

"Builder, can you give us a value for ln (4 million)?" the king asked.

Builder looked aghast, but a few minutes later he came back with the answer from his logarithm pools.

$$\ln (4 \text{ million}) = 15.2$$

$$C = 4 \times 15.2 = 60.8$$

"Now what does t equal when x is zero?" Recordis said. "That will tell us how long it will take Rutherford to scoop up all the beads."

$$t = -4 \ln x + 60.8$$

"Hold it," the professor protested. "We said that there was no such thing as $L(0)$ or $\log_{(\text{any base})} 0$. That means ln 0 doesn't exist."

"That means Rutherford can *never* pick up all the beads!" Trigonometeris said in dismay.

"Of course, that's right," the king stated. "The count said that in any time period Rutherford could scoop up only one fourth of the beads that were left. That way he could never get them all. Even if there was only 1 gram of beads left, he could pick up only one fourth of that gram."

"Don't be so picky!" Recordis said disrespectfully. "If Rutherford can sweep up all but 1 gram of beads, I will be glad to personally pick up the rest."

We easily calculated the value of t for $x = 1$:

$$t = -4 \ln 1 + 60.8 = 60.8 \text{ hours}$$

"That's 2½ days," Recordis said. "Not bad, considering how many beads there were to start with." Recordis went running off to take the answer to the count. Sure enough, the count did give us a nice gift for providing him with the answer. We were still amazed at the strange paths along which this investigation was taking us.

NOTE TO CHAPTER 9

The q in the integral defining the logarithm function could have been renamed g or q' or s'' or anything else without affecting the value of $L(x)$:

$$L(x) = \int_1^x \frac{1}{q}\, dq = \int_1^x \frac{1}{g}\, dg = \int_1^x \frac{1}{q'}\, dq' = \int_1^x \frac{1}{s''}\, ds''$$

See the note about dummy variables at the end of Chapter 8.

Exercises

1. On one occasion the gremlin tried to take over the kingdom with a bacteria dish. The bacteria multiplied at such a rate that $dn/dt = 3n$. At $t = 0$ hour, the number of bacteria was 10. If the gremlin had not been stopped, how long would it have been before n equaled 1,000?

Find y' for:

2. $y = \ln x^2$

3. $y = \ln (-x)$

4. $y = \ln \sin x$

5. $y = \ln \tan x$

6. $y = \ln (x^2 + 4)^{1/2}$

7. $y = \ln (ax^2 + bx + c)$

8. $y = \ln \left(x\sqrt{x+1} \right)$

9. Use the definition of the derivative to find y' for $y = \log_{10} x$.

10. Sketch the graph of the curve $y = (\ln x)/x$. Find any points with horizontal tangents. Find any points of inflection.

11. Use 10 inscribed rectangles to estimate ln 2. (For a helpful table of values, see Table 9–3.)

Table 9–3

x	$1/x$
1.1	0.9091
1.2	0.8333
1.3	0.7692
1.4	0.7143
1.5	0.6667
1.6	0.6250
1.7	0.5882
1.8	0.5556
1.9	0.5263
2.0	0.5000

12. For $\ln x = \int_1^x (1/t)\, dt$, show that $\ln (a/b) = \ln a - \ln b$.

13. For $\ln x = \int_1^x (1/t)\, dt$, show that $\ln (x^n) = n \ln x$.

Evaluate. In each case, make a substitution.

14. $y = \int \dfrac{1}{x+4}\, dx$ (let $u = x + 4$)

15. $y = \int \dfrac{1}{3x+4}\, dx$ (let $u = 3x + 4$)

16. $\int \dfrac{1}{x^2 + 2x + 1}\, dx$

17. $\int \dfrac{1}{ax + b}\, dx$

18. $\int \dfrac{1}{(ax + b)^2}\, dx$

19. $\int \dfrac{x}{3x^2 + 6}\, dx$

20. $\int \dfrac{2x + 4}{x^2 + 4x + 6}\, dx$

21. $\int \dfrac{\sin \theta}{\cos \theta}\, d\theta$

22. $\int \dfrac{\cos \theta}{\sin \theta + 4}\, d\theta$

23. Consider the function $y = f(x) = (1 + x)^{1/x}$. Find $f(x)$ for the following values of x: $10, 8, 6, 4, 2, 1, 0.5, 0.25, 0.1, -0.1, -0.25, -0.5, -0.6, -0.7, -0.8, -0.9$. Make a sketch of the curve. What is $f(0)$? What can you say about $\lim_{x \to 0} f(x)$? What is $\lim_{x \to \infty} f(x)$?

24. The *work* done in compressing a container of gas is given by $-\int_{V_1}^{V_2} P\,dV$, where P is the pressure of the gas, V_1 is the initial volume, and V_2 is the final volume. For a gas that obeys the ideal gas law (PV = constant), find the work done in compressing the gas from V_1 to V_2. What is the work if $V_1 = 2V_2$? What is the sign of the work if $V_1 > V_2$?

25. Find a formula for this integral:

$$\int_1^e t^n\,dt$$

where n approaches very close to -1. Then find a formula for e.

Exponential Functions
and Integration by Parts

Pal decided that the logarithm pool made a perfect adjustable swimming pool, and he had fun splashing around in the logarithms of different numbers. Meanwhile, Builder had finished the table of logarithms, and now we had to decide what to do with it.

"Maybe we should make a graph of the logarithm function," Trigonometeris suggested. "After we first discovered the trigonometric functions, we found that it helped to make graphs of them."

Builder got a giant glass plate out of his supply room and brought out his best etching equipment.

"The easiest point is $x = 1$," the king said. "We know that $y = \ln 1 = 0$."

"We also know that $\ln x$ is not defined for negative numbers," the professor added. "That means that we need to worry only about points that are to the right of the line $x = 0$."

"We can use what we did with derivatives," Recordis said. "Remember concave up and concave down?"

The derivatives were easy to calculate:

$$y = \ln x$$

$$y' = \frac{1}{x}$$

$$y'' = (-1)x^{-2} = \frac{-1}{x^2}$$

"We know that y' is never zero," the king said. "This means that the curve must never have a horizontal tangent. And y' is always positive, so the curve is sloping upward all the time."

"And y'' is always negative," Recordis said. "This means that the curve is always concave downward."

With this information, plus the results from the table of values, Builder was able to etch the curve on the glass plate. (See Figure 10–1.)

"Let's store this graph in the National Archives Room," the king suggested. Nobody else could carry the large plate, so Pal began to lift it. The glass was slippery at the edges, and he had barely picked it up when he started to slip. As he fell to the floor of the Main Conference Room, he tried to protect the glass plate. He was able to save it from breaking, but it landed on the ground backwards from the way it had been.

"Not fun!" Pal cried.

"Pal, what did you do to the graph?" the professor asked, looking at the backward view of the logarithm function through the glass plate. Pal began to cry, thinking that the professor was terribly angry at him.

"You made a whole new graph!" the professor said. "I've never seen a graph like this before!" She began to get excited, and she congratulated Pal. He smiled and jumped up and down.

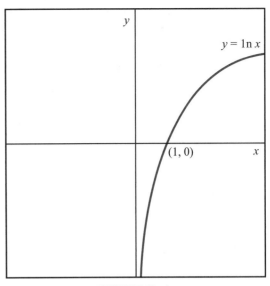

FIGURE 10–1.

"We have created an *inverse function*," Recordis said, after the kingdom had stopped shaking. "A long time ago we said that we could do that if we traded the x and y coordinates on a graph."

"Let's find out what the inverse function is," the king suggested. "We still have $y = \ln x$. Let's trade x and y around so that the horizontal axis is x again and the vertical axis is y again."

"That would make me feel less disoriented," the professor said. (Figure 10–2.)

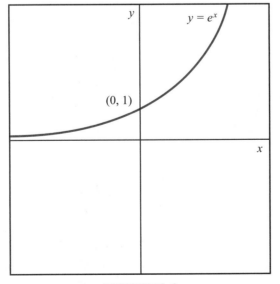

FIGURE 10–2.

We solved for y to determine the new function:

$$x = \ln y$$

$$e^x = e^{\ln y}$$

$$y = e^x$$

"Let's call it an *exponential function*," the professor said, "since we're raising x to an exponent."

"Is it good for anything?" Recordis asked.

"I think it's a good way to get big numbers fast," the professor replied thoughtfully. "Recordis, suppose I gave you a choice between having, after x days, either x^{20} or e^x dollars. Which would you choose?" Recordis, who knew something about algebraic functions, chose x^{20}, so the professor was left with e^x. We made a table of the number of dollars each would have on a given day.

x (days)	x^{20} (Recordis)	e^x (professor)
1	1	2.7
2	1,048,576	7.4
3	3.5 billion	20.1
4	1.1 trillion	54.6
5	9.5×10^{13}	148

"It looks as though you have no chance," Recordis said, feeling sorry for the professor. "Don't worry. I'll be generous."

"Let's wait a bit longer," the professor said.

10	1.0×10^{20}	2.2×10^4
20	1.0×10^{26}	4.9×10^8
50	9.5×10^{33}	5.2×10^{21}

"The professor is catching up!" the king remarked.

"Still, if I had that kind of money I could buy the whole world!" Recordis said.

100	1.0×10^{40}	2.7×10^{43}
200	1.0×10^{46}	7.2×10^{86}

"The professor was right!" the king said, surprised. "In less than a year she would have $7.2 \times 10^{40} = 72,000,000,000,000,000,000,000,000,000,000,000,000,000$ times as much money as Recordis would."

"I still wouldn't mind having 1.0×10^{46} dollars," Recordis replied defensively. "Besides, how do we know that the function e^x doesn't turn around and start to get smaller for larger values of x?"

"We can check the derivative and see whether it is ever negative," the professor said.

We used the power rule:

$$y = e^x$$

$$y' = xe^{x-1}$$

"That was easy," the professor said.

"It doesn't work, though," Recordis pointed out, anxious to prove that he wasn't as dumb as the professor had just made him look. "Look at $x = 0$."

$$x = 0$$

$$\left(\text{king's idea of } \frac{dy}{dx}\right) = 0$$

"But if you look at the graph (Figure 10–2) it is clear that when $x = 0$ the slope of the curve is not zero."

"What happened?" the professor said. We were all stunned. "The power rule always worked before."

Finally the king had a suggestion, "We proved the power rule using an exponent that was a constant. Maybe it doesn't work if you have a constant number raised to a variable exponent."

We tried to find the derivative of e^x directly, using the definition of the derivative, but a few minutes later we gave up. Finally Recordis came up with a suggestion.

"I remember some of the work we did with logarithms," he said. "We found that they were most useful because they could make some calculations a lot simpler. Naturally, I liked them very much. It seemed to me that, whenever we had numbers raised to powers, it helped to take the logarithm of both sides of the equation."

$$y = e^x$$

$$\ln y = \ln e^x = x$$

"Now we don't have an explicit function," the king said, "but maybe we can use implicit differentiation." (See Chapter 4.)

We applied d/dx to both sides:

$$\frac{d}{dx} \ln y = \frac{d}{dx} x$$

We could differentiate the left-hand side using the chain rule; and the right-hand side was obviously equal to 1:

$$\frac{1}{y} \frac{dy}{dx} = 1$$

$$\frac{dy}{dx} = y$$

$$= e^x$$

"That can't be right!" the professor objected. "You can't have a function whose derivative is itself!"

"Why not?" Recordis asked. The professor couldn't think of any reason why a function couldn't have a derivative equal to itself.

"That means you could take the second derivative and get the same thing!" the king said.

$$y = e^x$$

$$\frac{d^2 y}{dx^2} = e^x$$

"You could take the hundredth derivative and still get the same thing!" Recordis exclaimed.

$$y = e^x$$

$$\frac{d^{100} y}{dx^{100}} = e^x$$

"Amazing!" the professor murmured. "The function is indestructible! You could differentiate it forever and still not change it!"

"We can even work that backwards," Recordis said.

$$\int e^x \, dx = e^x + C$$

"Let's see what happens if we raise some other number to the x power," the professor suggested.

We used logarithms to find the derivative of $y = 2^x$:

$$\ln y = \ln 2^x = x \ln 2$$

$$\frac{1}{y} \frac{dy}{dx} = \ln 2$$

$$\frac{dy}{dx} = 2^x \ln 2$$

"At least its derivative isn't equal to itself," the professor remarked. We went through exactly the same calculations to make a general rule:

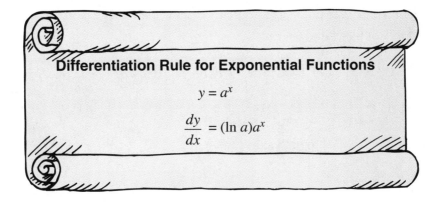

Differentiation Rule for Exponential Functions

$$y = a^x$$

$$\frac{dy}{dx} = (\ln a)a^x$$

"What if we have a variable in both the exponent and the base?" the professor asked. "How about $y = x^x$?" We used the same method, and it was only a little more complicated this time.

$$y = x^x$$

$$\ln y = x \ln x$$

Use the product rule for the right-hand side:

$$\frac{1}{y}\frac{dy}{dx} = x\frac{1}{x} + \ln x\left(\frac{dx}{dx}\right)$$

$$\frac{dy}{dx} = y(1 + \ln x)$$

"We should remember this method," Recordis said, always looking for simpler ways to do things. "We'll call it the method of *logarithmic implicit differentiation*."

"What other kinds of functions would you use it for?" the king asked.

"I have been having nightmares about complicated functions that I was afraid you would have me differentiate some day. Naturally, once you people have developed all the theory, I get stuck with all the hard work. Just imagine a function like . . . ," and Recordis wrote down the first complicated function that came into his head.

$$y = \sqrt{\frac{(x-1)(x+3)^2(x+4)^3}{(x+1)(x+2)(x-3)}}$$

"We could find dy/dx if we had to," the professor said, "using the power rule, the chain rule, and the product rule."

"But it will be a lot easier like this," Recordis said. "Look what happens if we take the logarithm of both sides."

$$\ln y = \ln\left[\frac{(x-1)(x+3)^2(x+4)^3}{(x+1)(x+2)(x-3)}\right]^{1/2}$$

Using the properties of logarithms that we knew, Recordis was able to simplify this expression quite a bit.

$$\ln y = \tfrac{1}{2}\{\ln [(x-1)(x+3)^2(x+4)^3] - \ln [(x+1)(x+2)(x-3)]\}$$

$$= \tfrac{1}{2}[\ln (x-1) + 2 \ln (x+3) + 3 \ln (x+4) - \ln (x+1) - \ln (x+2) - \ln (x-3)]$$

Now we applied d/dx to both sides:

$$\frac{1}{y}\frac{dy}{dx} = \frac{1}{2}\frac{d}{dx}[\ln (x-1) + 2 \ln (x+3) + 3 \ln (x+4) - \ln (x+1) - \ln (x+2) - \ln (x-3)]$$

$$= \frac{1}{2}\left(\frac{1}{x-1} + \frac{2}{x+3} + \frac{3}{x+4} - \frac{1}{x+1} - \frac{1}{x+2} - \frac{1}{x-3}\right)$$

$$\frac{dy}{dx} = \frac{1}{2}\sqrt{\frac{(x-1)(x+3)^2(x+4)^3}{(x+1)(x+2)(x-3)}} \times \left(\frac{1}{x-1} + \frac{2}{x+3} + \frac{3}{x+4} - \frac{1}{x+1} - \frac{1}{x+2} - \frac{1}{x-3}\right)$$

"That's still a complicated answer," the professor said.

"Of course the answer is just as complicated as it would be by any other method!" Recordis retorted. "It is just that the method involved much less work than we would have had if we had tried to find the answer by brute force. We definitely will remember this method when we have to differentiate complicated expressions involving powers or products. Now all I need is a logarithmic curve pool," Recordis went on excitedly. "After all, that is the function we started with before Pal inverted our graph."

$$y = \ln x \quad \text{from } x = 1 \text{ to } x = 5$$

"That will give us an expression for the area."

$$\int_1^5 \ln x \, dx =$$

Suddenly he stopped short.

"We don't know how to integrate the logarithm function," Recordis said slowly. His enthusiasm began to fade.

"Why did you decide to make a pool like that, anyway?" the professor asked him. "What if the gremlin comes back and repeats his fire-and-water threat, only we don't know the antiderivative function?"

We pondered for a long time, but we couldn't think of a function that would work.

"Maybe there isn't any function that works," Trigonometeris said.

"That's what we said last time," the king reminded him.

"But then we created a Mysterious Function to represent the area under the $1/x$ function, and we were able to find some simple properties. Suppose we create a Mysterious Function for the area under a logarithm curve, and it doesn't have any simple properties?" the professor asked.

"We can have Builder make another pool and get a table of values," the king said.

"But that could go on forever!" Recordis complained. "We'll make a pool for the integral of the logarithm, and then we'll need a pool for the integral of the integral of the logarithm and then a pool for the integral of the integral of the integral . . ."

"I see your point," the professor said. "We have developed rules that allow us to differentiate any function. But we still haven't developed rules that will allow us to integrate any function."

"That means we just need more integration methods," the king stated confidently.

"But how do we know that we will always be able to find rules that work? Suppose there are some functions that just don't have any kind of simple antiderivative. Then the only way we could integrate them would be to build a pool and measure the area directly." (See Chapter 17.)

"Maybe there are lonely functions like that—functions that lack antiderivatives," the king said. "But I'm sure a nice, simple function like ln x can't be one of them."

There was a long silence before the professor began to develop a carefully planned idea. She was anxious to prove that she was the one who came up with the ideas when they really counted.

"The only way to write an integration rule is to rewrite a differentiation rule backwards," she said slowly. "We did that for the power rule, and we did approximately the same thing for the differentiation chain rule when we developed the method of integration by substitution. But . . . we still have not developed an integral form for the differentiation product rule!"

Igor wrote down the rule:

$$\frac{d(uv)}{dx} = u\frac{dv}{dx} + v\frac{du}{dx}$$

"Remember what we said about differentials," Recordis reminded her. "We can simplify both sides by multiplying them by dx."

$$d(uv) = u\,dv + v\,du$$

"Let's integrate," the professor said.

$$\int d(uv) = \int u\,dv + \int v\,du$$
$$uv = \int u\,dv + \int v\,du$$
$$\int u\,dv = uv - \int v\,du$$

"How does that help?" the king asked.

"Why, u and v can be anything!" the professor said. "Once we define u and v, we can rewrite any integral we want in this form!"

"So what?" Recordis asked. "We still haven't gotten rid of the integral sign! We still have to evaluate the integral $\int v\, du$."

"But that integral might be simpler!" the professor said.

"But it might be more complicated!" Recordis retorted.

"Well, it *might* be simpler!" the professor said.

"There is only one way to decide," the king interrupted. "If this method results in a simpler integral, then we will use it. If it results in a more complicated integral, then we won't."

We went back to our problem: $\int \ln x\, dx$.

"Let's try these definitions," the professor said.

$$\text{Let } u = \ln x.$$

$$\text{Let } dv = dx.$$

"That's all right," Recordis said. "You're just changing the name of a variable. But now you need to find du and v."

$$u = \ln x$$

$$du = \frac{dx}{x}$$

$$dv = dx$$

$$v = x$$

We put these values into the professor's equation to see whether we ended up with a simpler integral:

$$\int u\, dv = uv - \int v\, du$$

$$\int \ln x\, dx = (\ln x)(x) - \int x\frac{1}{x}\, dx$$

$$= x \ln x - \int dx$$

"It works!" the professor said. "We did end up with a second integral that is much simpler than the first. We have found a new integration method!"

$$\int \ln x\, dx = x \ln x - x + C$$

"Let's differentiate this expression and make sure it works!" Recordis urged, before the professor got carried away.

$$\frac{d}{dx}(x \ln x - x + C) = ? \text{ (had better be } \ln x)$$

$$= \frac{d}{dx}(x \ln x) - \frac{dx}{dx} + \frac{dC}{dx}$$

$$= x \frac{d}{dx} \ln x + \ln x \frac{dx}{dx} - 1 + 0$$

$$= x \frac{1}{x} + \ln x - 1$$

$$\frac{d}{dx}(x \ln x - x + C) = \ln x$$

"It does work!" Recordis said. "We can find the area of the pool now."

$$A = \int_1^5 \ln x \, dx = (x \ln x - x) \Big|_1^5 = 5 \ln 5 - 5 - (1 \ln 1 - 1)$$

$$= 4.047$$

"What should we call this method?" Recordis asked.

"We'll call it the method of *integration by parts*," the king suggested. "The whole idea is to break the integral up into two parts: u and dv."

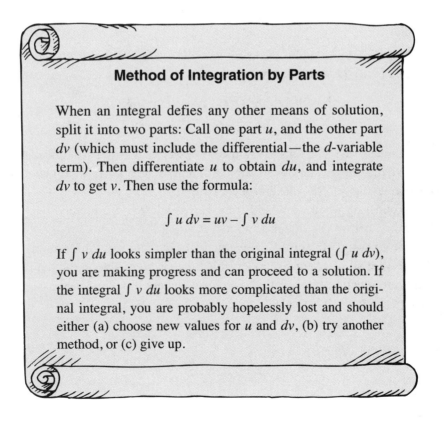

Method of Integration by Parts

When an integral defies any other means of solution, split it into two parts: Call one part u, and the other part dv (which must include the differential—the d-variable term). Then differentiate u to obtain du, and integrate dv to get v. Then use the formula:

$$\int u \, dv = uv - \int v \, du$$

If $\int v \, du$ looks simpler than the original integral ($\int u \, dv$), you are making progress and can proceed to a solution. If the integral $\int v \, du$ looks more complicated than the original integral, you are probably hopelessly lost and should either (a) choose new values for u and dv, (b) try another method, or (c) give up.

"Let's try a weird integral and see whether we can do it," the professor said. "How about $\int x \sin x \, dx$?"

"You don't want to do that!" Recordis objected. "That first x is a variable, rather than a constant, so you can't move it outside the integral sign."

We tried integration by parts:

$$\int x \sin x \, dx$$

Let $u = x \sin x$.

Let $dv = dx$.

Then: $du = (x \cos x + \sin x) \, dx$

and: $v = x$

$$\int u \, dv = uv - \int v \, du$$

$$\int x \sin x \, dx = (x \sin x)(x) - \int (x)(x \cos x + \sin x) \, dx$$

"That did not help!" Recordis said. "The new integral we came up with is *not* simpler than the original integral!"

The professor was stumped, but the king suggested, "Let's redefine u and dv."

Let $u = x$.

Let $dv = \sin x \, dx$.

Then: $du = dx$

and: $v = -\cos x$

Putting that into the formula, we found that:

$$\int x \sin x \, dx = (x)(-\cos x) - \int (-\cos x) \, dx$$

$$= -x \cos x + \sin x + C$$

"That's the answer!" the king exclaimed. We were easily able to differentiate this result to prove that the mysterious method of integration by parts had indeed produced the correct answer. (See Exercise 20.)

"I was wondering when we would come across a last-resort method like this," Recordis said. "I bet we can integrate almost anything!"

"Don't be so hasty!" the professor warned, afraid that the gremlin might be spying. Still, we were all in a happy mood as this time we were able to put the glass plate of the logarithm function safely into the archives, where it would stand as a monument to the progress we had made.

WORKSHEET

1. Evaluate this integral using integration by parts:

$$y = \int x \sin x \, dx$$

Let $u = x$, $dv = \sin x \, dx$.

$du =$ _____

$v = \int$ _____ $dx =$ _____

Fill in the expressions for u, du, v, and dv into the expression for integration by parts:

$$y = \int u \, dv = uv - \int v \, du$$

$$y = \underline{\hspace{1.5cm}} - \int \underline{\hspace{1.5cm}}$$

Evaluate the integral to find the final answer:

$$y = \underline{\hspace{1.5cm}}$$

Verify the result by finding the derivative.

2. Evaluate this integral using integration by parts:

$$y = \int x^2 \sin x \, dx$$

Let $u = x^2$, $dv = \sin x \, dx$.

$du =$ _____

$v = \int$ _____ $dx =$ _____

$$y = \int u \, dv = uv - \int v \, du = \underline{\hspace{1.5cm}} - \int \underline{\hspace{1.5cm}} dx$$

Let y_2 represent the integral on the right-hand side. Evaluate this by using integration by parts again.

Let $u = x$, $dv = \cos x \, dx$.

$du =$ _____

$v = \int$ _____ $dx =$ _____

$$y_2 = \int u \, dv = uv - \int v \, du = \underline{\hspace{1.5cm}} - \int \underline{\hspace{1.5cm}} dx$$

$$y_2 = \underline{\hspace{1.5cm}}$$

$$y = -x^2 \cos x + 2y_2$$

$$y = \int x^2 \sin x \, dx = \underline{\hspace{2cm}}$$

Verify the result by finding the derivative.

$$\frac{dy}{dx} = \underline{\hspace{2cm}}$$

Some of the terms will cancel out. The result is:

$$\frac{dy}{dx} = \underline{\hspace{2cm}}$$

3. Evaluate this integral using integration by parts:

$$y = \int x^3 \sin x \, dx$$

Let $u = x^3$, $dv = \sin x \, dx$

$$du = \underline{\hspace{2cm}}$$

$$v = \int \underline{\hspace{2cm}} dx = \underline{\hspace{2cm}}$$

$$y = \int u \, dv = uv - \int v \, du = \underline{\hspace{2cm}} - \int \underline{\hspace{2cm}} dx$$

Simplify:

$$y = \underline{\hspace{2cm}}$$

Let $y_3 = \int x^2 \cos x \, dx$. Evaluate this integral with integration by parts. Let $u = x^2$, $dv = \cos x \, dx$.

$$du = \underline{\hspace{2cm}}$$

$$v = \int \underline{\hspace{2cm}} dx = \underline{\hspace{2cm}}$$

$$y_3 = \int u \, dv = uv - \int v \, du = \underline{\hspace{2cm}} - \int \underline{\hspace{2cm}} dx$$

Simplify:

$$y_3 = \underline{\hspace{2cm}}$$

The answer to the last integral comes from question 1:

$$\int x \sin x \, dx = -x \cos x + \sin x$$

Therefore:

$$y_3 = \int x^2 \cos x \, dx = \underline{\hspace{2cm}} - 2(\underline{\hspace{2cm}})$$

Simplify:

$$y_3 = \underline{\hspace{2cm}}$$

Putting this result back in our original integral:

$$y = \int x^3 \sin x \, dx = \underline{\hspace{2cm}} + 3y_3$$

$$y = \int x^3 \sin x \, dx = \underline{\hspace{2cm}} + 3(\underline{\hspace{2cm}})$$

Simplify (the result will have four terms):

$$y = \int x^3 \sin x \, dx = \underline{\hspace{2cm}}$$

Find the derivative dy/dx to verify that we did the integration correctly. Use the product rule on the first three terms:

$$\frac{dy}{dx} = (\underline{\hspace{2cm}})(\underline{\hspace{2cm}})$$

$$+ (\underline{\hspace{2cm}})(\underline{\hspace{2cm}})$$

$$+ (\underline{\hspace{2cm}})(\underline{\hspace{2cm}})$$

$$+ (\underline{\hspace{2cm}})(\underline{\hspace{2cm}})$$

$$+ (\underline{\hspace{2cm}})(\underline{\hspace{2cm}})$$

$$+ (\underline{\hspace{2cm}})(\underline{\hspace{2cm}})$$

Six of the terms in the above expression will cancel out, leaving:

$$\frac{dy}{dx} = \underline{\hspace{2cm}}$$

NOTE TO CHAPTER 10

It turns out that the exponential function will always increase faster than any polynomial function for large values of x. This holds true no matter how big an exponent you use; for example, for large enough x, e^x is even bigger than $x^{100,000}$. Stating this as a theorem, we have

$$\lim_{x \to \infty} \frac{x^a}{e^x} = 0 \text{ for any } a, \text{ no matter how large}$$

Many functions lack antiderivatives that can be expressed in terms of elementary functions. The only way to integrate such a function is basically the same way in our country as it is in Carmorra. The method is known as *numerical integration*. Of course, in our country we would use a computer rather than construct a pool to measure the area. See Chapter 17.

Exercises

Find dy/dx for:

1. $y = e^{ax}$

2. $y = e^{-x^2}$

3. $y = e^{ax^2+b}$

4. $y = (e^{ax})^m$

5. $y = 10^x$

6. $y = e^{-ax} \sin x$

7. $y = a^{(x^x)}$

8. $y = \sqrt{(x-a)(x-b)}$

9. $y = \dfrac{x + 1}{(ax^2 + bx + c)^{3/2}}$

10. $y = \dfrac{x - 5}{(x + 2)(x + 3)}$

11. $y = \dfrac{1}{(2\pi)^{1/2}} e^{-[(x-\mu^2)/\sigma^2]}$

12. Use logarithmic implicit differentiation to verify the differentiation power rule.

13. Use logarithmic implicit differentiation to verify the differentiation product rule.

14. Use the definition of the derivative to show that $(d/dx)\,e^x = e^x$. Use a fact that we found later:

$$e^x = 1 + x + \frac{x^2}{2!} + \frac{x^3}{3!} + \cdots + \frac{x^n}{n!} + \cdots$$

Evaluate:

15. $y = \int e^{ax}\,dx$

16. $y = \int e^{ax+b}\,dx$

17. $y = \int x e^{x^2}\,dx$

18. $y = \int \dfrac{e^x}{e^x + 5}\,dx$

19. $y = \int \dfrac{\ln x}{x}\,dx$

20. Find dy/dx for $y = \sin x - x \cos x$ to verify the integration result from the chapter.

21. Find the area of Recordis' new exponential pool, which is bounded by $y = e^x$, $y = 0, x = 0$, and $x = 3$.

22. Find dy/dx for $y = x^2 e^{-x}$. Find points where the curve has horizontal tangents. Make a sketch of the curve.

23. The integral $y = \int \sin x \cos x\,dx$ may be evaluated three ways. (a) Find y by using integration by parts. (b) Find y by making the substitution $u = \sin x$. (c) Find y by using a trigonometric identity to simplify $\sin x \cos x$. Which method is easiest? Is the answer the same for all three methods?

24. Find out how many bacteria there will be at time t in the gremlin's bacteria dish $(dn/dt = 3n; t = 0$ when $n = 10)$. (Remember that t is measured in hours.) How many bacteria will there be after 5 hours?

25. Solve for y: $y = \int x \ln x\,dx$.

26. Verify the above result by differentiation.

27. Solve for y: $y = \int xe^x \, dx$.

28. Verify the above result by differentiation.

29. Solve for y: $y = \int x^2 e^x \, dx$.

30. Verify the above result by differentiation.

31. Solve for y: $y = \int x^2 \ln x \, dx$.

32. Solve for y: $y = \int x \sqrt{1+x} \, dx$.

33. Solve for y: $y = \int x^2 \sin x \, dx$.

34. Use integration by parts twice to evaluate $y = \int e^x \cos x \, dx$.

35. A *reduction formula* is a formula that tells how to express a complicated integral in terms of an integral that is slightly simpler. Use integration by parts to derive the reduction formula:

$$y = \int \sin^m x \, dx = -\frac{\sin^{m-1} x \cos x}{m} + \frac{m-1}{m} \int \sin^{m-2} x \, dx.$$

Integration by Trigonometric Substitution

"We should have a party!" Recordis said the next day. "Let's do something to celebrate our escape from the gremlin and all the other good events that have happened lately. We can invite all the children." The king agreed, so we quickly began to make plans for flowers, ribbons, refreshments, and rides.

The only person not enjoying the preparations was Trigonometeris, who was feeling left out again. I wanted to cheer him up, so I suggested that we join Recordis at the office of the Differentiation and Integration Business to see whether anything interesting was happening.

Recordis was talking to the Royal Gardener and National Park Superintendent. "We are planning flowers for the king's rose garden," the gardener was saying. "We want to fill the garden with roses that will bloom on the day of the party. The garden is shaped like an ellipse with length 20 units and width 10 units."

"The king always did like ellipses," Recordis said. "In this case it looks as though the ellipse has a semimajor axis of $a = 10$ and a semiminor axis of $b = 5$."

"We need to know how many rose bushes will be required to fill the garden," the gardener said. "And in order to do that, we need to know the area of the ellipse. Can you tell us what the area of the ellipse is?"

"You came to the right place," Recordis said, only this time he was not as overconfident as he had been when he was visited by his neighbor, Count Q. "We set up what we call a definite integral." He riffled through his notes to find the equation for an ellipse (which he had forgotten again).

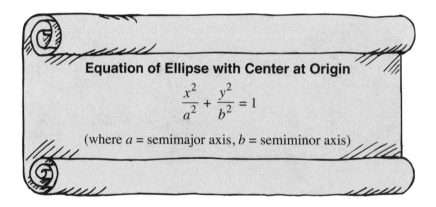

Equation of Ellipse with Center at Origin

$$\frac{x^2}{a^2} + \frac{y^2}{b^2} = 1$$

(where a = semimajor axis, b = semiminor axis)

"We can solve for y in terms of x, using algebra," Recordis said.

$$\frac{y^2}{b^2} = 1 - \frac{x^2}{a^2}$$

$$y^2 = b^2 \left(1 - \frac{x^2}{a^2} \right)$$

$$y = \pm b \sqrt{1 - x^2 / a^2}$$

We decided that we would first find the area of the quarter of the ellipse where x and y are both positive, so we used the plus sign in front of the square root.

$$y = b\sqrt{1 - x^2 / a^2}$$

"Is that the answer?" asked the gardener, who was getting impatient.
"No," Recordis said. "Now we set up the definite integral."

$$\text{(area of quarter-ellipse)} = b\int_0^a \sqrt{1 - x^2 / a^2}\; dx$$

$$\text{(area of whole ellipse)} = 4b\int_0^a \sqrt{1 - x^2 / a^2}\; dx$$

"Now, if you'll excuse us, we'll be right back after we figure out how to do the integral," Recordis said, and before the gardener knew what was happening Recordis had closed the business and was running through the palace shouting, "Help!"

"Do you always have problems?" the professor asked as we gathered in the Main Conference Room a few minutes later.

"This time it's the king's garden that is causing the problem," Recordis said. He had Igor display the integral that we didn't know how to do. "We can't use the power rule, because we don't have $\int x^n\, dx$. And we can't substitute like this: Let $u = 1 - x^2/a^2$, because then we would have $du = -2x\, dx/a^2$, and we don't have any way to turn dx into du (since we can't take that x outside the integral sign)."

"We could solve it if we had an extra x outside the square root sign," the king said.
"Suppose it was written as follows."

$$z = 4b\int_0^a x\sqrt{1 - x^2 / a^2}\; dx$$

"Then we could substitute."

$$u = 1 - \frac{x^2}{a^2}, \qquad du = \frac{-2x\, dx}{a^2}$$

"And," he continued, "we could turn dx into du by doing this."

$$z = 4b\int_{u=1}^{u=0} u^{1/2}\left(\frac{-a^2}{2}\right)\left(\frac{-2}{a^2}\right) x\, dx$$

$$= 4b\int_1^0 u^{1/2}\left(\frac{-a^2}{2}\right) du$$

$$= -2ba^2\,\frac{2}{3}\,u^{3/2}\,\Big|_1^0$$

$$= -\frac{4}{3}ba^2(0)^{3/2} - \left[-\frac{4}{3}ba^2(1)^{3/2}\right]$$

$$z = \frac{4ba^2}{3}$$

"Sure, that was easy," the professor said. "But it doesn't help us, because we have $4b \int_0^a \sqrt{1-x^2/a^2}\, dx$ without an extra x."

"If only we could get rid of the square root sign!" Recordis moaned. "I never did like square root signs anyway."

We stared at the integral for a long time. "We need to make a substitution like this," the professor said.

$$\sqrt{1-\text{variable}^2} = \text{something simple}$$

"Can anyone think of anything that we have *ever* done that looks like that?"

"I can," Trigonometeris said, suddenly becoming cheerful. "It's obvious. We have $\sqrt{1-\sin^2\theta} = \cos\theta$."

"I don't see any need to bring trigonometry into this!" Recordis objected.
"Try this," Trigonometeris said.

$$A = 4b \int_0^a \sqrt{1-x^2/a^2}\, dx$$

$$\text{Let } \frac{x}{a} = \sin\theta.$$

$$x = a\sin\theta$$

$$dx = a\cos\theta\, d\theta$$

$$\theta = \arcsin\left(\frac{x}{a}\right)$$

"What's an arcsin?" Recordis asked. "I remember the name, but I forget what it is."

"It's the inverse function for the sine function," Trigonometeris said. "In just the same way, the exponential function is the inverse function for the logarithm function. This is what we said."

$$\text{If } a = \sin b, \text{ then } b = \arcsin a.$$

"That means that $\sin(\arcsin a) = a$ for any a (provided that $-1 \le a \le 1$)."

"What happens when you put that into the integral?" the king asked. We made the change in the three places where we knew we had to: the integrand, the two limits of integration, and the differential dx term.

$$A = 4b \int_{\arcsin\,(0/a)}^{\arcsin\,(a/a)} \sqrt{1-\sin^2\theta}\; a\cos\theta\, d\theta$$

"And we know what $\sqrt{1-\sin^2\theta}$ is," Trigonometeris said.

"We do?" Recordis asked.

"It's cos θ!" Trigonometeris answered. "That's what I just told you!

"Maybe you had better review the list of trigonometric identities, just to make sure Recordis remembers them all," the professor said, trying to conceal the fact that she couldn't remember them all either.

Trigonometeris pulled a page from his book and displayed the following identities:

Trigonometric Identities (True for All *A*, *B*)

$\sin^2 A + \cos^2 A = 1$

$\sin^2 A = \frac{1}{2}(1 - \cos 2A)$

$\cos^2 A = \frac{1}{2}(1 + \cos 2A)$

$\sin(-A) = -\sin A$

$\cos(-A) = \cos A$

$\tan A = \sin A/\cos A$

$\operatorname{ctn} A = 1/\tan A$

$\sec A = 1/\cos A$

$\csc A = 1/\sin A$

$1 + \tan^2 A = \sec^2 A$

$1 + \operatorname{ctn}^2 A = \csc^2 A$

$\sin(A + B) = \sin A \cos B + \cos A \sin B$

$\cos(A + B) = \cos A \cos B - \sin A \sin B$

$\tan(A + B) = (\tan A + \tan B)/(1 - \tan A \tan B)$

$\sin 2A = 2 \sin A \cos A$

$\cos 2A = \cos^2 A - \sin^2 A = 1 - 2\sin^2 A = 2\cos^2 A - 1$

$\tan 2A = 2 \tan A/(1 - \tan^2 A)$

(There is a longer list of trigonometric formulas in Appendix 2 at the back of the book.)

We used the identity in the integral (also recall that arcsin $1 = \pi/2$ and arcsin $0 = 0$):

$$A = 4ba \int_0^{\pi/2} \sqrt{\cos^2 \theta} \, \cos \theta \, d\theta$$

"Now we can get rid of the square root sign," Recordis said, with sudden excitement. "Maybe this is a good method after all."

$$A = 4ab \int_0^{\pi/2} \cos^2 \theta \, d\theta$$

"How do we do that?" the king asked.

"Use another identity," the professor said.

$$\cos^2 \theta = \tfrac{1}{2}(1 + \cos 2\theta)$$

$$A = 4ab \int_0^{\pi/2} \tfrac{1}{2}(1 + \cos 2\theta) \, d\theta$$

$$= 2ab \int_0^{\pi/2} d\theta + 2ab \int_0^{\pi/2} \cos 2\theta \, d\theta$$

$$= 2ab \, \theta \, \Big|_0^{\pi/2} + 2ab\tfrac{1}{2}\sin 2\theta \, \Big|_0^{\pi/2}$$

$$= ab\pi - 2ab \times 0 + ab(\sin \pi - \sin 0)$$

Since $\sin \pi = 0$:

$$A = \pi ab$$

"That's a pretty simple answer," Recordis said.

"It makes sense," Trigonometeris told him. "If you have a rectangle (Figure 11–1), its area is $4ab$. It stands to reason that the inscribed ellipse would have an area of about $3ab$."

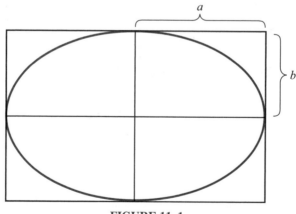

FIGURE 11–1.

"The formula also works for a circle," the professor pointed out, "We said that a circle is the same as an ellipse with $a = b$. In that case the area would be πa^2, where a is the radius. We already know that that's the correct answer for the area of a circle."

We easily found the area of the rose garden:

$$A = (3.14)(10)(5) = 157$$

"We'll call this method the method of *integration by trigonometric substitution*." Everyone protested that Trigonometeris was naming the method after himself, but I couldn't think of any better name.

"Is the method very generally applicable?" the king asked.

"I'm sure it is," Trigonometeris answered. "I can think of lots of integrals where you would use it."

"Name one," Recordis challenged.

Trigonometeris gave us the first complicated integral that came into his head:

$$z = \int (a^2 + b^2 x^2)^{-1/2}\, dx$$

"How can you do that?" Recordis asked. "It's an indefinite, rather than a definite, integral, so you can't change the bounds after you make the substitution."

"We can do the same thing," Trigonometeris told him. "At the end we will have to substitute backwards and write x in terms of θ." Trigonometeris thought that the key identity was $\tan^2 \theta + 1 = \sec^2 \theta$. So we made the substitution $bx/a = \tan \theta$:

$$z = \int (a^2 + b^2 x^2)^{-1/2}\, dx$$

$$= \int \frac{1}{\sqrt{a^2 \left(1 + \dfrac{b^2 x^2}{a^2}\right)}}\, dx$$

$$= \frac{1}{a} \int \frac{1}{\sqrt{1 + b^2 x^2 / a^2}}\, dx$$

$$= \frac{1}{a} \int \frac{1}{\sqrt{1 + \tan^2 \theta}}\, dx$$

Now we need to write dx in terms of $d\theta$:

$$x = \frac{a \tan \theta}{b}$$

The derivative of $\tan \theta$ is $\sec^2 \theta$. (See Chapter 5.)

$$dx = \frac{a}{b} \sec^2 \theta\, d\theta$$

$$z = \frac{1}{a} \int \frac{1}{\sqrt{\sec^2 \theta}} \frac{a}{b} \sec^2 \theta\, d\theta$$

$$= \frac{1}{b} \int \sec \theta\, d\theta$$

"That is a lot simpler," Recordis admitted. "We don't know how to integrate the secant function, though." Trigonometeris spent a long time trying to figure that one out. Nothing he tried worked. Finally he said, "I'll have the answer tomorrow, I promise."

"I don't suppose the method of trigonometric substitution is much good for any other type of integral," Recordis needled Trigonometeris. "It surely didn't help much for that one."

"It will help if you have an expression like $(1 + x^2)$ raised to a negative power," Trigonometeris retorted defensively. He was deeply embarrassed that he had been stuck with an integral of one of his own functions that he could not do. "Suppose you had $(1 + x^2)^{-1}$. If you had $(1 + x^2)$ raised to a power that is a positive integer, you could multiply out the result quite easily: $(1 + x^2)^2 = 1 + 2x^2 + x^4$."

"Let's try the other one you mentioned," the professor said.

$$z = \int \frac{1}{1 + x^2}\, dx$$

We decided to use the secant-tangent identity again, so we made the substitution:

Let $x = \tan \theta$.

$$\theta = \arctan x$$

$$dx = \sec^2 \theta\, d\theta$$

$$z = \int \frac{1}{1 + \tan^2 \theta}\, \sec^2 \theta\, d\theta$$

$$= \int \frac{1}{\sec^2 \theta}\, \sec^2 \theta\, d\theta$$

$$= \int d\theta$$

"That's a perfect integral!" the professor exclaimed.

$$z = \theta$$

"We have to substitute back in for θ," Recordis said.

$$\int \frac{1}{1 + x^2}\, dx = \arctan x + C$$

"That integrates into a standard function!" the king pointed out in surprise.
"But if that's the case, then we can do this," the professor said.

$$\frac{d(\arctan x)}{dx} = \frac{d}{dx} \int \frac{1}{1 + x^2}\, dx = \frac{1}{1 + x^2}$$

Trigonometeris was stunned. "This equation tells us how to differentiate the arctangent function," he said slowly. "How could I have forgotten to look for the derivatives of the inverse functions when we looked for the derivatives of the other trigonometric functions?" He kicked himself (rather hard). "They were right under my nose. How could I have overlooked them?"

We decided to find the derivatives of the other inverse trigonometric functions. After a couple of guesses that didn't work, we tried to evaluate the integral:

$$z = \int \frac{1}{\sqrt{1 - x^2}}\, dx$$

We realized that this type of integral called for the sine-cosine identity:

$$\text{Let } x = \sin \theta.$$

$$\theta = \arcsin x$$

$$dx = \cos \theta\, d\theta$$

$$z = \int \frac{1}{\sqrt{1 - \sin^2 \theta}} \cos \theta\, d\theta$$

$$= \int \frac{\cos \theta\, d\theta}{\cos \theta}$$

$$= \int d\theta$$

$$= \theta$$

$$z = \arcsin x$$

$$\frac{d \arcsin x}{dx} = \frac{1}{\sqrt{1 - x^2}}$$

"That even fits," the king said. "Remember that arcsin x is defined only for $-1 \le x \le 1$ which are the same values of x where $(1 - x^2)^{-1/2}$ is defined."

"The derivative of $y = \arccos x$ should be just the negative of the derivative of $y = \arcsin x$," Recordis guessed. We were able to establish that

$$\frac{d \arccos x}{dx} = -\frac{1}{\sqrt{1 - x^2}}$$

As Recordis started to make a list of the integrals we had done, Trigonometeris got up to leave the room. "Not so fast," Recordis said. "Remember that you still have to tell us how to integrate the secant function. We'll be waiting for your answer tomorrow."

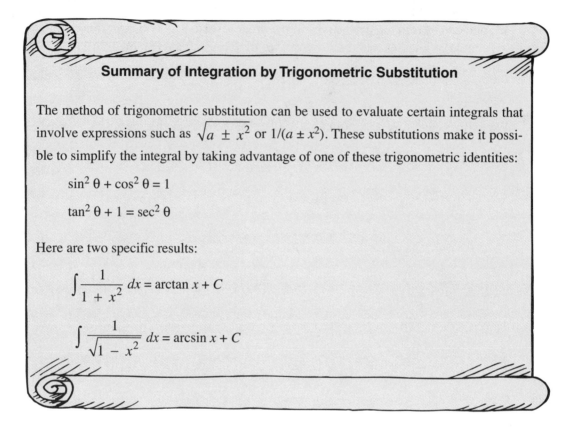

Summary of Integration by Trigonometric Substitution

The method of trigonometric substitution can be used to evaluate certain integrals that involve expressions such as $\sqrt{a \pm x^2}$ or $1/(a \pm x^2)$. These substitutions make it possible to simplify the integral by taking advantage of one of these trigonometric identities:

$$\sin^2 \theta + \cos^2 \theta = 1$$

$$\tan^2 \theta + 1 = \sec^2 \theta$$

Here are two specific results:

$$\int \frac{1}{1 + x^2} \, dx = \arctan x + C$$

$$\int \frac{1}{\sqrt{1 - x^2}} \, dx = \arcsin x + C$$

NOTE TO CHAPTER 11

When the functions arcsin x and arccos x are used, it is assumed that the principal values are taken:

$$-\pi/2 \leq \arcsin x \leq \pi/2$$

$$0 \leq \arccos x \leq \pi$$

See Figure 11–2. The two functions are connected by the formula

$$\arcsin x = \pi/2 - \arccos x$$

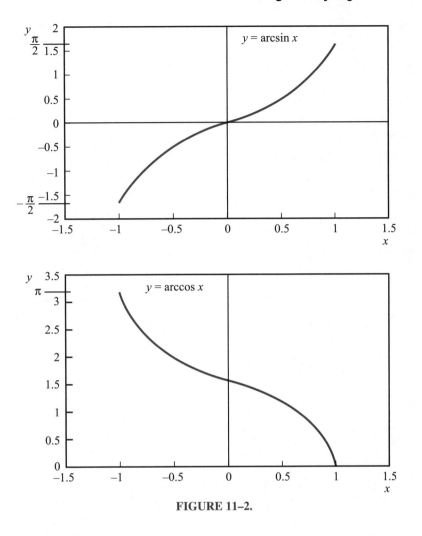

FIGURE 11–2.

WORKSHEET

Use trigonometric substitution to evaluate the integral. Fill in the blanks for each step.

$$y = \int \frac{1}{\sqrt{m^2 - n^2 x^2}} \, dx$$

Factor m^2 out of the expression in the square root sign:

$$y = \int \frac{1}{\sqrt{m^2 \left(1 - \left(\dfrac{\underline{\quad} x}{\underline{\quad}}\right)^2\right)}} \, dx$$

Move the m^2 outside the square root sign, then outside the integral:

$$y = \left(\frac{1}{\underline{}}\right) \int \frac{1}{\sqrt{1 - \left(\dfrac{x}{\underline{}}\right)^2}} \, dx$$

Let $u = nx/m$. Then find:

$$\frac{du}{dx} = \underline{}$$

$du = \underline{}$

$dx = \underline{}$

Use the above expressions to substitute into the integral:

$$y = \left(\frac{1}{\underline{}}\right) \int \frac{1}{\sqrt{1 - \underline{}}} \, (\underline{}) \, du$$

Move the constants in front of du outside the integral:

$$y = (\underline{}) \int \frac{1}{\sqrt{1 - \underline{}}} \, du$$

Let $u = \sin \theta$. Find:

$$\frac{du}{d\theta} = \underline{}$$

$du = \underline{} \, d\theta$

Substitute the expressions for u and du into the integral:

$$y = (\underline{}) \int \frac{1}{\sqrt{1 - \underline{}}} \, \underline{} \, d\theta$$

Simplify the expression under the square root sign:

$$y = (\underline{}) \int \frac{1}{\sqrt{\underline{}}} \, \cos \theta \, d\theta$$

The expression for the integral becomes very simple:

$$y = (\underline{}) \int \underline{}$$

Rewrite the expression without the integral sign:

$$y = \underline{}$$

Write an expression for θ in terms of u. (See the definition of θ above.)

$$\theta = \underline{\qquad}$$

Write an expression for θ in terms of x. (See the definition of u above.)

$$\theta = \underline{\qquad}$$

Put this expression for θ back into the expression for the integral, and there is the final answer:

$$y = \int \frac{1}{\sqrt{m^2 - n^2 x^2}}\ dx = \underline{\qquad}$$

For the final step, verify the result by finding $\dfrac{dy}{dx}$.

Exercises

Solve for y:

1. $y = \displaystyle\int \dfrac{dx}{x^2 \sqrt{1 + x^2}}$

2. $y = \displaystyle\int \dfrac{dx}{x\sqrt{x^2 - 1}}$

3. $y = \displaystyle\int \dfrac{e^x}{1 + e^{2x}}\ dx$

4. Evaluate: $\displaystyle\int_0^{a/b} \sqrt{a^2 - b^2 x^2}\ dx$

5. Evaluate these two expressions: $y_1 = \int x \sqrt{1 - x^2}\ dx$ and $y_2 = x \int \sqrt{1 - x^2}\ dx$. Compare the results.

6. Find the area of a circle of radius r.

7. One strip of pink roses will be planted at the tip of the rose garden (Figure 11–3). Find the area of the strip of pink roses.

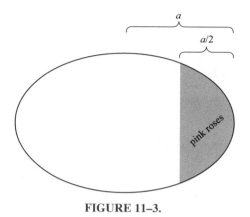

FIGURE 11–3.

8. Evaluate this indefinite integral: $z = 4b \int \sqrt{1 - x^2 / a^2} \, dx$. (This is the same as the ellipse area integral.) Verify the result by differentiation.

Solve for y:

9. $y = \int \dfrac{1}{(x/a)^2 + 1} \, dx$

10. $y = \int \dfrac{1}{x^2 + a^2} \, dx$

11. $y = \int \dfrac{1}{x^2 + 25} \, dx$

12. $y = \int \dfrac{1}{x^2 + 4x + 13} \, dx$ (Let $u = x + 2$.)

13. $y = \int \dfrac{1}{x^2 + 2bx + c} \, dx$ (Let $u = x + b$. Assume that $c > b$.)

14. $y = \int \dfrac{1}{x^2 + x + 5} \, dx$

15. Solve for $\int \sqrt{1 - x^2} \, dx$ using (a) $x = \sin \theta$ and (b) $x = \cos \theta$. Do the two results agree? Which method is easier?

16. Evaluate $y = \int \arcsin x \, dx$. Verify the result by differentiation.

17. Evaluate $y = \int \arctan x \, dx$. Verify the result by differentiation.

Integration by Partial Fractions

"I'd like to see how Trigonometeris weasels his way out of this one," Recordis said when we met the next morning. Trigonometeris had not joined us yet. "Maybe there isn't any way to integrate the secant function. In my opinion, once you've introduced trigonometry to a situation you've created worse problems than you've solved."

The professor wanted to make a summary of all the integration techniques we had developed. She said that we had finished with all the theory that we needed to know for calculus and that all that remained were the menial bookkeeping tasks (meaning that Recordis would have to do the rest of the work). Recordis etched our results on a large plate so that we could have them permanently.

$$\int x^n\,dx = \frac{1}{n+1}\,x^{n+1} + C \qquad (n \neq -1)$$

$$= \ln|x| + C \qquad (n = -1)$$

$$\int f(u)\,\frac{du}{dx}\,dx = \int f(u)\,du$$

$$\int \sin x\,dx = -\cos x + C$$

$$\int \cos x\,dx = \sin x + C$$

$$\int \tan x\,dx = \ln|\sec x| + C \text{ (see Exercise 9–21)}$$

$$\int \ln x\,dx = x \ln x - x + C$$

$$\int e^x\,dx = e^x + C$$

"We can add what we did yesterday," Recordis said. "We discovered what to do with a quadratic term in the denominator of an integrand." He etched on the plate:

$$\int \frac{1}{1+x^2}\,dx = ?$$

$$\int \frac{1}{1-x^2}\,dx = ?$$

He flipped through his notes to find the answer. "This is easy," he said.

$$\int \frac{1}{1+x^2}\,dx = \arctan x + C$$

Suddenly he turned pale. He stood in front of the last integral that he had written so that we couldn't see it. "That's all we need to do for now," he said.

"Wait a minute," the professor objected. "What about that last one? The one with $\int (1-x^2)^{-1}\,dx$?"

"What last one?" Recordis said, turning around. "Oh, that last one. It looks as though we didn't do it yesterday, so I think we should forget it."

"But you already wrote it on the plate, so we can't erase it," the king said. "We should try to find out what it is. Let's try this substitution: $x = \sin \theta$, $dx = \cos \theta \, d\theta$."

"I'm warning you: This is only going to cause trouble," Recordis muttered.

$$\int \frac{1}{1 - x^2} \, dx = \int \frac{1}{1 - \sin^2 \theta} \cos \theta \, d\theta$$

$$= \int \frac{\cos \theta}{\cos^2 \theta} \, d\theta$$

$$= \int \frac{1}{\cos \theta} \, d\theta$$

"Oh no!" Recordis moaned.

$$\int \frac{1}{1 - x^2} \, dx = \int \sec \theta \, d\theta$$

"That's exactly what we're trying to find the answer for," the professor said. "I hope Trigonometeris gets here soon with that answer."

"I bet there is no answer, and that's why Trigonometeris isn't here yet," Recordis said. "There must be a way to do this integral without trigonometry. For one thing, we can factor the denominator, since it is the dfference of two squares."

$$\int \frac{1}{1 - x^2} \, dx = \int \frac{1}{(1 - x)(1 + x)} \, dx$$

"Does that help?" the king asked.

We were trying to figure out what to do with that integral when Builder came in, carrying two bottles of liquids.

"Maybe you can help me," he said. "I'm working on the fireworks for the party. Your work on figuring out the speed of things might help. I have 10 grams of this red juice (call it x) and 9 grams of this yellow juice (call it y). If you mix them together, they form another liquid, called z. It takes exactly 3 grams of x to combine with exactly 2 grams of y to produce exactly 5 grams of z. When there is more x and y available, the reaction goes faster; but when there is more z around, the reaction goes slower. In fact, the speed of the reaction is proportional to the product of the amounts of x and y and inversely proportional to the amount of z."

$$\frac{dz}{dt} = \frac{xy}{z}$$

"Can you figure out how much z will have been produced at a particular time after I start mixing them together?"

"We'll at least set up the integral," the professor said. We realized that we would have to figure out the amounts of x and y that would be present at a given time, and it was clear that the amounts would be equal to the starting amounts (10 and 9, respectively) minus the amounts that had gone into the formation of the z:

$$x = 10 - 0.6z$$

$$y = 9 - 0.4z$$

Putting these expressions into Builder's rate equation, we were able to set up the integral:

$$\frac{dz}{dt} = \frac{xy}{z}$$

$$\frac{z}{xy}\,dz = dt$$

$$\frac{z}{(10 - 0.6z)(9 - 0.4z)}\,dz = dt$$

$$t = \int \frac{z}{(10 - 0.6z)(9 - 0.4z)}\,dz$$

"We can't do that one either," the professor said. "That's basically the same as the other problem we're stuck on. We have two factors in the denominator of a fraction."

$$\int \frac{1}{(1 - x)(1 + x)}\,dx = ?$$

"We need a new method to tell us what to do when we have a bunch of factors in the denominator like that," the king stated. "If only we had a bunch of fractions *added* together, instead of one giant fraction!"

"It's so easy to take a bunch of fractions that are added together and turn them into one giant fraction," Recordis moaned. "It's too bad you can't do it the other way."

"What did you say was easy?" the professor asked.

"Adding a bunch of fractions together," Recordis said. "Suppose you had this."

$$y = \frac{1}{x - a} + \frac{1}{x - b}$$

"You can easily add these together and turn them into one giant fraction. All you need to do is find the common denominator."

$$y = \frac{1}{x-a} \frac{x-b}{x-b} + \frac{1}{x-b} \frac{x-a}{x-a}$$

$$= \frac{x-b+x-a}{(x-a)(x-b)}$$

$$= \frac{2x-(b+a)}{(x-a)(x-b)}$$

"It's too bad you can't do that procedure backwards," Recordis said. "Then you could start with a fraction having a bunch of factors in the denominator and turn it into a sum of fractions."

"Why can't you do it backwards?" the professor asked.

Recordis didn't say anything for a while. "I thought there was a reason why you couldn't do it backwards," he finally said.

"An equation must work in both directions," the king pointed out. "It may be a little more tricky to start with a single fraction and end up with a sum of fractions, but we should be able to do it. Take for example, Builder's problem."

$$\frac{z}{(10-0.6z)(9-0.4z)}$$

"That expression must be the sum of two partial fractions, one with denominator $(10-0.6z)$ and one with denominator $(9-0.4z)$."

"We don't know what the numerators are, though," Recordis said.

We decided to call the numerators A and B while we tried to solve for them.

$$\frac{z}{(10-0.6z)(9-0.4z)} = \frac{A}{10-0.6z} + \frac{B}{9-0.4z}$$

"We don't know what z is," Recordis remarked.

"But this equation must hold for all values of z," the king said.

"That means we call it an *identity*," the professor stated.

We found the common denominator for the right-hand side:

$$\frac{z}{(10-0.6z)(9-0.4z)} = \frac{A(9-0.4z)+B(10-0.6z)}{(10-0.6z)(9-0.4z)}$$

"Since these two fractions are equal, their numerators must be equal," Recordis said.

$$z = A(9-0.4z)+B(10-0.6z)$$
$$z+0 = 9A-0.4Az+10B-0.6Bz$$
$$(1 \times z)+0 = (-0.4A-0.6B)z+(9A+10B)$$

"Now what?" Recordis asked.

"Since this equation must hold for all values of z, the coefficient of z on the left-hand side (which is 1) must equal the coefficient of z on the right-hand side, and the constant term on the left-hand side (which is 0) must equal the constant term on the right-hand side." (You can easily verify that, if these two conditions are not met, you can find a value of z such that the equation will not hold.)

coefficients of z:

$$1 = -0.4A - 0.6B$$

constant terms:

$$0 = 9A + 10B$$

"That's easy," Recordis said. "That's just two equations in two unknowns." From the second equation:

$$9A = -10B$$

$$A = \frac{-10}{9}B$$

Now substitute this formula for A into the first equation:

$$1 = -(0.4)\left(\frac{-10}{9}\right)B - 0.6B$$

$$= \frac{4}{9}B - \frac{3}{5}B$$

$$= -\left(\frac{7}{45}\right)B$$

$$B = \frac{-45}{7} = -6.4$$

$$A = \left(\frac{-10}{9}\right)(-6.4) = 7.1$$

"So that means . . ."

$$\frac{z}{(10 - 0.6z)(9 - 0.4z)} = \frac{7.1}{10 - 0.6z} - \frac{6.4}{9 - 0.4z}$$

Recordis was still skeptical, so he double-checked to make sure that the expression on the right did indeed equal the expression on the left. (See Exercise 12–6.)

"Now it's easy!" Recordis said. "We can break the original integral into two integrals."

$$t = \int \frac{z}{(10 - 0.6z)(9 - 0.4z)} \, dz = \int \frac{7.1dz}{10 - 0.6z} - \int \frac{6.4dz}{9 - 0.4z}$$

$$= 7.1 \int (10 - 0.6z)^{-1} \, dz - 6.4 \int (9 - 0.4z)^{-1} \, dz$$

Let $u = 10 - 0.6z$; $du = -0.6 \, dz$; $dz = -du/0.6$.
Let $v = 9 - 0.4z$; $dv = -0.4 \, dz$; $dz = -dv/0.4$.

$$t = 7.1 \int u^{-1} \left(\frac{-du}{0.6} \right) - 6.4 \int v^{-1} \left(\frac{-dv}{0.4} \right)$$

$$= -\frac{7.1}{0.6} \ln u + \frac{6.4}{0.4} \ln v + C$$

$$= -11.833 \ln u + 16 \ln v + C$$

$$= \ln |10 - 0.6z|^{-11.833} + \ln |9 - 0.4z|^{16} + C$$

$$t = \ln \left| \frac{(9 - 0.4z)^{16}}{(10 - 0.6z)^{11.833}} \right| + C$$

To solve for the arbitrary constant C, use the initial condition $z = 0$ when $t = 0$:

$$0 = \ln \left| \frac{9^{16}}{10^{11.833}} \right| + C$$

$$C = -7.9$$

$$t = \ln \left| \frac{(9 - 0.4z)^{16}}{(10 - 0.6z)^{11.833}} \right| - 7.9$$

To check this result, we found the derivative dt/dz of this expression:

$$t = \ln |10 - 0.6z|^{-11.833} + \ln |9 - 0.4z|^{16} + C$$

$$\frac{dt}{dz} = \frac{-11.833 \times (-0.6)}{10 - 0.6z} + \frac{16 \times (-0.4)}{9 - 0.4z}$$

$$\frac{dt}{dz} = \frac{7.1}{10 - 0.6z} + \frac{-6.4}{9 - 0.4z}$$

Finding the common denominator and adding the fractions:

$$\frac{dt}{dz} = \frac{7.1(9 - 0.4z)}{(10 - 0.6z)(9 - 0.4z)} + \frac{-6.4(10 - 0.6z)}{(10 - 0.6z)(9 - 0.4z)}$$

$$\frac{dt}{dz} = \frac{7.1(9 - 0.4z) - 6.4(10 - 0.6z)}{(10 - 0.6z)(9 - 0.4z)}$$

$$\frac{dt}{dz} = \frac{64 - 2.84z - 64 + 3.84z}{(10 - 0.6z)(9 - 0.4z)}$$

$$\frac{dt}{dz} = \frac{z}{(10 - 0.6z)(9 - 0.4z)}$$

The derivative dz/dt is just the reciprocal of dt/dz:

$$\frac{dz}{dt} = \frac{(10 - 0.6z)(9 - 0.4z)}{z}$$

This is the original formula for dz/dt that we started with, so we were reassured that our integration method worked because we could reverse the process by taking the derivative and get back to where we had started from.

We realized that it would be pretty hard to solve for z as an explicit function of t, but Builder was satisfied because we could make a table of values and create a graph.

"We must think of a name for this method," Recordis said.

"The key seems to be to break the fraction into a sum of partial fractions," the king offered.

"Let's call it the method of *partial fractions*," the professor said.

"Now we can solve this other integral," Recordis said.

$$\int \frac{1}{(1 - x)(1 + x)}\, dx$$

"There are two factors in the denominator, so that means that there should be two partial fractions, with denominators $(1 - x)$ and $(1 + x)$."

$$\frac{1}{(1 - x)(1 + x)} = \frac{A}{1 - x} + \frac{B}{1 + x}$$

Now it was just a matter of algebra to solve for A and B:

$$\frac{1}{(1 - x)(1 + x)} = \frac{A(1 + x)}{(1 - x)(1 + x)} + \frac{B(1 - x)}{(1 - x)(1 + x)}$$

$$\frac{1}{(1 - x)(1 + x)} = \frac{A(1 + x) + B(1 - x)}{(1 - x)(1 + x)}$$

Setting the numerators equal gave us:

$$1 = A + Ax + B - Bx$$

$$(0 \times x) + 1 = (A - B)x + (A + B)$$

Since the equation must hold for every value of x, we could form two equations:

coefficients of z:

$$0 = A - B$$

constant terms:

$$1 = A + B$$

These two equations imply:

$$A = B$$

$$1 = A + A$$

$$= 2A$$

$$A = \frac{1}{2}$$

$$B = \frac{1}{2}$$

"Now we can do the integral easily," Recordis said.

$$\int \frac{1}{(1 - x)(1 + x)} \, dx = \int \left[\frac{1}{2(1 - x)} + \frac{1}{2(1 + x)} \right] dx$$

$$= -\frac{1}{2} \ln|1 - x| + \frac{1}{2} \ln|1 + x| + C$$

$$\int \frac{1}{1 - x^2} \, dx = \frac{1}{2} \ln \left| \frac{1 + x}{1 - x} \right| + C$$

At that moment Trigonometeris walked in. He looked terrible, as if he hadn't had any sleep the night before.

"I'll have it for you tomorrow, I promise," he said weakly.

"We don't need you anymore," Recordis said unkindly. "We tried a trigonometric substitution on this integral: $\int (1 - x^2)^{-1} \, dx$, and got:

$$x = \sin \theta$$

$$\int \frac{1}{1 - x^2} \, dx = \int \sec \theta \, d\theta$$

"That is exactly the same problem that you haven't been able to find the answer to. But we found another way of getting the answer, using algebra without a trace of trigonometry."

$$\int \frac{1}{1 - x^2}\, dx = \frac{1}{2}\ln\left|\frac{1 + x}{1 - x}\right| + C$$

"We called it the *method of partial fractions*."

"Wait!" the king said. "This means that we can find the integral of the secant function!"

"We can?" Trigonometeris asked.

"Look at this!"

If $x = \sin\theta$ then $\int \sec\theta\, d\theta = \int \frac{1}{1 - x^2}\, dx = \frac{1}{2}\ln\left|\frac{1 + x}{1 - x}\right| + C$

"That means we have the following."

$$\int \sec\theta\, d\theta = \frac{1}{2}\ln\left|\frac{1 + \sin\theta}{1 - \sin\theta}\right| + C$$

"Of course!" Recordis said. "The old finding-the-integral-of-the-secant-function-by-the-method-of-partial-fractions trick!"

Trigonometeris wanted to contribute something to the discussion, so he found a few ways to simplify this expression for us:

$$\int \sec\theta\, d\theta = \frac{1}{2}\ln\left|\frac{(1 + \sin\theta)\,(1 + \sin\theta)}{(1 - \sin\theta)\,(1 + \sin\theta)}\right| + C$$

$$= \frac{1}{2}\ln\left|\frac{(1 + \sin\theta)^2}{1 - \sin^2\theta}\right| + C$$

$$= \frac{1}{2}\ln\left|\frac{1 + \sin\theta}{\cos\theta}\right|^2 + C$$

$$= \ln\left|\frac{1 + \sin\theta}{\cos\theta}\right| + C$$

$$= \ln\left|\frac{1}{\cos\theta} + \frac{\sin\theta}{\cos\theta}\right| + C$$

$$\int \sec\theta\, d\theta = \ln|\sec\theta + \tan\theta| + C$$

"It is amazing that we could get an answer so complicated, yet so simple," the king remarked.

WORKSHEET

Use the method of partial fractions to split each fraction into two fractions. In each case you will need to solve a two-equation, two-variable system. There is more than one way to do this and you may use whichever method you prefer. One method is *Cramer's rule*, which says that the solution of the equation system

$$a_1 H + b_1 K = c_1$$

$$a_2 H + b_2 K = c_2$$

where H and K are unknown can be found from these formulas:

$$H = \frac{c_1 b_2 - b_1 c_2}{a_1 b_2 - b_1 a_2}$$

$$K = \frac{a_1 c_2 - c_1 a_2}{a_1 b_2 - b_1 a_2}$$

1.

$$\frac{102x + 126}{(3x+9)(21x+18)}$$

$$= \frac{H}{3x+9} + \frac{K}{21x+18}$$

$$\frac{H(\underline{\hphantom{xx}}x + \underline{\hphantom{xx}}) + K(\underline{\hphantom{xx}}x + \underline{\hphantom{xx}})}{(\underline{\hphantom{xx}}x + \underline{\hphantom{xx}})(\underline{\hphantom{xx}}x + \underline{\hphantom{xx}})}$$

Multiply across the parentheses for the terms in the numerator:

$$\frac{\underline{\hphantom{xx}}xH + \underline{\hphantom{xx}}H + \underline{\hphantom{xx}}xK + \underline{\hphantom{xx}}K}{(\underline{\hphantom{xx}}x + \underline{\hphantom{xx}})(\underline{\hphantom{xx}}x + \underline{\hphantom{xx}})}$$

Collect the terms with x:

$$\frac{x(\underline{\hphantom{xx}}H + \underline{\hphantom{xx}}K) + (\underline{\hphantom{xx}}H + \underline{\hphantom{xx}}K)}{(\underline{\hphantom{xx}}x + \underline{\hphantom{xx}})(\underline{\hphantom{xx}}x + \underline{\hphantom{xx}})}$$

Look only at the numerators:

$$102x + 126 = x(\underline{\hphantom{xx}}H + \underline{\hphantom{xx}}K) + (\underline{\hphantom{xx}}H + \underline{\hphantom{xx}}K)$$

This equation needs to hold for all values of x, so the coefficients of x must be equal, and the constant terms must be equal. The result is two equations:

$$102 = \underline{\hphantom{xx}}H + \underline{\hphantom{xx}}K$$

$$126 = \underline{\hphantom{xx}}H + \underline{\hphantom{xx}}K$$

Solve these two equations for H and K.

The final result is:

$$\frac{102x+126}{(3x+9)(21x+18)} = \frac{H}{3x+9} + \frac{K}{21x+18} = \frac{}{3x+9} + \frac{}{21x+18}$$

2.

$$\frac{36x+57}{(2x+4)(6x+9)}$$

$$= \frac{H}{2x+4} + \frac{K}{6x+9}$$

$$\frac{H(\underline{\hspace{0.6cm}}x+\underline{\hspace{0.6cm}})+K(\underline{\hspace{0.6cm}}x+\underline{\hspace{0.6cm}})}{(\underline{\hspace{0.6cm}}x+\underline{\hspace{0.6cm}})(\underline{\hspace{0.6cm}}x+\underline{\hspace{0.6cm}})}$$

Multiply across the parentheses for the terms in the numerator:

$$\frac{\underline{\hspace{0.6cm}}xH+\underline{\hspace{0.6cm}}H+\underline{\hspace{0.6cm}}xK+\underline{\hspace{0.6cm}}K}{(\underline{\hspace{0.6cm}}x+\underline{\hspace{0.6cm}})(\underline{\hspace{0.6cm}}x+\underline{\hspace{0.6cm}})}$$

Collect the terms with x:

$$\frac{x(\underline{\hspace{0.6cm}}H+\underline{\hspace{0.6cm}}K)+(\underline{\hspace{0.6cm}}H+\underline{\hspace{0.6cm}}K)}{(\underline{\hspace{0.6cm}}x+\underline{\hspace{0.6cm}})(\underline{\hspace{0.6cm}}x+\underline{\hspace{0.6cm}})}$$

Look only at the numerators:

$$36x + 57 = x(\underline{\hspace{0.6cm}}H + \underline{\hspace{0.6cm}}K) + (\underline{\hspace{0.6cm}}H + \underline{\hspace{0.6cm}}K)$$

This equation needs to hold for all values of x, so the coefficients of x must be equal, and the constant terms must be equal. The result is two equations:

$$36 = \underline{\hspace{0.6cm}}H + \underline{\hspace{0.6cm}}K$$

$$57 = \underline{\hspace{0.6cm}}H + \underline{\hspace{0.6cm}}K$$

Solve these two equations for H and K.

The final result is:

$$\frac{36x+57}{(2x+4)(6x+9)} = \frac{H}{2x+4} + \frac{K}{6x+9} = \frac{}{2x+4} + \frac{}{6x+9}$$

3.

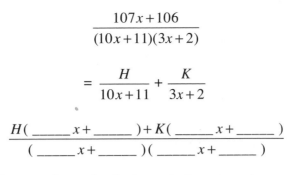

$$\frac{107x+106}{(10x+11)(3x+2)}$$

$$= \frac{H}{10x+11} + \frac{K}{3x+2}$$

$$\frac{H(\underline{\quad}x+\underline{\quad})+K(\underline{\quad}x+\underline{\quad})}{(\underline{\quad}x+\underline{\quad})(\underline{\quad}x+\underline{\quad})}$$

Multiply across the parentheses for the terms in the numerator:

$$\frac{\underline{\quad}xH+\underline{\quad}H+\underline{\quad}xK+\underline{\quad}K}{(\underline{\quad}x+\underline{\quad})(\underline{\quad}x+\underline{\quad})}$$

Collect the terms with x:

$$\frac{x(\underline{\quad}H+\underline{\quad}K)+(\underline{\quad}H+\underline{\quad}K)}{(\underline{\quad}x+\underline{\quad})(\underline{\quad}x+\underline{\quad})}$$

Look only at the numerators:

$$107x + 106 = x(\underline{\quad}H+\underline{\quad}K) + (\underline{\quad}H+\underline{\quad}K)$$

This equation needs to hold for all values of x, so the coefficients of x must be equal, and the constant terms must be equal. The result is two equations:

$$107 = \underline{\quad}H+\underline{\quad}K$$

$$106 = \underline{\quad}H+\underline{\quad}K$$

Solve these two equations for H and K.

The final result is:

$$\frac{107x+106}{(10x+11)(3x+2)} = \frac{H}{10x+11} + \frac{K}{3x+2} = \frac{}{10x+11} + \frac{}{3x+2}$$

4.

$$\frac{408x+208}{(20x+15)(24x+4)}$$

$$= \frac{H}{20x+15} + \frac{K}{24x+4}$$

$$\frac{H(\underline{\quad}x+\underline{\quad})+K(\underline{\quad}x+\underline{\quad})}{(\underline{\quad}x+\underline{\quad})(\underline{\quad}x+\underline{\quad})}$$

Multiply across the parentheses for the terms in the numerator:

$$\frac{\underline{\hspace{1cm}}xH + \underline{\hspace{1cm}}H + \underline{\hspace{1cm}}xK + \underline{\hspace{1cm}}K}{(\underline{\hspace{1cm}}x + \underline{\hspace{1cm}})(\underline{\hspace{1cm}}x + \underline{\hspace{1cm}})}$$

Collect the terms with x:

$$\frac{x(\underline{\hspace{1cm}}H + \underline{\hspace{1cm}}K) + (\underline{\hspace{1cm}}H + \underline{\hspace{1cm}}K)}{(\underline{\hspace{1cm}}x + \underline{\hspace{1cm}})(\underline{\hspace{1cm}}x + \underline{\hspace{1cm}})}$$

Look only at the numerators:

$$408x + 208 = x(\underline{\hspace{1cm}}H + \underline{\hspace{1cm}}K) + (\underline{\hspace{1cm}}H + \underline{\hspace{1cm}}K)$$

This equation needs to hold for all values of x, so the coefficients of x must be equal, and the constant terms must be equal. The result is two equations:

$$408 = \underline{\hspace{1cm}}H + \underline{\hspace{1cm}}K$$

$$208 = \underline{\hspace{1cm}}H + \underline{\hspace{1cm}}K$$

Solve these two equations for H and K.

The final result is:

$$\frac{408x + 208}{(20x + 15)(24x + 4)} = \frac{H}{20x + 15} + \frac{K}{24x + 4} = \frac{\underline{\hspace{1cm}}}{20x + 15} + \frac{\underline{\hspace{1cm}}}{24x + 4}$$

"We must make a general plan for the method of partial fractions," the professor said. "This method will help whenever we have two polynomials in a fraction, such as $N(x)/D(x)$. Here N stands for numerator and D stands for denominator. If we want to, we can say that the degree of $N(x)$ is less than the degree of $D(x)$. That's what we used to call a *proper rational function*."

"Why can we say that?" Trigonometeris asked.

"If there was a higher power of x in the numerator than in the denominator, we could use algebraic division to express the fraction as the sum of a polynomial and a proper fraction," the professor answered. "I always thought it was fascinating the way algebraic division worked."

"It is fascinating the way algebraic division works, but it is a lot of work," Recordis said. "Anyway, we know that we can factor $D(x)$ into a bunch of linear factors. Then we know that each of those linear factors will be in the denominator of one of the partial fractions."

"Are you sure you can factor every polynomial into linear factors?" the king said. "I thought . . ."

"Yes, you can factor anything," Recordis said. "Take my word for it. Except . . ." Recordis suddenly began to shiver, and then he trembled violently. "No—no—not those numbers!" he screamed, and he fell to the floor.

The king rushed to his assistance. "What happened to him?"

We were all puzzled, until the professor realized what had happened. "Recordis gets like this whenever he sees an *imaginary number*, such as $i = \sqrt{-1}$. You can factor any polynomial into linear factors, but some of the factors may contain terms that are not real numbers. For example, $(x^2 + 1) = (x - \sqrt{-1})(x + \sqrt{-1}) = (x - i)(x + i)$."

"We can't lose Recordis like this!" the king said. "We must find some way to avoid imaginary numbers."

"We could factor any polynomial into some linear factors and some quadratic factors," I suggested. "That way we will be assured that we won't have to deal with any imaginary numbers."

After repeated assurances that we would use only real numbers, Recordis finally calmed down, got up, and returned to normal. "This plan will create other problems, though," he said. "Suppose we have one linear factor and one quadratic factor in the denominator:

$$z = \frac{n_1 x^2 + n_2 x + n_3}{(a_1 x - b_1)(a_2 x^2 + b_2 x + c_2)}$$

"And you try partial fractions on that."

$$z = \frac{A}{a_1 x + b_1} + \frac{B}{a_2 x^2 + b_2 x + c_2}$$

$$n_1 x^2 + n_2 x + n_3 = A(a_2 x^2 + b_2 x + c_2) + B(a_1 x + b_1)$$
$$= Aa_2 x^2 + (Ab_2 + Ba_1)x = (Ac_2 + Bb_1)$$

coefficients of x^2:

$$n_1 = Aa_2$$

coefficients of x:

$$n_2 = Ab_2 + Ba_1$$

constant terms:

$$n_3 + Ac_2 + Bb_1$$

"Now we're in real trouble," Recordis said. "We have three equations but only two unknowns (A and B). There is no way we can find a solution."

We realized he had a good point. "There doesn't seem to be any way of getting rid of an equation," the professor said. "So it looks as though our only choice will be to find some place where we can add another variable to go along with A and B." We experimented with a few possibilities until we were able to establish that, whenever there was a partial fraction with a quadratic denominator, the numerator would have to look like ($Ax + B$), with A and B both unknown.

"Let's see if it works," the professor said.

We tried an example:

$$y = \int \frac{3x^2 + 2x + 5}{(x - 1)(x^2 + 2x + 2)}\, dx$$

Setting up the partial fractions, we obtained:

$$\frac{3x^2 + 2x + 5}{(x - 1)(x^2 + 2x + 2)} = \frac{A}{x - 1} + \frac{Bx + C}{x^2 + 2x + 2}$$

$$= \frac{A(x^2 + 2x + 2) + (Bx + C)(x - 1)}{(x - 1)(x^2 + 2x + 2)}$$

Setting the numerators equal gave us:

$$3x^2 + 2x + 5 = Ax^2 + 2Ax + 2A + Bx^2 - Bx + Cx - C$$
$$= (A + B)x^2 + (2A - B + C)x + (2A - C)$$

coefficients of x^2:

$$3 = A + B$$

coefficients of x:

$$2 = 2A - B + C$$

constant terms:

$$5 = 2A - C$$

A bit of substitution made it possible to solve this three-equation, three-unknown system:

$$B = 3 - A$$
$$C = 2A - 5$$
$$\overline{2 = 2A - (3 - A) + (2A - 5)}$$
$$= 2A - 3 + A + 2A - 5$$
$$= 5A - 8$$
$$10 = 5A$$
$$A = 2$$
$$B = 1$$
$$C = -1$$

The result became:

$$\frac{3x^2 + 2x + 5}{(x - 1)(x^2 + 2x + 2)} = \frac{2}{x - 1} + \frac{x - 1}{x^2 + 2x + 2}$$

$$y = \int \frac{2}{x - 1}\, dx + \int \frac{x - 1}{x^2 + 2x + 2}\, dx$$

$$= 2 \ln |x - 1| + \int \frac{x - 1}{x^2 + 2x + 2}\, dx$$

"I hate to disillusion you," Recordis said, "but we still don't know how to do that last integral."

"I hate the expression $x^2 + 2x + 2$," Recordis moaned. "At least I hate having it in the denominator in an integral. If only there was some kind of substitution that would get rid of the $2x$ term. Then it would fit in better with the kind of integrals we have been doing lately."

"Let's consider the general case where we have an expression of the form $ax^2 + bx + c$, where a, b, and c are given," the king suggested. "Recordis has the right idea—there must be some substitution that will help."

234   Integration by Partial Fractions

We tried the substitution $u = x + k$, or $x = u - k$. We realized this substitution had the huge advantage that $dx = du$ (making it easier to use this substitution in an integral). Putting $x = u - k$ into $ax^2 + bx + c$ gave us:

$$a(u - k)^2 + b(u - k) + c = a(u^2 - 2uk + k^2) + b(u - k) + c$$
$$= au^2 - 2auk + ak^2 + bu - bk + c$$
$$= au^2 + (-2ak + b)u + ak^2 - bk + c$$

"If only we could choose the value of k," Recordis said. "Then let $k = b/2a$, and the term involving u disappears."

"We *can* choose k!" the king said. He issued the proclamation:

When the expression $ax^2 + bx + c$ appears in an integral, it can be simplified with the substitution $x = u - b/2a$, which gives us the expression:

$$au^2 + \frac{b^2}{4a} - \frac{b^2}{2a} + c = au^2 - \frac{b^2}{4a} + c = \frac{4a^2u^2 + (4ac - b^2)}{4a}$$

The new expression is easier to deal with, since it does not contain a term involving u to the first power.

For our expression $(x^2 + 2x + 2)$, we made the substitution $x = u - 2/2 = u - 1$.

$$\int \frac{x - 1}{x^2 + 2x + 2} \, dx = \int \frac{u - 2}{(u - 1)^2 + 2(u - 1) + 2} \, du$$

$$= \int \frac{u - 2}{u^2 - 2u + 1 + 2u - 2 + 2} \, du$$

$$= \int \frac{u - 2}{u^2 + 1} \, du$$

Trigonometeris, in turn, realized that a trigonometric substitution would help for this integral: Let $u = \tan\theta$, $du = \sec^2\theta \, d\theta$.

$$\int \frac{x - 1}{x^2 + 2x + 2} \, dx = \int \frac{(\tan\theta - 2)\sec^2\theta}{\tan^2\theta + 1} \, d\theta$$

$$= \int \frac{(\tan\theta - 2)\sec^2\theta}{\sec^2\theta} \, d\theta$$

$$= \int (\tan\theta - 2) \, d\theta$$

$$\int \frac{x - 1}{x^2 + 2x + 2} \, dx = \ln|\sec\theta| - 2\theta + C$$

We had performed two substitutions in evaluating this integral: First we had substituted u for x, and then we had substituted θ for u. That meant that we had to make two reverse substitutions to get our answer back in terms of x:

$$\int \frac{x-1}{x^2+2x+2}\,dx = \ln\sqrt{1+\tan^2\theta} - 2\theta$$

$$= \ln\sqrt{1+u^2} - 2\arctan u$$

$$= \tfrac{1}{2}\ln\left|[1+(x+1)^2]\right| - 2\arctan(x+1)$$

$$= \tfrac{1}{2}\ln\left|1+(x^2+2x+1)\right| - 2\arctan(x+1)$$

$$\int \frac{x-1}{x^2+2x+2}\,dx = \tfrac{1}{2}\ln\left|x^2+2x+2\right| - 2\arctan(x+1)$$

After doing all this work, we still had to write the final answer to the original integral.

$$y = \int \frac{3x^2+2x+5}{(x-1)(x^2+2x+2)}\,dx$$

$$= 2\ln|x-1| + \tfrac{1}{2}\ln\left|x^2+2x+2\right| - 2\arctan(x+1) + C$$

"I surely hope we never run into a denominator where we have to use a quadratic factor," Recordis said, gasping for breath. "Partial fractions was a nice method before we discovered that complication."

"I just thought of another case where we might not have enough unknowns," the king said. "Suppose that one of the factors in the denominator is repeated."

$$y = \frac{n_1 x^2 + n_2 x + n_3}{(x-a_1)(x-a_1)(x-a_2)}$$

"Then it won't work to set up the partial fractions as follows."

$$y = \frac{A}{x-a_1} + \frac{B}{x-a_1} + \frac{C}{x-a_2}$$

"A and B aren't really separate constants in this problem, because they both have the same denominator."

We realized that we had to make another adjustment in the partial fractions formula. It turned out that, whenever a factor in the denominator is repeated, one of the partial fractions has to have the *square* of that factor in the denominator:

$$\frac{n_1 x^2 + n_2 x + n_3}{(x - a_1)^2 (x - a_2)} = \frac{A}{x - a_1} + \frac{B}{x - a_1} + \frac{C}{x - a_2}$$

"We'd better make a summary of the method now, before anyone thinks of any more complications," Recordis said. (See page 238.)

"That wraps up everything for today," Trigonometeris said, looking greatly relieved.

"You're not off the hook yet," Recordis told him. "You still need to solve your problem from yesterday. We started with this."

$$z = \int \frac{1}{\sqrt{a^2 + b^2 x^2}} \, dx$$

"And," he continued, "we ended up as follows."

$$x = \frac{a \tan \theta}{b}$$

$$z = \frac{1}{b} \int \sec \theta \, d\theta$$

"Let's use our result for the integral of the secant function," Trigonometeris said.

$$z = \frac{1}{b} \ln | \sec \theta + \tan \theta | + C$$

"Now we need to write the result in terms of x," Recordis said. "I know an expression for $\tan \theta$."

$$\tan \theta = \frac{bx}{a}$$

"How do we find $\sec \theta$?"

"We'll set up a triangle to find sec θ (Figure 12–1)," Trigonometeris said.

$$\tan \theta = \frac{\text{(opposite side)}}{\text{(adjacent side)}} = \frac{bx}{a}$$

$$\sec \theta = \frac{\text{(hypotenuse)}}{\text{(adjacent side)}} = \left(1 + \frac{b^2 x^2}{a^2}\right)^{1/2}$$

$$z = \int (a^2 + b^2 x^2)^{-1/2} \, dx$$

$$= \frac{1}{b} \ln \left| \left(1 + \frac{b^2 x^2}{a^2}\right)^{1/2} + \frac{bx}{a} \right| + C$$

$$\sqrt{1 + b^2 x^2 / a^2}$$

bx/a

θ

1

FIGURE 12–1.

All of us, especially Trigonometeris, were greatly relieved when we finally finished the work for that day. One result was that Recordis and Trigonometeris realized they were very dependent on each other. After that day Recordis was less likely to make snide remarks about trigonometry, and Trigonometeris was less likely to belittle algebra.

NOTE TO CHAPTER 12

The *degree* of a polynomial is the highest power that appears in the polynomial. For example, $(x + 1)$ is a first-degree polynomial, $(x^2 + 2x + 3)$ is a second-degree polynomial, and $(x^3 + 3x^2 + 4)$ is a third-degree polynomial. A *rational function* is a fraction in which both the numerator and denominator are polynomials. If the degree of the numerator is less than the degree of the denominator, the rational function is called a *proper rational function*. If the degree of the numerator is greater than the degree of the denominator, the function is called an *improper rational function*. An improper rational function can always be written as the sum of a polynomial plus a proper rational function. Note the analogy with real numbers. A fraction n/d is called a *proper fraction* if $n < d$, and an *improper fraction* if $n > d$. An improper fraction can always be written as the sum of an integer and a proper fraction. (For example, ³⁄₂ can be written as 1 + ½.) See a book on algebra for more information on polynomials.

Evaluating Integrals by Partial Fractions

The method of partial fractions is useful when the integrand contains a rational function, that is, a fraction with a polynomial in the numerator of degree less than the degree of the polynomial in the denominator:

$$\int \frac{a_m x^m + a_{m-1} x^{m-1} + a_{m-2} x^{m-2} + \cdots a_1 x + a_0}{b_n x^n + b_{n-1} x^{n-1} + b_{n-2} x^{n-2} + \cdots b_1 x + b_0} \, dx \qquad (m < n)$$

The goal of the method is to break the integrand into a sum of fractions that are much simpler. The first step is to factor the denominator. The result will be a product of some linear factors and some quadratic factors. All the numbers that result will be real, but there is no guarantee that they will be rational. (Of course, if the denominator comes to you already factored, you will be saved a *lot* of work.) The integrand can then be resolved as a sum of partial fractions as follows:

1. If a linear factor (such as $ax + b$) occurs once in the denominator, then there will be a partial fraction of the form $A/(ax + b)$.
2. If the linear factor $(ax + b)$ occurs k times in the denominator, then there are k partial fractions of the form $A_1/(ax + b), A_2/(ax + b)^2, \ldots, A_k/(ax + b)^k$.
3. If a quadratic factor (such as, $(ax^2 + bx + c)$ occurs once in the denominator, then there is a partial fraction of the form $(Ax + B)/(ax^2 + bx + c)$. (Note that $(b^2 - 4ac)$ is negative in this case. Otherwise, the quadratic factor can be broken into a product of two linear factors.)
4. If the quadratic factor $(ax^2 + bx + c)$ occurs j times in the denominator, then there are j partial fractions of the form $(A_1 x + B_1)/(ax^2 + bx + c)$,

    $(A_2 x + B_2)/ax^2 + bx + c)^2, \ldots, (A_j x + B_j)/(ax^2 + bx + c)^j$.

The numerators of the partial fractions must be solved for next. Once the integrand has been broken up into partial fractions, each integral can be solved individually. The integrals with linear denominators can be solved with logarithms, and the integrals with quadratic denominators can be solved with trigonometric substitution, using the secant-tangent identity.

Exercises

Express each of these fractions as a sum of partial fractions:

1. $\dfrac{x+4}{(x-3)(x-2)}$

2. $\dfrac{3x}{(x+\frac{1}{2})(x-\frac{1}{2})}$

3. $\dfrac{2x-4}{(2x-2)(4x+3)}$

4. $\dfrac{1}{x(x+3)}$

5. $\dfrac{x^2+x+1}{x^3+3x^2+3x+1}$

Combine into single fractions:

6. $\dfrac{7.1}{10-0.6z}-\dfrac{6.4}{9-0.4z}$

7. $\dfrac{1}{2-2x}+\dfrac{1}{2+2x}$

8. $\dfrac{2}{x-1}+\dfrac{x-1}{x^2+2x+2}$

Solve for y:

9. $\dfrac{dy}{dx}=\dfrac{2x-7}{(x-4)(x-3)}$

10. $\dfrac{dy}{dx}=\dfrac{2x+9}{(\frac{1}{2}x+5)(x-1)}$

11. $\dfrac{dy}{dx} = \dfrac{(1-\sqrt{3})x-4}{(x-\sqrt{3})(x-1)}$

12. $\dfrac{dy}{dx} = \dfrac{x}{x^2-1}$

Differentiate each function to verify an integration result from the chapter:

13. $y = \dfrac{1}{2}\ln\left(\dfrac{1+x}{1-x}\right)$

14. $y = \ln(\sec x + \tan x)$

15. $y = -\ln\cos x$

16. $y = \dfrac{1}{2}\ln\mid(x^2+2x+2)\mid - 2\arctan(x+1) + 2\ln\mid x-1\mid$

17.   Evaluate: $y = \int \sec^3 x\, dx$. Verify the result by differentiation.

18.   Find the area of the section of the hyperbola $x^2/a^2 - y^2/b^2 = 1$ that is bounded by the curve and the line $x = 2a$. (See Figure 12–2.)

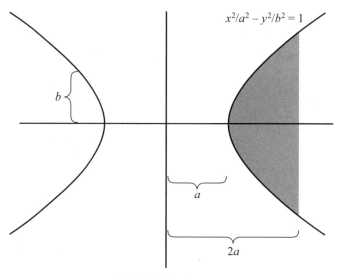

FIGURE 12–2.

Evaluate:

19. $y = \int \sqrt{1+x^2}\, dx$. (Use the result from Exercise 17.)

20. $y = \int (1+x^2)^{-1/2}\, dx$

21. $y = \displaystyle\int \frac{\sqrt{1-x^2}}{x}\, dx$

22. $y = \displaystyle\int \frac{x}{x\sqrt{1+x^2}}\, dx$

23. $y = \displaystyle\int \frac{x^2}{\sqrt{1+x^2}}\, dx$

Express each improper rational function as the sum of a polynomial and a proper rational function:

24. $\dfrac{2x^4 + 5x^3 + 7x^2 + 5x}{2x^2 + 3x + 4}$

25. $\dfrac{15x^4 - 15x + 1}{3x^3 - 3}$

26. $\dfrac{5x^4 + 4x^3 - 5x^2 - 4x + 1}{x^2 - 1}$

27. $\dfrac{3x^4 - x^3 - 33x^2 - 25x - 61}{x^2 - x - 12}$

28. Find a general formula for the integral

$$y = \int \frac{a}{ax^2 + bx + c}\, dx$$

Consider two cases: (a) $b^2 - 4ac > 0$. Define a new constant D so that $D = \sqrt{b^2 - 4ac}$. (b) $b^2 - 4ac < 0$. Let $D = \sqrt{4ac - b^2}$.

Evaluate each of the following integrals. The result from Exercise 28 will help. Another useful result, which can be found in a table of integrals, is:

$$\int \frac{Ax + B}{ax^2 + bx + c}\, dx = \frac{A}{2a} \ln|ax^2 + bx + c| + \left(\frac{B - Ab}{2a}\right) \int \frac{1}{ax^2 + bx + c}\, dx$$

29. $\displaystyle\int \frac{1}{x^2 + 2x + 2}\, dx$

30. $\displaystyle\int \frac{1}{x^2 - 3x - 1}\, dx$

31. $\int \dfrac{1}{x^2+x+1}\, dx$

32. $y = \int \dfrac{1}{x^2-x+1}\, dx$

33. $y = \int \dfrac{1}{x^3-1}\, dx$

34. $y = \int \dfrac{1}{x^3+1}\, dx$

35. $y = \int \dfrac{1}{2x^2+2x}\, dx$

36. $y = \int \dfrac{1}{3x^2+5x+1}\, dx$

37. $y = \int \dfrac{1}{x^4-1}\, dx$

38. $y = \int \dfrac{1}{(x-1)(x^2+1)}\, dx$

39. Verify the formula from Exercise 28 by differentiation.

40. Find a general formula for the integral:

$$Z = \int \dfrac{x}{\sqrt{ax^2+bx+c}}\, dx$$

Assume that $a < 0$, and $b^2 - 4ac > 0$. Hint: First perform a substitution to eliminate the term involving x to the first power.

41. Find a general formula for the integral:

$$Z = \int \dfrac{x}{\sqrt{2ax-x^2}}\, dx$$

Hint: Make the substitution $x = a(1 - \cos \theta)$.

Finding Volumes with Integrals

Plans for the party were proceeding smoothly, but there were still a lot of arrangements to be made. Every day Recordis or Builder came up with a new problem to be solved. "There are so many little things that need doing," Recordis kept saying.

The professor was working on her outline of calculus. She was convinced that we were completely finished with the subject. The king, though, was becoming uneasy. "I'm sure there's something we're missing," he said. "Definite integrals must be good for more than just finding area."

"No way!" the professor said. "We know that integrals are good for two things—finding antiderivatives and finding areas."

At that moment Builder came into the room in a very frustrated state of mind. "I need to know the volume of Column Mountain so I can make a model for the playground. I know the mountain is shaped like a paraboloid." (See Figure 13–1.)

"I remember what a paraboloid is," the professor said. "You make one by taking a regular two-dimensional parabola and rotating it in three dimensions."

FIGURE 13–1.

"It's too bad we can't help you," Recordis said. "Of course, if you needed to know the area of something . . ."

"We should be able to do something," the king protested, and he started to call the others.

"No, we can't do anything," Recordis said. "I just flipped through all my old tables. We don't have any formula for the volume of a paraboloid."

Builder gave us the exact measurements of the mountain anyway. The cross section of the mountain obeyed the relationship $y = -\frac{1}{2}x^2$, and we needed to know the volume from $y = 0$ to $y = -100$.

"I know how we can approximate the volume," the king said. "Let's start in the same way that we started to find the area under a curve. (See Chapter 8.) We can divide the mountain up into a series of little cylinders." (Figure 13–2.)

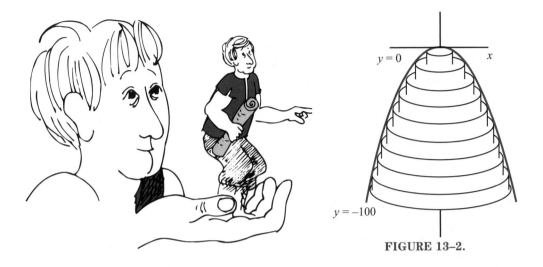

$y = 0$ x

$y = -100$

FIGURE 13–2.

"Pancakes! Pancakes!" Pal chanted.

("If he wants to call them pancakes, that's what we'd better call them," Recordis whispered.)

"Now all we need is the volume of each pancake," the professor said.

"That's easy," Recordis said. "The volume of a pancake is as follows."

$$\text{(base area)} \times \text{(height)} = \pi(\text{radius})^2(\text{height})$$

"In this case x is the radius," the professor said.

$$\text{(volume of pancake)} = \pi x^2(\text{height})$$

"What are we going to call the height?" Recordis asked.

"What are we going to call the volume?" the king asked.

"We could call the height Δy," I suggested. "And we could call the volume of each little pancake Δv_i, where the little i stands for the ith pancake."

$$\Delta v_i = \pi x_i^2 \, \Delta y$$

"Then to get the approximate volume of the paraboloid we just add up the volumes of all the pancakes," the professor said.

$$\text{(volume of paraboloid)} \simeq \sum_{i=1}^{n} \Delta v_i = \sum_{i=1}^{n} \pi x_i^2 \, \Delta y$$

"We're still missing a lot of volume," Recordis objected. "Look at how much volume is part of the paraboloid that isn't part of any of the pancakes."

"Then we can add more and more pancakes, so the approximate volume comes closer to the total volume," the professor said.

Recordis was about to protest that he would have to add together all of the pancakes, but before he could the king cried out, "I know how we can get the exact volume of the paraboloid! We take the limit as the number of pancakes goes to infinity!"

$$(\text{volume of paraboloid}) = \lim_{n\to\infty, \Delta y\to 0} \sum_{i=1}^{n} \pi x_i^2 \, \Delta y$$

"We could call that the *continuous sum*," the professor said. "Before we take the limit we can say that we have a *discrete sum*."

We were all impressed that we had proceeded this far, but we realized that we did not know where to go next.

"This won't work!" Recordis said emphatically. "In fact, we have already proved that this doesn't work! This is exactly like the first method we tried to find the area under a curve, and we know that didn't work at all!" (See Chapter 8.)

"That's it!" the professor exclaimed suddenly.

"What's it?" Recordis said.

"What you just said!" the professor went on excitedly. "We already found that the area under a curve is equal to the continuous sum."

$$(\text{area under curve}) = \lim_{\Delta x\to 0} \sum_{i=1}^{n} f(x_i) \, \Delta x$$

"I know that!" Recordis said. "But we don't know how to calculate a continuous sum."

"But we also found that the area under a curve is equal to the definite integral."

$$(\text{area under curve}) \int_{a}^{b} f(x) \, dx$$

"I already know that," Recordis said.

"If both the continuous sum and the definite integral equal the area, then they must be equal to each other!" the professor proclaimed triumphantly.

$$\lim_{\Delta x\to 0} \sum_{i=1}^{n} f(x_i) \, \Delta x = (\text{continuous sum}) = (\text{area})$$

$$= (\text{definite integral}) = \int_{a}^{b} f(x) \, dx$$

Therefore:

$$\lim_{\Delta x\to 0} \sum_{i=1}^{n} f(x_i) \, \Delta x = \int_{a}^{b} f(x) \, dx$$

"Now we can evaluate the continuous sum," the king said. "We just need to solve the definite integral, and we know how to do that. And the continuous sum can represent anything, instead of just an area. It can represent a volume or even something else."

"There is an uncanny similarity between the two notations," the professor said. "Notice how this part, the crooked s, Σ, corresponds to the curvy part, which is almost like a letter s: $\int$. Notice also how the $f(x_i)$ corresponds to the $f(x)$ part, and the Δx part corresponds to the dx part."

$$\lim_{\Delta x \to 0} \sum_{i=1}^{n} f(x_i)\,\Delta x$$

$$\int_{a}^{b} f(x)\,dx$$

"That's amazing," the king said. "I wonder how that happened?"

"It almost looks as though the integral sign was set up to stand for a continuous sum," Recordis remarked, looking at me suspiciously.

"Can you solve the problem now?" Builder asked.

We wrote our expression for the volume of the paraboloid as a continuous sum:

$$V = \lim_{\Delta y \to 0} \sum_{i=1}^{n} \pi x^2\,\Delta y$$

"Now it's easy to turn this into a definite integral," the king said.

$$V = \int \pi x^2\,dy$$

"What about the limits of integration?" Recordis asked.

"The limits must be in terms of y, since we are integrating along y," the king said. "We need to set up the limits so that our pancakes will cover all of the volume that we want."

"It looks as though we start integrating where $y = -100$ and keep integrating until $y = 0$," the professor said. (See Figure 13–2.)

$$V = \pi \int_{-100}^{0} x^2\,dy$$

"Now we're stuck," Recordis said. "We have the integrand written in terms of x, but the d-variable term is in terms of y."

Builder told him, "I already told you that $y = -\tfrac{1}{2}x^2$."

We put that expression into the integral, and then we found everything was quite straightforward:

$$x^2 = -2y$$

$$V = \pi \int_{-100}^{0} (-2y)\, dy$$

$$= (-2\pi)(\tfrac{1}{2}y^2)\bigg|_{-100}^{0}$$

$$= -\pi(0)^2 - [-\pi(100)^2]$$

$$= 10{,}000\pi$$

"That's a lot of dirt," Builder whistled. "I'll take your word for it."

"I don't take your word for it," Recordis disagreed. "I'm not sure that this pancake method really works. Let's try it for something that we know the volume of already, such as a sphere of radius r."

"We know that the volume of a sphere is $(4/3)\pi r^3$," the king said.

The professor started to divide the sphere into pancakes. "Let's find the volume of one hemisphere first," she suggested. "Then we can multiply by 2 to get the volume of the whole sphere." (Figure 13–3.)

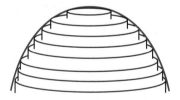

FIGURE 13–3.

I suggested that we call the volume of a single pancake dV, to make it easier to set up the integral $\int dV = V$. We decided to call the height of each pancake dy, because we would be integrating along y. Then the volume of each pancake would be $dV = \pi x^2\, dy$. We set up the integral:

$$V = \int dV = \int \pi x^2\, dy$$

We decided that the limits of integration, which had to be written in terms of y, would be $y = 0$ to $y = r$:

$$V = \int_{0}^{r} \pi x^2\, dy$$

Now we had to solve for x^2 in terms of y. Recordis recognized an algebra problem when he saw it. Since the cross section of a sphere is a circle, he used the equation of a circle:

$$x^2 + y^2 = r^2$$

$$x^2 = r^2 - y^2$$

Substitute this formula for x^2 into the integral:

$$V = \pi \int_0^r (r^2 - y^2)\, dy$$

Now it was just a matter of using the methods we had developed to solve definite integrals:

$$V = \pi \int_0^r r^2\, dy - \pi \int_0^r y^2\, dy$$

Note that r is constant:

$$V = \pi r^2 y \Big|_0^r - \pi (\tfrac{1}{3}) y^3 \Big|_0^r$$

$$= \pi r^2 (r - 0) - \pi (\tfrac{1}{3})(r^3 - 0^3)$$

$$= \pi r^3 - \tfrac{1}{3}\pi r^3$$

$$V = \tfrac{2}{3}\pi r^3$$

"Just what we wanted!" the professor said with relief. "If the volume of a hemisphere is $2/3\pi r^3$, then the volume of a whole sphere is $4/3\pi r^3$."

"So this method does work!" Recordis rejoiced. "I have a whole bunch of party arrangements that I'm working on that this method could help with." He flipped to a page in one of his notebooks.

"Here is a good example. We designed special vases to hold flowers at the party."

The curved vases were formed by rotating the curve $x = e^y - 1$ about the y axis, from $y = 0$ to $y = h$ (Figure 13–4). (We used h to stand for the height of the vases because Recordis had forgotten what the height was.)

"We can use pancakes easily," the professor said. "We will be integrating along y, from $y = 0$ to $y = h$."

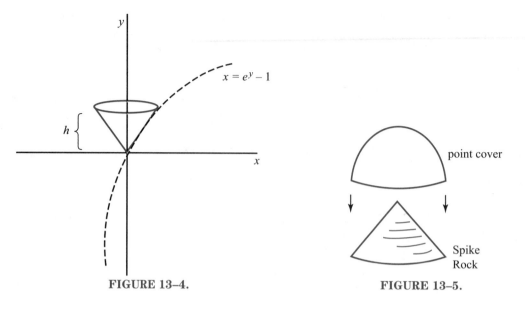

FIGURE 13–4.   **FIGURE 13–5.**

The volume of each pancake was given by:

$$dV = \pi x^2\, dy$$

$$x = e^y - 1$$

$$dV = \pi (e^y - 1)^2\, dy$$

$$V = \pi \int_0^h (e^{2y} - 2e^y + 1)\, dy$$

$$= \pi \left(\tfrac{1}{2}e^{2y} - 2e^y + y\right)\Big|_{y=0}^{h}$$

$$= \pi \left(\tfrac{1}{2}e^{2h} - 2e^h + h - \tfrac{1}{2}e^0 + 2e^0 - 0\right)$$

$$V = \pi \left(\tfrac{1}{2}e^{2h} - 2e^h + h + \tfrac{3}{2}\right)$$

"I have another problem we should be able to solve now," the king said. "The gardener mentioned it to me. Remember Spike Rock, the rock with the perfect conical top next to the rose garden (Figure 13–5)? I am very much afraid that one of the children at the party could be hurt unless we do something to cover up the spike. So Builder and I figured out how to build a round point-cover out of clay to put on the spike. If we can figure out the volume of the point-cover, we will know how much clay we need to buy."

It turned out that the point-cover was formed by taking the region bounded by the two curves $y = x$ and $y = x^2$ and rotating that region about the y-axis (Figure 13–6). (Notice that we turned the point-cover upside down to make it a bit easier to figure out the volume.)

"I hate to disillusion you," Recordis said, "but there is no way that we can fit any pancakes into a shape like that."

"Do we have to use pancakes?" the king asked. "Maybe we could use another shape."

"How?" Recordis inquired.

"All we need to do is set up the continuous sum," the professor agreed. "We might be able to find another shape that works."

I suggested that we try using little cylindrical shells (Figure 13–7). "If the shell is thin enough, we can say that the volume of the shell is approximately equal to $dV = (2\pi r)(h)(dx)$." (Note that $2\pi r$ is the circumference of the cylinder, so $2\pi rh$ is the outer surface area of the cylinder. The volume of the shell is then equal to the surface area times the thickness.)

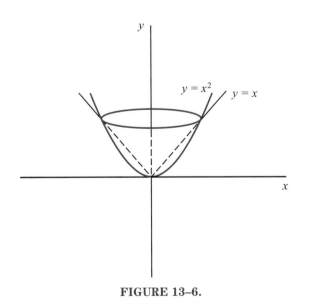

FIGURE 13–6.

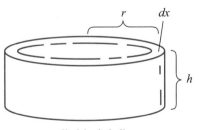

cylindrical shell
$dV = 2\pi rh\, dx$

FIGURE 13–7.

"You had better make sure that you don't let the shells become too thick," Recordis said. "If the shells are too thick, the approximate formula you are using for the volume won't work."

"That shouldn't be any problem," the professor told him. "Remember that we are going to take the limit where the thickness of the shells goes to zero."

Next, we had to fit a representative cylindrical shell into the point-cover (Figure 13–8).

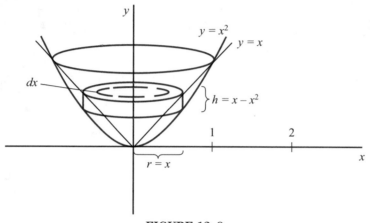

FIGURE 13–8.

"Now we need the radius of the shell and the height of the shell," the professor said. The radius turned out to be simply equal to the x coordinate of the shell, and the height of the shell was equal to the distance between the two curves: $h = x - x^2$.

"Now we have to put all this together into a sum," the king stated.

$$V = \lim_{\Delta x \to 0} \sum_{i=1}^{n} 2\pi r_i h_i \, \Delta x$$

$$= \lim_{\Delta x \to 0} 2\pi \sum_{i=1}^{n} x_i (x_i - x_i^2) \, \Delta x$$

$$= 2\pi \int_0^1 x(x - x^2) \, dx$$

"That's an easy integral," Recordis said.

$$V = 2\pi \int_0^1 (x^2 - x^3) \, dx$$

$$= 2\pi \left(\tfrac{1}{3} x^3 \Big|_0^1 - \tfrac{1}{4} x^4 \Big|_0^1 \right)$$

$$= 2\pi \left(\tfrac{1}{3} - \tfrac{1}{4} \right)$$

$$V = \frac{\pi}{6}$$

WORKSHEET

1. Find the volume of a box which has height h, and a square base with side s.
   (Integrate from $y = 0$ to $y = h$).

$$\int \underline{\hspace{1cm}} dy = \underline{\hspace{1cm}}$$

2. Find the volume of a cylinder that has height h, and a circular base with radius
   r. (Integrate from $y = 0$ to $y = h$).

$$\int \underline{\hspace{1cm}} dy = \underline{\hspace{1cm}}$$

3. Find the volume of an upside-down cone. Put the vertex of the cone at $y = 0$,
   and integrate from $y = 0$ to $y = h$. The volume of each pancake is $\pi x^2\, dy$,
   where x is the radius of each pancake. Find the general formula for x, noting
   that $x = 0$ when $y = 0$, and $x = R$ when $y = h$.

$$x = \underline{\hspace{1cm}}$$

$$\int_0^h \pi x^2\, dy = \int_0^h \pi (\underline{\hspace{1cm}})^2\, dy = \underline{\hspace{1cm}}$$

4. Find the volume of an upside-down pyramid with height h and a square base
   with area s^2. (Follow a similar method as in Question 3.)

$$\int \underline{\hspace{1cm}} dy = \underline{\hspace{1cm}}$$

5. Evaluate this integral:

$$I = \int_0^R 2\pi x \sqrt{R^2 - x^2}\, dx$$

Use the substitution $u = R^2 - x^2$, $du = \underline{\hspace{1cm}} dx$, $dx = \underline{\hspace{1cm}} du$. For what shape are
you finding the volume?

"It's a good thing that we just invented the method of cylindrical shells," Recordis
said. "It will help with something else I need to do for the party. We're giving out
doughnuts to the children. I have been trying to figure out what the volume of a
doughnut is, so that I will know how much doughnut mix I need to order."

The doughnuts were formed by rotating the circle $x^2 + y^2 = 1$ about the line $x = 2$
(Figure 13–9). (The geometric term for a doughnut-shaped figure is *toroid*.)

Igor drew a representative cylindrical shell in the doughnut. The next problem was
to figure out the height and radius of the shell. "The radius must be $(2 - x)$," the professor said.

"The height must be $2y$," the king added (Figure 13–10).

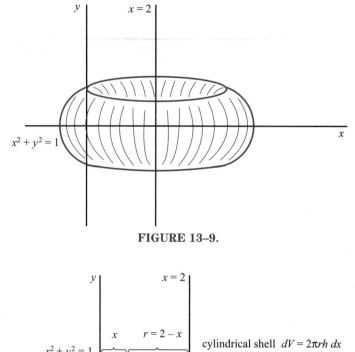

FIGURE 13–9.

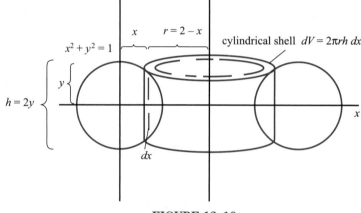

FIGURE 13–10.

We set up the equation for the volume of the cylindrical shell:

$$dV = 2\pi\, rh\, dx$$

$$r = 2 - x$$

$$h = 2y$$

"We can solve for y," Recordis noted.

$$x^2 + y^2 = 1$$

$$y^2 = 1 - x^2$$

$$y = \sqrt{1 - x^2}$$

$$dV = 2\pi\,(2 - x)2\sqrt{1 - x^2}\ dx$$

"Now we need to figure out the limits of integration," the professor said. "They must be in terms of x, since we are integrating along x." After looking closely at the diagram, we decided that we could catch all of the volume of the doughnut with our cylindrical shells if we integrated from $x = -1$ to $x = 1$.

$$V = \int_{-1}^{1} 2\pi(2-x)2\sqrt{1-x^2}\ dx$$

$$= 4\pi \int_{-1}^{1} (2-x)\sqrt{1-x^2}\ dx$$

$$= 4\pi \int_{-1}^{1} 2\sqrt{1-x^2}\ dx - 4\pi \int_{-1}^{1} x\sqrt{1-x^2}\ dx$$

We made the substitution $u = 1 - x^2$, $du = -2x\ dx$ for the second integral, with the result:

$$V = 8\pi \int_{-1}^{1} \sqrt{1-x^2}\ dx + 2\pi \int_{0}^{0} u^{1/2}\ du$$

"That second integral can't be right!" Recordis exclaimed. "That's equal to zero!"

After we had looked at the integral for a while, the professor said, "Of course it must be zero. Look at a graph of the integrand (Figure 13–11). Half of the graph is above the x-axis, so its area is positive; but exactly half of the graph is below the x-axis, so its area is negative. The two areas cancel out, so the total value of the definite integral must be zero."

"Anyway, we know that the volume of the doughnut can't be zero," Recordis said.

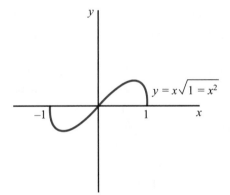

FIGURE 13–11.

"It's lucky for us that we still have the first integral to evaluate."

$$V = 8\pi \int_{-1}^{1} (1-x^2)^{1/2}\ dx$$

We made the trigonometric substitution $x = \sin\theta$, $dx = \cos\theta\ d\theta$.

$$V = 8\pi \int_{-\pi/2}^{\pi/2} \cos^2\theta\ d\theta$$

We had done the integral of $\cos^2\theta$ before (see Chapter 11), so the final answer for the volume turned out to be:

$$V = 8\pi\left(\frac{\pi}{2}\right)$$

$$= 4\pi^2$$

"I have another problem related to the party," Builder said. "We want to build a grandstand in one corner of the auditorium (Figure 13–12). The height of the grandstand is h. The base of the grandstand is an isosceles right triangle, with each side equal to s. I'm sure you can figure out the volume under the grandstand."

"We can?" Recordis asked. "That's not a figure of revolution, so it's not like anything else we've done. There is no way that you could fit either pancakes or cylindrical shells into a pyramid like that. In fact, the only shape that you could fit in there would be little triangles."

"That's a good idea," the professor said. "We'll use little triangles." (Figure 13–13.)

The volume of each little triangular segment was easy to figure out, since it was just the area of the triangle times the thickness. "With a right triangle it's easy to find the area," Recordis noted. "You just multiply together the two legs and divide by 2." In our case the two legs were equal, so we just needed to find one of the legs. We called x the length of one leg for our representative triangle, and we called z the distance down from the top of the pyramid-shaped grandstand (Figure 13–14).

"I know how we can find x," the professor said. "We can use a similar triangle relationship to establish that $z/h = x/s$.

$$x = \frac{zs}{h}$$

We wrote down the volume of the differential triangular element:

$$dV = \frac{1}{2}x^2\,dz = \frac{s^2}{2h^2}z^2\,dz$$

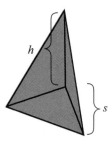

FIGURE 13–12.

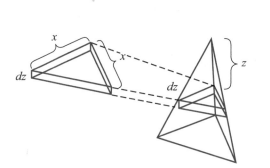

FIGURE 13–13.

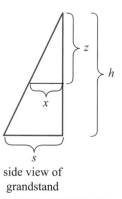

side view of
grandstand

FIGURE 13–14.

We integrated along z, from $z = 0$ to $z = h$:

$$V = \int_0^h \frac{s^2 z^2}{2h^2}\, dz$$

Since s and h are constants, they slide across the integral sign:

$$V = \frac{s^2}{2h^2} \int_0^h z^2\, dz$$

$$= \frac{s^2}{2h^2} \frac{1}{3} z^3 \Big|_0^h$$

$$= \frac{s^2 h^3}{6h^2}$$

$$= \frac{s^2 h}{6}$$

"Now we can work out the volume of the new swimming pool for the party," Builder said. "Then we'll know how much water it will take to fill it."

"As easy as floating in the shallow end," the professor said. "The volume of the pool is area times depth."

"That only works if the pool has the same depth everywhere!" the king exclaimed. "Just like you said, the pool should have a shallow end."

Builder said the depth of the pool at a particular coordinate (x, y) was

$$z(x, y) = D - ax - by$$

where $z(x, y)$ is the depth of the pool, which is a function of the two variables x and y, D is the depth at the deepest corner of the pool, and a and b are constants that determine the slant of the bottom of the pool. For convenience, we can locate the origin of the coordinate system at the deepest corner. (See Figure 13–15.)

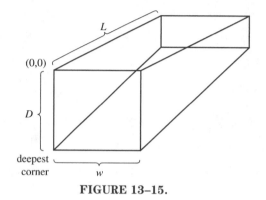

FIGURE 13–15.

"That's not a figure of revolution, so we can't treat it like the other problems," the professor said worriedly.

"The problem is that the depth of the pool changes at each point," the king realized. "If only we could look at a very small square at the bottom of the pool, then the depth wouldn't change very much over that square, and so the volume of the column of water above the square would equal the area of the square multiplied by the depth of the pool at that point."

The professor developed the idea further. "Call the lengths of the sides of the square Δx and Δy. Let z be the depth of the pool at that point, and the volume (Δv) of the column of water above that square is $z\, \Delta x\, \Delta y$."

"We'll need a lot of these columns to fill the volume of the pool," the king said. "We'll need to use subscripts to identify each square: x_i for the x value and y_j for the y value for the square in row i, column j. (See Figure 13–16.)

$$\text{volume of column} = \Delta v_{ij} \approx z(x_i, y_j)\, \Delta y\, \Delta x$$

We let w represent the width along the x dimension and L the length along the y dimension. We arranged to have n rectangles of width $\Delta x = w/n$ along the x dimension, and m rectangles of width $\Delta y = L/m$ along the y dimension.

"To approximate the total volume (V) of the pool, we'll have to add up all of these columns," Builder realized.

$$
\begin{aligned}
V \approx\ & [z(x_1, y_1) + z(x_1, y_2) + z(x_1, y_3) + \cdots + z(x_1, y_m) \\
& + z(x_2, y_1) + z(x_2, y_2) + z(x_2, y_3) + \cdots + z(x_2, y_m) \\
& + z(x_3, y_1) + z(x_3, y_2) + z(x_3, y_3) + \cdots + z(x_3, y_m) \\
& \cdots \\
& + z(x_n, y_1) + z(x_n, y_2) + z(x_n, y_3) + \cdots + z(x_n, y_m)]\, \Delta y\, \Delta x
\end{aligned}
$$

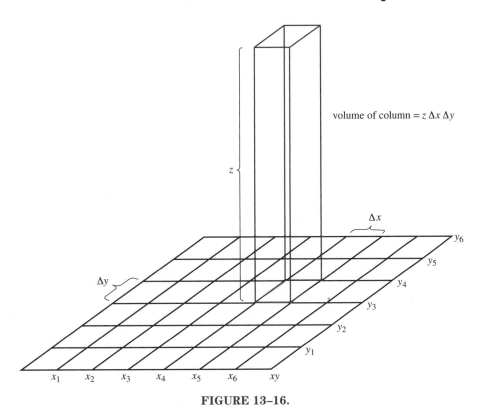

volume of column = $z \, \Delta x \, \Delta y$

FIGURE 13–16.

Recordis' eyes glazed over. "We have to use summation notation for this expression."

$$V \approx \sum_{i=1}^{n} \sum_{j=1}^{m} z(x_i, y_j) \, \Delta y \, \Delta x$$

"To get the exact volume, we'll have to take the limit as Δx and Δy go to zero, and m and n go to infinity," Builder said.

$$V = \lim_{\Delta x \to 0, \Delta y \to 0} \sum_{i=1}^{n} \sum_{j=1}^{m} z(x_i, y_j) \, \Delta y \, \Delta x$$

Instantly we recognized that the summations turned to integrals:

$$V = \int_{x=0}^{w} \int_{y=0}^{L} z(x, y) \, dy \, dx$$

"We haven't had two integrals on one line before," Recordis was skeptical. "Although I suppose this just means we work out the inner integral first, and then the outer integral."

We inserted our formula for z:

$$V = \int_{x=0}^{w} \int_{y=0}^{L} (D - ax - by) \, dy \, dx$$

Initially we pretended that we only had one integral involving y:

$$\int_{y=0}^{L} (D - ax - by)\, dy$$

"When we're doing the integral for y, I'm going to assert that x will be constant. We'll deal with the variability of x later," the professor said.

Treating D, a, x, and b as constants, the integral was not hard:

$$\int_{y=0}^{L} (D - ax - by)\, dy = (Dy - axy - 0.5by^2)\Big|_{y=0}^{L} = DL - axL - 0.5bL^2$$

Now we put this result for the integral for y into the integral for x (the outer integral from our original pair):

$$V = \int_{x=0}^{w} (DL - axL - 0.5bL^2)\, dx$$

We worked out the integral (remembering that all letters other than x are constants:

$$V = (DLx - 0.5aLx^2 - 0.5bL^2x)\Big|_{x=0}^{w}$$

$$V = DLw - 0.5aLw^2 - 0.5bL^2w$$

$$V = DLw - 0.5Lw(aw + bL)$$

We were satisfied with this result, and Builder made it even easier when he decided that the pool would only slope in the x direction, so b would be zero. Then the volume became

$$V = w(DL - 0.5aLw)$$

The professor decided to write a general description of what she called a *double integral* of a two-variable function $f(x, y)$, which represents the volume under the surface $z = f(x, y)$ and above the xy plane in a specified region.

For example,

$$\int_{x=a}^{x=b} \int_{y=c}^{y=d} f(x, y)\, dy\, dx$$

represents the volume under the surface $z = f(x, y)$ over the rectangle from $x = a$ to $x = b$ and $y = c$ to $y = d$. (See Figure 13–17.)

This assumes that $f(x, y)$ is nonnegative everywhere within the limits of integration. If $f(x, y)$ is negative, then the double integral will give the negative of the volume above the surface and below the x, y plane.

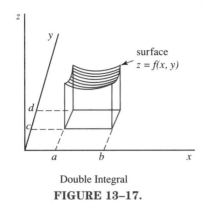

Double Integral
FIGURE 13–17.

"I never thought an integral could be so versatile," Recordis said. "This will be the best party ever, and we've made calculus do all the work. And I have an eerie feeling that my problem closet is filled with similar problems that we can now solve."

Exercises

1. Find the volume of the solid formed by rotating the ellipse $x^2/a^2 + y^2/b^2 = 1$ about the x axis.

2. The base of a solid is the region bounded by the parabola $y = 1/2x^2$ and the line $y = 2$. Each plane section of the solid perpendicular to the y axis is an equilateral triangle. Find the volume of the solid.

3. A paraboloid dish (cross section $y = x^2$) is 8 units deep. It is filled with water up to a height of 4 units. How much water must be added to the dish to fill it completely?

4. Write the integral that represents the volume of the solid formed by rotating the region bounded by $y = f(x)$, $x = a$, $x = b$, and $y = 0$ about (a) the x axis; (b) the line $x = b$; and (c) the line $x = c$, where $c > a, b$.

5. Consider the solid formed by rotating the curve $y = f(x)$ from $x = a$ to $x = b$ about the x axis. Let $V(x)$ be the function whose value is the volume of the solid between $x = a$ and $x = x$. (a) What is $V(a)$? (b) For some small Δx, what is $V(x + \Delta x) - V(x)$? (c) Using the definition of the derivative, find the definite integral that represents the total volume of the solid. (Let $F(x)$ be a function such that $dF/dx = \pi[f(x)]^2$.)

6. Find the volume of the top quarter of a sphere:

$$V = \int_{r/2}^{r} \pi x^2 \, dy$$

7. Find the volume of the sphere by integrating from $y = -r$ to $y = r$. Compare the result with the one arrived at by taking twice the integral from $y = 0$ to $y = r$.

8. Find the volume of a pyramid with a square base and sides that are equilateral triangles.

9. Use pancakes to find the volume of the hole in the doughnut described in the chapter.

10. One day Recordis made a mistake while making doughnuts. The doughnuts turned out exactly the same as the doughnuts described in the chapter except that the hole was missing. Use pancakes to find the total volume of one of these doughnuts.

11. Use the method of cylindrical shells to find the volume of a sphere.

12. Use the method of cylindrical shells to find the volume of the paraboloid with cross section $y = -\frac{1}{2}x^2$, from $y = 0$ to $y = -100$.

13. The volume of a hemisphere can be found with the double integral:

$$\int_{x=-r}^{x=r} \int_{y=-\sqrt{r^2-x^2}}^{y=\sqrt{r^2-x^2}} \sqrt{r^2 - x^2 - y^2} \, dy \, dx$$

First, evaluate the inner integral involving y (treating x as a constant while evaluating this integral):

$$\int_{y=-\sqrt{r^2-x^2}}^{y=\sqrt{r^2-x^2}} \sqrt{r^2 - x^2 - y^2} \, dy$$

Define $A = \sqrt{r^2 - x^2}$; then use the trigonometric substitution $y = A \sin \theta$; $dy = A \cos \theta \, d\theta$; $\theta = \arcsin (y/A)$.

After you've evaluated the inner integral, you'll need to evaluate the outer integral involving x, which becomes

$$\int_{-r}^{r} \frac{\pi}{2} (r^2 - x^2) \, dx$$

Arc Lengths, Surface Areas, and the Center of Mass

The next day we were in the courtyard preparing to cut ribbons to use as streamers around the party site. Builder had already built poles throughout the grounds from which he could hang the ribbons. The professor unrolled the ribbons while Recordis stood next to the roll, prepared to cut them.

"How long should we make the ribbons?" Recordis called out to the professor.

The professor thought for a moment. "We had better not cut them too long, or else they will hang down too far and will get in the way when the adults walk under them."

"But we can't cut them too short," Recordis said. "We want the ribbons to hang down a little bit, so they make a nice curve."

"We should be able to figure out the exact length," the king told them. "Builder, how far apart are the poles?"

"They're all 4 meters apart, and each pole is 2.2 meters tall," Builder answered.

"The lowest part of the hanging ribbon should be 2 meters high," Trigonometeris said. "That will be high enough so that all the adult heads will fit under it."

"We don't know the equation of a hanging ribbon, and we don't know how to figure out the length of a curve even if we do know its equation," Recordis pointed out.

"Builder, you should be able to figure out the shape of a hanging ribbon," the king said. Builder nodded hesitantly, as if he knew that it would be a lot of work to find the equation of a hanging ribbon. "We should be able to find the length of a curve," the king went on. "We'll use what we did yesterday. We'll make an approximation for the length of the curve in terms of a sum, and then turn the sum into a definite integral."

The professor was irritated because this meant we were not completely done with calculus yet, and Recordis was irritated because it sounded as if calculating lengths would be hard, but we all prudently agreed to the king's plan anyway. Builder took some string and returned to his workroom. The rest of us went into the Main Conference Room to set up a definite integral to represent the length of a curve.

"We don't know how to find the length of anything curved, except circles," the professor said.

"Then we'll have to approximate the length of a curve in terms of straight lines," the king suggested. "We know the length of straight lines."

Igor drew a series of straight lines that approximately traced out the curve (Figure 14–1).

We decided to call the length of each little line segment ds. It was clear that the length of the whole curve (which we called S) was approximately equal to the sum of all the little lengths ds:

$$S \simeq \sum_{i=1}^{n} ds_i$$

"We can get the exact length by taking the limit as ds goes to zero," the professor noted.

$$S \simeq \lim_{ds \to 0} \sum_{i=1}^{n} ds_i$$

FIGURE 14–1.

"And that, of course, is equal to the definite integral."

$$S = \int ds$$

"We can't do very much with that integral when it is written in terms of s," Recordis said. "For one thing, we don't know what the limits of integration are in terms of s. We need to express the integral in terms of x."

Igor displayed a close-up version of one of the little segments ds (Figure 14–2).

"We can call the distance that ds goes up dy, and the distance that it goes sideways dx," the professor said. "That makes use of differential notation."

"We can use the Pythagorean theorem to express ds in terms of dx and dy," the king suggested.

$$ds^2 = dx^2 + dy^2$$
$$ds = \sqrt{dx^2 + dy^2}$$

"Do we have to bring the Pythagorean theorem into this?" Recordis complained. "Whenever you use that theorem, you end up with a square root sign. Wouldn't it be easier to just make the approximation that, as ds goes to zero, ds and dx are approximately equal? Then we could just integrate dx to get the length of the curve." Recordis' method certainly sounded simpler, until we realized that it would give the same result for the lengths of all the curves in Figure 14–3. Sadly we went back to the method that had the square root sign.

$$(\text{length}) = S = \int ds = \int \sqrt{dx^2 + dy^2}$$

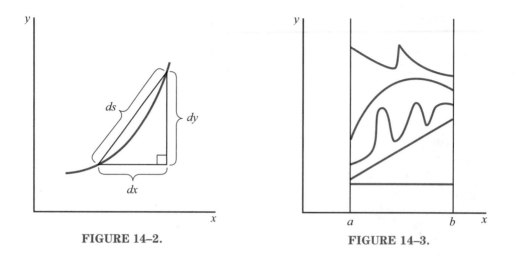

FIGURE 14–2.　　　　　　　　　**FIGURE 14–3.**

The professor suggested a way to modify the square root so that we could integrate along x:

$$S = \int \sqrt{(dx^2)[1+(dy/dx)^2]}$$

$$= \int \sqrt{1+(dy/dx)^2}\ dx$$

"We'd better make sure that we remember the difference between the second derivative (d^2y/dx^2) and the square of the first derivative $(dy/dx)^2$," Recordis cautioned.

"That's the integral we need!" the king said. "Now all we have to do is write the limits in terms of x.

$$(\text{length of curve from } x = a \text{ to } x = b) = S = \int_a^b \sqrt{1+(dy/dx)^2}\ dx$$

"With this integral we can find the length of any curve, as long as we know dy/dx."

Just then Builder came into the room with the answer to the hanging ribbon problem. "This curve turns out to be a weird one, but I guarantee this is the right curve. The general shape of a hanging ribbon is given by this expression."

$$y = \frac{1}{2}a\ (e^{x/a} + e^{-x/a})$$

(This curve is known as a *catenary*. It can also be expressed using a function known as the *hyperbolic cosine*, defined by $\cosh x = \frac{1}{2}(e^x + e^{-x})$.)

"That number e again!" Recordis said. "I am always amazed how many times that one number turns up."

"In your particular case $a = 10$," Builder told him. "Your specific curve is given by:

$$y = 5(e^{x/10} + e^{-x/10}) - 8$$

"The poles are 4 meters apart, which means that one pole is at $x = -2$ and the other pole is at $x = 2$."

We tried our length integral on the general curve, with distance d between the ends:

$$S = \int_{-d/2}^{d/2} \sqrt{1 + (dy/dx)^2}\ dx$$

$$y = \frac{1}{2}a\,(e^{x/a} + e^{-x/a})$$

$$\frac{dy}{dx} = \frac{1}{2}a\left(\frac{1}{a}e^{x/a} - \frac{1}{a}e^{-x/a}\right)$$

$$= \frac{1}{2}(e^{x/a} - e^{-x/a})$$

$$\left(\frac{dy}{dx}\right)^2 = \frac{1}{4}(e^{2x/a} - 2e^{x/a}e^{-x/a} + e^{-2x/a})$$

$$= \frac{1}{4}(e^{2x/a} - 2 + e^{-2x/a})$$

$$S = \int_{-d/2}^{d/2} \sqrt{1 + \frac{1}{4}(e^{2x/a} - 2 + e^{-2x/a})}\ dx$$

$$= \int_{-d/2}^{d/2} \sqrt{1 + \frac{1}{4}e^{2x/a} - \frac{1}{2} + \frac{1}{4}e^{-2x/a}}\ dx$$

$$= \int_{-d/2}^{d/2} \sqrt{\frac{1}{4}e^{2x/a} + \frac{1}{2} + \frac{1}{4}e^{-2x/a}}\ dx$$

"If only we could get rid of the square root sign!" Recordis said.
"We can," the professor assured him. "Watch this."

$$S = \int_{-d/2}^{d/2} \frac{1}{2}\sqrt{e^{2x/a} + 2 + e^{-2x/a}}\ dx$$

$$(e^{x/a} + e^{-x/a})^2 = e^{2x/a} + 2 + e^{-2x/a}$$

$$S = \int_{-d/2}^{d/2} \frac{1}{2}\sqrt{(e^{2x/a} + e^{-2x/a})^2}\ dx$$

$$= \frac{1}{2}\int_{-d/2}^{d/2} (e^{x/a} + e^{-x/a})\ dx$$

$$= \frac{1}{2}ae^{x/a}\Big|_{-d/2}^{d/2} - \frac{1}{2}ae^{-x/a}\Big|_{-d/2}^{d/2}$$

$$= \frac{1}{2}ae^{d/2a} - \frac{1}{2}ae^{-d/2a} - \frac{1}{2}ae^{-d/2a} + \frac{1}{2}ae^{d/2a}$$

$$S = ae^{d/2a} - ae^{-d/2a}$$

"And in our case $d = 4$ and $a = 10$," the professor said.

We calculated the length of our ribbons:

$$S = (10)(e^{4/20} - e^{-4/20})$$

$$= 4.03$$

"That certainly looks about right," Recordis said. "We know that the ribbons must be a little bit longer than 4, but they can't be very much longer than 4. I'm sure that this is a nice method, but I would feel more comfortable if we used the method to find the length of something that we already know, such as a circle."

We all agreed to that plan. The professor was nervous, though. "After we've come this far, it would be terrible to find that something we've done in calculus is inconsistent."

We set up the equation of a circle:

$$x^2 + y^2 = r^2$$

Using implicit differentiation, we found dy/dx:

$$2x + 2y \frac{dy}{dx} = 0$$

$$\frac{dy}{dx} = \frac{-x}{y}$$

$$\left(\frac{dy}{dx}\right)^2 = \frac{x^2}{y^2}$$

$$= \frac{x^2}{r^2 - x^2}$$

We decided to call the length of an arc equal to one eighth of the circumference of the circle S (Figure 14–4). Next we set up the integral for S:

$$S = \int_{x=0}^{x=r/\sqrt{2}} \left[1 + \left(\frac{dy}{dx}\right)^2\right]^{1/2} dx$$

$$= \int_0^{r\sqrt{2}} \sqrt{1 + [x^2/(r^2 - x^2)]}\ dx$$

$$= \int_0^{r/\sqrt{2}} \sqrt{(r^2 - x^2 + x^2)/(r^2 - x^2)}\ dx$$

$$= \int_0^{r/\sqrt{2}} \sqrt{r^2/(r^2 - x^2)}\ dx$$

$$S = r \int_0^{r/\sqrt{2}} \frac{1}{\sqrt{r^2 - x^2}}\ dx$$

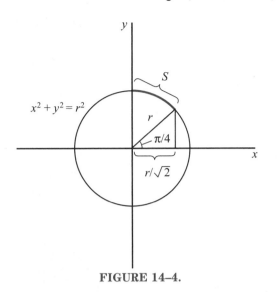

FIGURE 14–4.

"This is a trigonometric substitution integral," Trigonometeris said.

$$x = r \sin \theta$$

$$dx = r \cos \theta \, d\theta$$

$$\theta = \arcsin\left(\frac{x}{r}\right)$$

We calculated the two limits of integration in terms of θ:

$$\arcsin\left(\frac{1}{\sqrt{2}}\right) = \frac{\pi}{4}$$

$$\arcsin 0 = 0$$

$$S = r \int_0^{\pi/4} \frac{1}{\sqrt{r^2 - r^2 \sin^2 \theta}} \, r \cos \theta \, d\theta$$

$$= r \int_0^{\pi/4} \frac{r \cos \theta}{r \cos \theta} \, d\theta$$

$$= r \int_0^{\pi/4} d\theta$$

$$= r \, \theta \, \Big|_0^{\pi/4}$$

$$= r \left(\frac{\pi}{4} - 0\right)$$

$$S = \frac{\pi r}{4}$$

"Whew! That's right," the professor said. "We know that C, the circumference of the whole circle, is given by $C = 8S$, since S is the length of one eighth of the circle. That means $C = 8(\pi r/4) = 2\pi r$. We already know that that is the correct answer."

We went out into the courtyard to cut the ribbons. While we were doing this, Builder came into the courtyard carrying the parabolically shaped clay point-cover, which he prepared to put over the top of Spike Rock.

"You aren't putting it on there like that, are you?" Recordis asked. "That is such an ugly clay color. Don't you think you should paint it first?"

"If you insist," Builder said, although it was clear that he had a lot of other responsibilities as the date of the party approached. "First I have to know how much paint I need."

"It's a good thing he didn't suggest that we should figure out how much paint he needs," Recordis remarked, as he returned to ribbon-cutting.

"We should be able to," the king said. "We would have to calculate the surface area of the outside of the paraboloid. Builder is so busy with all the other arrangements that I think we should do this for him."

"We don't know how to calculate surface areas!" Recordis protested. "I hope you aren't about to suggest that we set up a sum that approximates the surface area, and then turn the sum into a definite integral!"

That was exactly what the king had in mind, so we all returned to the Main Conference Room and had Igor draw a picture of a paraboloid (Figure 14–5).

"Now we need to fit some little shapes in there so we can set up a sum," the professor said.

We decided to fill the paraboloid with a series of sections cut from cones. "I remember what those sections of cones are," Trigonometeris said. "We called them *frustums*. They were very frustrating shapes." (Figure 14–6.)

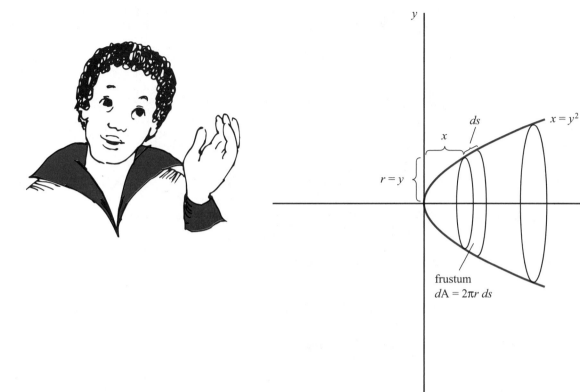

FIGURE 14–5.

Recordis looked in his book and found a formula for the surface area of a frustum:

$$(\text{surface area}) = A = 2\pi r \,(\text{slant height})$$

We decided to call the slant height of each frustum ds:

$$dA = 2\pi r \, ds$$

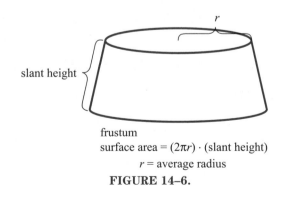

frustum
surface area $= (2\pi r) \cdot (\text{slant height})$
$r =$ average radius
FIGURE 14–6.

"We know what ds is," the professor said. "That's the same ds that we used when we calculated the length of an arc of the curve."

$$ds = \sqrt{1+(dy/dx)^2}\ dx$$

$$dA = 2\pi r\ \sqrt{1+(dy/dx)^2}\ dx$$

"And in the case of the paraboloid the radius of each frustum is simply equal to y," the king said.

$$dA = 2\pi y\ \sqrt{1+(dy/dx)^2}\ dx$$

The cross section of the paraboloid obeyed the relationship $y = x^{1/2}$, so we could calculate y and dy/dx in terms of x:

$$y = x^{1/2}$$

$$\frac{dy}{dx} = \frac{1}{2}x^{-1/2}$$

$$\left(\frac{dy}{dx}\right)^2 = \frac{1}{4x}$$

We calculated the surface area for a general paraboloid, having a height of a:

$$A = \int_0^a 2\pi\ \sqrt{x}\sqrt{1 + \frac{1}{4x}}\ dx$$

$$= 2\pi \int_0^a \sqrt{x + \frac{1}{4}}\ dx$$

Let $u = x + \frac{1}{4}$; $du = dx$.

$$A = 2\pi \int_{1/4}^{a+1/4} u^{1/2}\ du$$

$$= 2\pi\ \frac{2}{3}u^{3/2}\ \Big|_{1/4}^{a+1/4}$$

$$= \frac{4\pi}{3}[(a + \tfrac{1}{4})^{3/2} - (\tfrac{1}{4})^{3/2}]$$

"In our case $a = 1$," the king said.

$$A = \frac{4\pi}{3}(1.25^{3/2} - 0.25^{3/2})$$

$$= 5.33$$

"I suppose you want us to calculate the surface area of a sphere, just to make sure that the method works," the professor said to Recordis.

We set up the integral for the surface area of a hemisphere: (Figure 14–7).

$$dA = 2\pi y \left[1 + \left(\frac{dy}{dx} \right)^2 \right]^{1/2} dx$$

$$\left(\frac{dy}{dx} \right)^2 = \frac{x^2}{r^2 - x^2} \text{ (See page 268.)}$$

$$A = 2\pi \int_0^r \sqrt{r^2 - x^2} \sqrt{\frac{r^2 - x^2 + x^2}{r^2 - x^2}} \, dx$$

$$= 2\pi \int_0^r \sqrt{r^2} \, dx$$

$$= 2\pi r \int_0^r dx$$

$$= 2\pi r \left. x \right|_0^r$$

$$= 2\pi r \, (r - 0)$$

$$A = 2\pi r^2$$

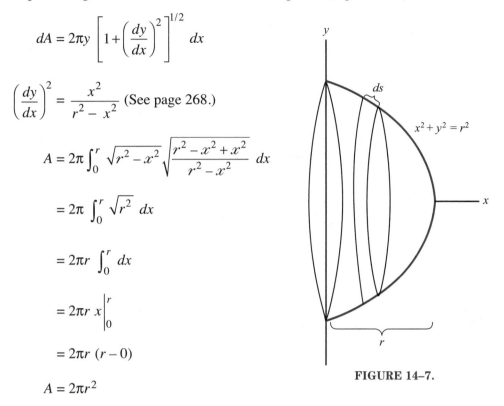

FIGURE 14–7.

"Therefore the surface area of a whole sphere is $4\pi r^2$, which is what we knew," the professor said.

Even Recordis was convinced by now that the method for finding surface areas worked.

We were so busy with all the party preparations we almost forgot to order the ice cream cones. We realized that we still hadn't solved Mr. Carrier's problem about the optimal shape for the ice cream cone (see Chapter 6). Then it hit us: now that we could find volumes and surface areas we could solve the problem and find the shape of the cone with the greatest volume for a fixed amount of surface area. If a cone has base radius r and height h, we worked out that its volume would be:

$$V = \frac{\pi r^2 h}{3}$$

(See Chapter 13, page 253, Worksheet Question 3.) The surface area s of the cone is:

$$s = \pi r \sqrt{h^2 + r^2}$$

(See Exercise 9, and rewrite the answer substituting $b = h$ and $a = r/h$.) The sight of the square root sign sent chills up our spines, as we realized it would be no easy matter to solve for h or r. However, Recordis found a way to avoid the square root sign. The square of the surface area is:

$$C = s^2 = \pi^2 r^2(h^2 + r^2) = BRH + BR^2$$

If s is constant, then C will also be constant. We also defined $R = r^2$, where r is the radius of the base of the cone; $H = h^2$, where h is the height of the cone; $A = \pi^2/9$; and $B = \pi^2$. The square of the volume is:

$$V = v^2 = \frac{\pi^2}{9}r^4h^2 = AR^2H$$

Maximizing the square of the volume while holding the square of the surface area constant leads to the same answer as maximizing the volume while holding the surface area constant.

The new simpler version of the problem required us to maximize V:

$$V = AR^2H$$

but this equation must be true:

$$C = BRH + BR^2$$

(Note: A, B, and C are constants.)

This problem is the same general form as the fence problems in Chapter 6. Solve for H in the second equation:

$$C = BRH + BR^2$$

$$C - BR^2 = BRH$$

$$\frac{C - BR^2}{BR} = H$$

Substitute this expression for H into the formula for volume:

$$V = AR^2H$$

$$V = AR^2\left(\frac{C - BR^2}{BR}\right)$$

Cancel an R:

$$V = AR\left(\frac{C - BR^2}{B}\right)$$

$$V = \left(\frac{AC}{B}\right)R - \left(\frac{AB}{B}\right)R^3$$

$$V = \left(\frac{AC}{B}\right)R - AR^3$$

Find the derivative:

$$\frac{dV}{dR} = \frac{AC}{B} - 3AR^2$$

Set the derivative equal to zero:

$$0 = \frac{AC}{B} - 3AR^2$$

Solve this equation for R:

$$3AR^2 = \frac{AC}{B}$$

$$3R^2 = \frac{C}{B}$$

$$R^2 = \frac{C}{3B}$$

$$R = \sqrt{\frac{C}{3B}}$$

Recall $R = r^2$ and $B = \pi^2$:

$$r^2 = \sqrt{\frac{C}{3\pi^2}}$$

Suppose that $r = 1$. Then

$$C = 3\pi^2$$

Since

$$C = s^2 = \pi^2 r^2 (h^2 + r^2)$$

solve for h:

$$C = \pi^2(h^2 + 1)$$

$$h^2 = \frac{C}{\pi^2} - 1$$

$$h^2 = \frac{3\pi^2}{\pi^2} - 1$$

$$h^2 = 3 - 1$$

$$h^2 = 2$$

$$h = \sqrt{2}$$

The result tells us that for the cone to have maximum volume for given surface area, the ratio of the height to the base radius should be $\sqrt{2}$.

Just before Builder went out to paint the parabolic point-cover, he asked us whether we could help him with another problem where he thought a continuous sum could be useful.

"I have a large metal half-circle that will be the new concert stage for the party. I want to know the exact point where it will balance so I can best design the supporting structure to hold it together (Figure 14–8). It is obvious that the balancing point (I call it the *center of mass*) must lie along the center line of the semicircle, because the semicircle is symmetric about the center line. I don't know where to balance it along that line, though."

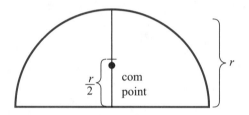

FIGURE 14–8.

"How can we help?" the professor asked. "I don't see how that has any connection with what we have been doing."

"We can use your work with discrete sums and continuous sums," Builder said. "It's easy to find the center of mass if you have a certain number of metal bars strung out along a rod." (Figure 14–9.)

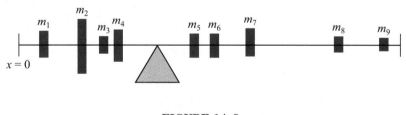

FIGURE 14–9.

"It doesn't look very easy," the king said. "Suppose each mass was dffferent, or suppose the bars weren't arranged symmetrically."

"You find the center of mass by calculating a *weighted average*."

$$(\text{position of center of mass}) = x_{\text{com}} = \frac{\sum\limits_{i=1}^{n} x_i m_i}{m}$$

"I use m to stand for the total mass of all the bars, and m_i for the mass of each individual bar."

"That tells you where the semicircle balances?" the professor asked, puzzled. (The professor was not very skillful in dealing with practical matters.)

"You bet it does," Builder said.

"How can we figure out the center of mass of the half-circle?" Recordis asked.

"We can imagine that the circle is made up of a bunch of tiny bars," Builder said. (See Figure 14–10.) "That means that the center of mass is approximately as follows."

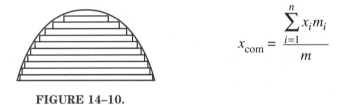

$$x_{\text{com}} = \frac{\sum\limits_{i=1}^{n} x_i m_i}{m}$$

FIGURE 14–10.

"I see!" the professor said suddenly. "To get the exact center of mass you take the limit as the number of bars goes to infinity and the mass of each bar goes to zero!"

$$x_{\text{com}} = \lim_{\Delta m \to 0} \frac{\sum\limits_{i=1}^{n \to \infty} x_i \Delta m}{m}$$

"We can write that as an integral!" the king realized.

$$x_{\text{com}} = \frac{\int x \, dm}{m}$$

"What's dm?" Recordis asked.

"That must be $dm = \rho \, dV$, where ρ is the density of the plate and dV is the volume of each little bar."

$$x_{\text{com}} = \int \frac{x\rho \, dV}{m}$$

"What's dV?" Recordis asked.

"That's just the volume of the little rectangular bar," the king said. We decided to call the thickness of each rectangular bar h (which is also the thickness of the semi-circular plate). The length of each bar, we decided, was $2y$, and the width of each bar was dx (Figure 14–11). That made the volume of each bar equal to $dV = 2yh\, dx$.

$$x_{\text{com}} = \int \frac{x\rho h\, 2y\, dx}{m}$$

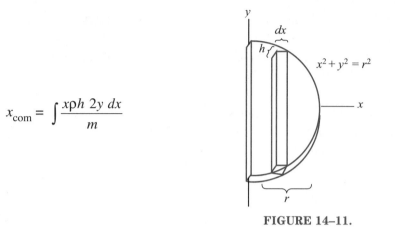

FIGURE 14–11.

"What's y?" Recordis asked.

"We can get that from the equation of a circle," the king said.

$$y = \sqrt{r^2 - x^2}$$

$$x_{\text{com}} = 2\rho h \int_0^r \frac{x\sqrt{r^2 - x^2}\, dx}{m}$$

"What's m?" Recordis inquired. It was the only other thing he could think of to ask.

"That's easy," Builder said. "It's equal to (volume) × (density)."

$$m = \frac{\rho h\, \pi r^2}{2}$$

$$x_{\text{com}} = \frac{2\rho h}{\pi\rho h r^2 / 2} \int_0^r x\sqrt{r^2 - x^2}\, dx$$

"I can take it from here," Recordis interrupted. "I was wondering what was taking you so long."

$$x_{\text{com}} = \frac{4}{\pi r^2} \int_0^r x\sqrt{r^2 - x^2}\, dx$$

Let $u = r^2 - x^2$.

$$du = -2x \, dx$$

$$x_{com} = \frac{4}{\pi r^2} \left(-\frac{1}{2}\right) \int_{r^2}^{0} u^{1/2} \, du$$

$$= -\frac{2}{\pi r^2} \frac{2}{3} u^{3/2} \Big|_{r^2}^{0}$$

$$= -\frac{4}{3\pi r^2} [0^{3/2} - (r^2)^{3/2}]$$

$$= \frac{4r^3}{3\pi r^2}$$

$$= \frac{4r}{3\pi}$$

$$x_{com} = 0.424r$$

"That looks about right," Builder said. "I would have predicted that the semicircle would have balanced at slightly less than half the radius from the edge."

We went back to work on preparations for the party. That evening the king and I were walking together along the grounds where we had hung the ribbons. "I never would have guessed it," the king said. "We started out trying to solve one simple problem—how fast the train was going at a given instant. Who would have known that we would find a new way to solve optimum-value problems, and a new way for calculating related rates and areas? And now we have the most mysterious part of all—we find that we can add together an infinite number of infinitesimal things."

Exercises

1. Find the length of the arc of the curve $y = x^{3/2}$ from $x = 1$ to $x = 2$.

2. Find the length of the parabola $y = 1/2x^2$ from $x = 0$ to $x = a$. (For a helpful integration result, see Chapter 12, Exercise 19.) Find a numerical result when $a = 2$.

3. Find the length of the arc of the curve $y = \ln(\sec x)$ from $x = 0$ to $x = \pi/4$. What is the length of the curve from $x = 0$ to $x = \pi/2$?

4. Find the length of the curve $y = ax$ from $x = 0$ to $x = b$.

Set up the integrals for the following arc lengths:

5. $y = x^{1/2}$ from $x = 0$ to $x = a$ $(a > 0)$

6. $y = \ln x$ from $x = 1$ to $x = a$ $(a > 1)$

7. $y = \sin x$ from $x = 0$ to $x = \pi$

8. Find the surface area of the paraboloid $y = x^2$ between $y = 3/4$ and $y = 15/4$.

9. Find the surface area of the cone formed by rotating the line $y = ax$ about the x-axis from $x = 0$ to $x = b$.

10. Find the surface area of the solid formed by rotating one arch of the curve $y = \sin x$ about the x-axis. (For a helpful integration result, see Exercise 2.)

11. Find the surface area of the solid formed by rotating the curve $y = 1/3x^3$ about the x-axis from $x = 0$ to $x = a$. Find a numerical result when $a = 2$.

12. Find the surface area of the solid formed by rotating the curve $y = e^x$ about the x-axis from $x = 0$ to $x = a$.

13. Find the surface area of the top quarter of the sphere of radius r:

$$A = 2\pi \int_{r/2}^{r} x \sqrt{1 + \frac{y^2}{x^2}}\, dy$$

14. The ceremonial shield of Carmorra has a thickness of h units and a density of ρ, and it is shaped like a parabola ($y = x^2$). Find the y coordinate of the center of mass (y_{com}). The center of mass is located along the line $x = 0$.

15. Find the y coordinate of the center of mass of the cone with cross section $y = ax$ and height b.

16. Write the expression for the center of mass of the solid generated by rotating the curve $y = f(x)$ from $x = a$ to $x = b$ about the x axis.

17. The professor has a long rod that is heavier at one end. The cross section of the rod has area A, the total length of the rod is L, and the density of the rod is given by $\rho = \sqrt{x} + 2$, where x is the distance from the light end of the rod. Find x_{com}.

18. The *moment of inertia* of a solid about a particular axis of rotation is given by $I = \int x^2 \, dm$, where x is the distance from the axis of rotation. (Note that the moment of inertia for the same solid is different when different axes of rotation are considered.) For example, a thin rod with cross-sectional area A, density ρ, and length L has $dm = \rho A \, dx$. Its moment of inertia about an axis through its center perpendicular to the rod (Figure 14–12) is given by:

$$I = \int_{-L/2}^{L/2} x^2 \rho A \, dx = \rho A \int_{-L/2}^{L/2} x^2 \, dx = \rho A (\tfrac{1}{3}) x^3 \Big|_{-L/2}^{L/2}$$

$$= \frac{\rho A L^3}{12}$$

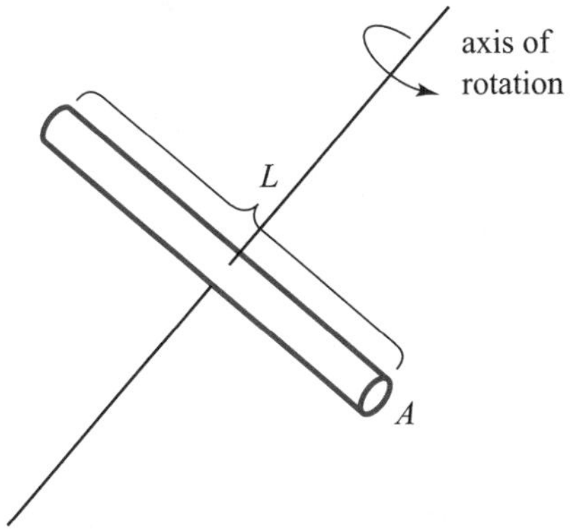

FIGURE 14–12.

Note that $AL = V$, where V is the volume of the rod, so $\rho AL = M$, where M is the total mass of the rod. Therefore $I = ML^2/12$. Find the moment of inertia of the rod about an axis of rotation perpendicular to the rod that passes through one end of the rod.

19. The moment of inertia of a disk (thickness h, density ρ, radius R), for rotations about the axis perpendicular to the disk through its center, is given by:

$$I = \int_{r=0}^{r=R} r^2 \, dm$$

where $dm = 2\pi r h \rho \, dr$. Solve the integral to find the moment of inertia. What is M, the total mass of the disk? Express the result for I in terms of M and R. (One reason that the moment of inertia is important is that the kinetic energy of a body rotating with angular speed ω is given by K.E. $= 1/2 I\omega^2$.)

20. Find the moment of inertia of a sphere (radius R, density ρ) with respect to rotations about an axis through its center. (Use cylindrical shells.)

21. Find the moment of inertia of Pal's conical-shaped top with respect to the central axis of the cone (height $= h$, base radius $= R$).

Introduction to Differential Equations

Builder was making the final arrangements for the party. "I just thought of something," he said one day. "I have a few old parts in my workroom. I could make them into a nice ride that the children would like. I've figured out the acceleration of the ride, and I'd like to know what its motion looks like."

"How do you know the acceleration?" the professor asked.

"Any object moves according to the equation $F = ma$, where F is the force acting on the object, m is its mass, and a is the acceleration ($a = d^2x/dt^2$). With my ride the force is given by $F = -kx$, and the mass of the ride is m. So that makes the final equation of motion as follows."

$$m \frac{d^2x}{dt^2} = -kx$$

"Can you tell me what x is as a function of t?"

"We can't do that!" Recordis said. "There's no way to turn that into an integral, and we can solve problems like that only if they're written as integrals."

"If only the equation were $d^2x/dt^2 = -kt$," the professor said. "Then we could integrate the function of t like this."

$$\frac{dx}{dt} = \int (-kt)\, dt = -\frac{1}{2}kt^2 + C_1$$

$$x = \int \left[-\frac{1}{2}kt^2 + C_1 \right] dt = -\frac{1}{6}kt^3 + C_1 t + C_2$$

"If only we had $dx/dt = -kx$," the king pointed out. "Then we could take the x over to the other side."

$$\frac{dx}{x} = -k\, dt$$

$$\ln x = -kt + C$$

$$x = e^{-kt + C}$$

"We could call an equation with derivatives in it a *differential equation*," the professor said.

"No way!" Recordis objected. "Take my word for it—as soon as we start applying functions to derivatives, we are really going to be sunk."

"We already know how to solve one type of differential equation," the king said helpfully.

$$\frac{dx}{dt} = f(t)$$

"When we have an equation of that form, we know that the solution is as follows."

$$x = \int f(t)\, dt$$

"But that means that these differential equation things are worse than integrals!" Recordis said. "You're telling me that first we have to solve the differential equation and then we have to solve the integral. I didn't think that anything could be harder than integrals."

"Recordis does have a point," the professor agreed. "We had better put some restrictions on the differential equations before we look for a way to solve them. It looks to me as if it would be very difficult to solve an equation if we are allowed to write an arbitrary function of lots of variables and their derivatives."

"First, we had better say that we are considering only one dependent variable (in this case x) and its derivative with respect to one independent variable (in this case t)," the king suggested. (A differential equation with one dependent variable and its derivatives with respect to one independent variable is known as an *ordinary differential equation*. If a differential equation involves derivatives with respect to more than one independent variable, it is called a *partial differential equation*.)

"We had better place some more restrictions," the professor said.

"Let's agree not to apply any functions to the derivatives," Recordis said. "We'll be in real trouble if we allow expressions like $(d^2x/dt^2)^2$ or $(d^2x/dt^2)(dx/dt)$. While we're at it, let's make sure that we don't have any functions of x, such as x^2, or any terms with an x multiplied by its derivative."

We agreed to these restrictions. They meant that the differential equation could be written in the form:

$$\frac{d^n x}{dt^n} + f_{n-1}(t)\,\frac{d^{n-1}x}{dt^{n-1}} + f_{n-2}(t)\,\frac{d^{n-2}x}{dt^{n-2}} + \cdots + f_0(t)x = f(t)$$

"That looks like a reasonable sort of equation," the professor said. "We have one term with no x, one term with one x, one term with the first derivative, one term with the second derivative, etc. Each f represents a function of t, but note that we never apply a function to x or one of its derivatives."

Recordis decided to write the equation in a shorter fashion:

$$\left[\frac{d^n}{dt^n} + f_{n-1}(t)\,\frac{d^{n-1}}{dt^{n-1}} + f_{n-2}(t)\,\frac{d^{n-2}}{dt^{n-2}} + \cdots + f_0(t)\right]x = f(t)$$

"This way, we have to write x only once. Also, we can save a lot of writing if we refer to this big, long expression with all the derivatives by some letter, say T. We can call it a *differential operator*."

$$T = \frac{d^n}{dt^n} + f_{n-1}(t)\,\frac{d^{n-1}}{dt^{n-1}} + f_{n-2}(t)\,\frac{d^{n-2}}{dt^{n-2}} + \cdots + f_0(t)$$

That meant we could write the equation in a much shorter form:

$$Tx = f(t)$$

The king was worried that this notation would confuse us by making it look like T was being multiplied by x, but Recordis was insistent so we agreed to use this notation for differential operators.

"I see what an operator is," the professor said. "It's almost like a function. You apply a function to a *number*; the result is another number. You apply an operator to a *function* (in this case x); the result is another function (in this case Tx)."

"I know what a differential operator is," the king added. "It means you have a *linear combination* of derivative operators." ("A 'linear combination' is just a fancy way of saying that you set out your list of derivatives, multiply each one by its own coefficient, and then add the whole thing up," Recordis whispered to me.)

We established that any differential operator satisfied the following properties:

1. $T(x_1 + x_2) = Tx_1 + Tx_2$

   (x_1 and x_2 are functions of t.) This follows from the sum rule for derivatives:

$$\frac{d}{dt}(x_1 + x_2) = \frac{dx_1}{dt} + \frac{dx_2}{dt}$$

2. $T(ax_1) = aTx_1$,

   where a is a constant.

   Combining these two properties, we could say:

$$T(ax_1 + bx_2) = aTx_1 + bTx_2$$

where x_1 and x_2 are two functions and a and b are two constants. We invented the term *linear operator* for an operator satisfying that property.

"Let's call an equation that can be written in this form a *linear differential equation*," the professor said.

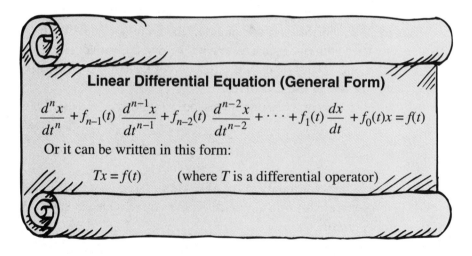

Linear Differential Equation (General Form)

$$\frac{d^n x}{dt^n} + f_{n-1}(t)\,\frac{d^{n-1}x}{dt^{n-1}} + f_{n-2}(t)\,\frac{d^{n-2}x}{dt^{n-2}} + \cdots + f_1(t)\,\frac{dx}{dt} + f_0(t)x = f(t)$$

Or it can be written in this form:

$$Tx = f(t) \qquad \text{(where } T \text{ is a differential operator)}$$

(Note: Since we can multiply both sides of the equation by a constant without changing the solution, we can conveniently arrange for the coefficient of the first term to be one.)

"It would help if we did not have that $f(t)$," the king suggested. "Then we would have an equation where each term contained either x or one of its derivatives."

I suggested that we call this type of equation a *homogeneous linear differential equation*.

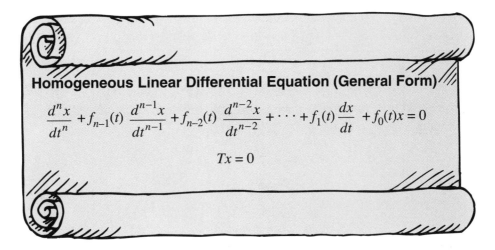

Homogeneous Linear Differential Equation (General Form)

$$\frac{d^n x}{dt^n} + f_{n-1}(t)\frac{d^{n-1}x}{dt^{n-1}} + f_{n-2}(t)\frac{d^{n-2}x}{dt^{n-2}} + \cdots + f_1(t)\frac{dx}{dt} + f_0(t)x = 0$$

$$Tx = 0$$

"That still doesn't look very encouraging," Recordis said. "But it does look as if Builder's equation is linear and homogeneous."

$$\frac{d^2 x}{dt^2} + \frac{k}{m}x = 0$$

We spent hours trying to solve this problem. Finally Trigonometeris had an idea. "We could guess," he said.

"We can't guess!" Recordis told him sternly. "This is Math. This is Serious Business. We don't have time for guessing games."

"Where did you get the parts that you are using to build this ride?" Trigonometeris asked Builder.

Builder thought a moment. "I remember," he said. "I'm using the giant spring that was originally part of the gremlin's oscillating chicken-scaring machine."

"Of course!"' Trigonometeris said. "Now I know what the answer must be. The chicken-scaring machine moved like a sine wave, so I bet that the answer will be as follows."

$$x = A \sin t$$

"We could try," the professor said. We differentiated Trigonometeris' guess twice to see whether it fit the equation:

$$x = A \sin t$$

$$\frac{dx}{dt} = A \cos t$$

$$\frac{d^2x}{dt^2} = -A \sin t$$

$$= -x$$

"It almost works," the king admitted. He suggested that we try multiplying the t in the middle of the sine function by some number ω: (ω is the Greek lowercase letter omega.)

$$x = A \sin \omega t$$

$$\frac{dx}{dt} = A\omega \cos \omega t$$

$$\frac{d^2x}{dt^2} = -A\omega^2 \sin \omega t$$

$$= -\omega^2 x$$

Putting this expression for x back in the original differential equation:

$$\frac{d^2x}{dt^2} + \frac{k}{m}x = 0$$

gave us:

$$-\omega^2 x + \frac{k}{m}x = 0$$

"Let $\omega = \sqrt{k/m}$," Recordis said. "Then we have $d^2x/dt^2 = (-k/m)x$. Maybe it is possible to solve differential equations after all."

Builder gave us some initial conditions for the ride:

$$x = 1 \text{ and } \frac{dx}{dt} = 4 \text{ when } t = 0$$

We put in the first initial condition to solve for the arbitrary constant A:

$$1 = A \sin 0$$

"It won't work!" Recordis moaned. "When $t = 0$ our guess for x will always be zero."

The king had an idea. "Every time we did one integral, we ended up with one arbitrary constant, and we needed one initial condition. And every time we've done two integrals, we've ended up with two arbitrary constants and we've needed two initial conditions. (See the exercises for Chapter 7.) Builder gave us two initial conditions in this case, so I bet that we need to put two arbitrary constants into the solution."

"We put one constant, A, in front of the sine function," the professor said. "Where could we put another one?"

"We could put another constant in the middle of the sine," Trigonometeris suggested.

$$x = A \sin(\omega t + B)$$

We tried this solution and found that we could use our two initial conditions to solve for the two arbitrary constants:

$$1 = A \sin B$$

$$4 = A\omega \cos B$$

$$\frac{\sin B}{\cos B} = \tan B = \frac{\omega}{4}$$

$$B = \arctan\left(\frac{\omega}{4}\right)$$

Now that we found B, we had to look for A.

$$A = \frac{1}{\sin B}$$

Since $\tan B = \omega/4$, we could find $\sin B$:

$$\sin B = \frac{\sin B / \cos B}{1 / \cos B} = \frac{\tan B}{\sec B} = \frac{\tan B}{\sqrt{1 + \tan^2 B}}$$

$$\sin B = \frac{\omega/4}{\sqrt{1 + \omega^2/4^2}}$$

We could simplify this expression to find:

$$A = \frac{\sqrt{16 + \omega^2}}{\omega}$$

Recordis summarized our result:

The solution to the differential equation:

$$\frac{d^2 x}{dt^2} + \omega^2 x = 0$$

with the initial conditions $x = 1$ and $dx/dt = 4$ when $t = 0$ can be found from the formula

$$x = \left(\frac{\sqrt{16 + \omega^2}}{\omega} \right) \sin \left[\omega t + \arctan(\omega/4) \right]$$

Builder was satisfied with this answer, so he left to begin work on the ride.

Encouraged by our success, we decided to look for more ways to solve linear homogeneous differential equations. "We can't guess all the time," the professor said.

$$\frac{d^n x}{dt^n} + f_{n-1}(t) \frac{d^{n-1} x}{dt^{n-1}} + \cdots + f_2(t) \frac{d^2 x}{dt^2} + f_1(t) \frac{dx}{dt} + f_0(t)x = 0$$

"I know another restriction that would help a lot," Recordis offered.

"We have all those functions of t in the equation: $f_{n-1}(t)$, etc. It would be much easier if we restricted each derivative to have a constant coefficient. After all, the derivatives in Builder's equation had constant coefficients."

We added that restriction, and then had the general form for a linear homogeneous differential equation with constant coefficients:

$$\frac{d^n x}{dt^n} + c_{n-1} \frac{d^{n-1} x}{dt^{n-1}} + c_{n-2} \frac{d^{n-2} x}{dt^{n-2}} + \cdots + c_2 \frac{d^2 x}{dt^2} + c_1 \frac{dx}{dt} + c_0 x = 0$$

"I know a function that looks as though it might fit!" the king said. "The exponential function!"

"Of course!" the professor agreed. "The indestructible exponential function! We know that, if $x = e^t$, then $d^n x/dt^n = x$."

We guessed that the solution to our equation might look like this:

$$x = e^{rt}$$

Then we established that:

$$\frac{dx}{dt} = re^{rt}$$

$$\frac{d^2 x}{dt^2} = r^2 e^{rt}$$

$$\frac{d^3 x}{dt^3} = r^3 e^{rt}$$

$$\vdots$$

$$\frac{d^n x}{dt^n} = r^n e^{rt}$$

We put that solution back into our equation:

$$r^n e^{rt} + c_{n-1} r^{n-1} e^{rt} + c_{n-2} r^{n-2} e^{rt} + \cdots + c_1 r e^{rt} + c_0 e^{rt} = 0$$

"We can factor out the e^{rt}," Recordis noted.

$$e^{rt}(r^n + c_{n-1} r^{n-1} + c_{n-2} r^{n-2} + \cdots + c_2 r^2 + c_1 r + c_0) = 0$$

"And no matter what t is, e^{rt} will never be zero," the professor said. "That means that we can divide both sides by e^{rt}."

$$r^n + c_{n-1} r^{n-1} + c_{n-2} r^{n-2} + \cdots + c_2 r^2 + c_1 r + c_0 = 0$$

"That's just an algebra equation!" Recordis said with relief. "As much as I like calculus, it is nice to see a regular old algebra equation. Now we can solve for r. Of course," he thought a minute more, "it won't be *easy* to solve for r, but I'm sure we can do it somehow."

"We could start by restricting our attention to an equation with only two derivatives," the professor said. "That would simplify matters."

We decided that we would call the *order* of a differential equation the highest order derivative that appeared anywhere in the equation. For example, $dx/dt = t^2$ is a first-order equation, $d^2x/dt^2 = -5x$ is a second-order equation, and

$$\frac{d^n x}{dt^n} + f_{n-1}(t) \frac{d^{n-1} x}{dt^{n-1}} + \cdots + f_1(t) \frac{dx}{dt} + f_0(t)x = 0$$

is an equation of order n. Igor wrote down the general form for a second-order linear homogeneous differential equation with constant coefficients:

$$\frac{d^2 x}{dt^2} + c_1 \frac{dx}{dt} + c_0 x = 0$$

After making the test solution $x = e^{rt}$, we found that r must satisfy this equation:

$$r^2 + c_1 r + c_0 = 0$$

At that moment Recordis began to tremble. Trigonometeris said, "We're forgetting something. Builder's equation was a second-order linear homogeneous differential equation with constant coefficients, but our answer didn't involve an exponent at all. It involved a sine function."

"Trigonometeris is right," the king said. "This is a mystery."

We realized that we could solve for the two values of r by using the quadratic formula:

$$r^2 + c_1 r + c_0 = 0$$

$$r = \frac{-c_1 \pm \sqrt{c_1^2 - 4c_0}}{2}$$

"Let's hope that $c_1^2 > 4c_0$," Recordis said. "Then we get two answers:

$$r_1 = \frac{-c_1 - \sqrt{c_1^2 - 4c_0}}{2}$$

$$r_2 = \frac{-c_1 + \sqrt{c_1^2 - 4c_0}}{2}$$

"This way we won't have to use any trigonometry, and we won't have to use any of those . . . those . . . you know, those numbers," Recordis still couldn't bring himself to mention the term "imaginary number."

"But we did get a trigonometric answer, rather than an exponential answer, to Builder's equation," the king said.

"Maybe trigonometric answers and exponential answers are really the same," the professor suggested.

"But there is a huge difference," the king said. "An exponential answer gets bigger all the time, whereas a trigonometric answer just oscillates back and forth. We must be able to figure out why we get different kinds of answers."

We looked at Builder's equation again:

$$\frac{d^2x}{dt^2} + \omega^2 x = 0$$

(Remember that $\omega^2 = k/m$.) We tried an exponential solution:

$$x = e^{rt}$$

Putting that solution back in the differential equation:

$$r^2 e^{rt} + \omega^2 e^{rt} = 0$$

$$r^2 + \omega^2 = 0$$

We could solve for r using the quadratic formula, with $c_1 = 0$ and $c_0 = \omega^2$:

$$r = \frac{0 \pm \sqrt{0 - 4\omega^2}}{2}$$

$$= \pm \sqrt{-\omega^2}$$

$$r = i\omega \text{ or } r = -i\omega$$

$$x = e^{i\omega t} \text{ or } x = e^{-i\omega t}$$

"No!" Recordis screamed, and fainted. Trigonometeris fanned him absent-mindedly.

"But we can't use that answer, because we don't know what it means when we take the exponential of an imaginary number," the king said. We were able to revive

Recordis after we promised him that we would not use any imaginary numbers. (After advancing a bit farther in calculus, we found that it is possible to make a consistent definition for the exponential of an imaginary number: $e^{i\theta} = \cos\theta + i\sin\theta$.)

"We can use the imaginary root as a signal that it is time to use a trigonometric answer," the professor suggested. "It looks as though a purely imaginary number will result only when we have an equation of this form."

$$\frac{d^2x}{dt^2} = -c_0 x$$

"If c_0 is negative, then we want a function that is proportional to its second derivative with the same sign as its second derivative. That means we use an exponential function. If c_0 is positive, that means we need a function that is proportional to its second derivative but has the opposite sign as the second derivative. The one function that satisfies that condition is the function $x = \sin t$.

"Actually there are two functions," Trigonometeris said. "If $x = \cos t$, then $d^2x/dt^2 = -x$."

"All right, we have two functions that work," the professor agreed.

"The main thing is that we know we use a trigonometric answer."

"We could use any function similar in form to this one," the king said.

$$x = B_1 \sin\omega t + B_2 \cos\omega t$$

"Remember that the complicated differential operator is a linear operator, so that, if we have two functions, x_1 and x_2, that satisfy the equation ($Tx_1 = 0$ and $Tx_2 = 0$), we know that any linear combination of the two functions will also be a solution."

$$T(B_1x_1 + B_2x_2) = B_1Tx_1 + B_2Tx_2 = B_10 + B_20 = 0$$

"That's just what we want," the professor said. "We have two arbitrary constants, B_1 and B_2."

"This is getting even more confusing," Recordis said. "When we solved Builder's problem, we said that the answer was $x = A\sin(\omega t + B)$, not $x = B_1\sin\omega t + B_2\cos\omega t$.

We puzzled over this mystery for a few minutes before Trigonometeris realized that the two answers are equivalent:

$$x = A\sin(\omega t + B)$$
$$= A\cos B\sin\omega t + A\sin B\cos\omega t$$

(See the formula for the sine of a sum in Appendix 2.)

Let $A\cos B = B_1$,
$A\sin B = B_2$.

Then

$$x = A\sin(\omega t + B) = B_1\sin\omega t + B_2\cos\omega t$$

"We now have the hang of second-order linear homogeneous differential equations with constant coefficients," Recordis said.

$$\frac{d^2x}{dt^2} + c_1 \frac{dx}{dt} + c_0 x = 0$$

"For an equation like that, we set up this equation."

$$r^2 + c_1 r + c_0 = 0$$

"We could call that the *characteristic equation*," the professor suggested.

"Then we solve for r," Recordis went on. "If we get two real values of r (call them r_1 and r_2), we set up this solution."

$$x = B_1 e^{r_1 t} + B_2 e^{r_2 t}$$

"Then we use the two initial conditions to solve for the two arbitrary constants, B_1 and B_2. If we get two imaginary values for r (call them $i\omega$ and $-i\omega$), we just ignore the fact that the numbers are imaginary and set up the solution for x."

$$x = B_1 \sin \omega t + B_2 \cos \omega t \quad \text{or} \quad x = A \sin (\omega t + B)$$

"We can choose whichever of these two forms we like better. That should take care of everything except . . ." Recordis almost fainted again. "What if we have a complex solution, such as $r = r_0 + i\omega$?"

At that moment Builder came into the room. "I left out an important term in the equation I gave you," he said. "When I was building the ride, I found that there is a certain amount of friction that acts to slow the ride down. The faster the ride goes, the greater is the force of the friction. The equation of motion for the ride should be as follows."

$$\frac{d^2x}{dt^2} = -b \frac{dx}{dt} - \omega^2 x$$

"Here b is a constant that measures the friction," Builder finished. We rewrote the equation:

$$\frac{d^2x}{dt^2} + b \frac{dx}{dt} + \omega^2 x = 0$$

Then we set up the characteristic equation:

$$r^2 + br + \omega^2 = 0$$

$$r = \frac{-b \pm \sqrt{b^2 - 4\omega^2}}{2}$$

Builder gave us some numerical values: $b = 3$, $\omega^2 = 5$.

$$r = \frac{-3 \pm \sqrt{9 - 20}}{2}$$

$$= \frac{-3}{2} \pm \frac{1}{2}\sqrt{-11}$$

$$= \frac{-3}{2} \pm i\,\frac{1}{2}\sqrt{11}$$

$$r_1 = -1.5 + 1.66i$$

$$r_2 = -1.5 - 1.66i$$

"We're in real trouble now!" Recordis said. "If we had a purely real number, we could use an exponential solution. If we had a purely imaginary number, we could use a trigonometric solution. But we have a complex number: partly real and partly imaginary."

"We could try both kinds of solutions," the king guessed. "We could put the real part in an exponential solution, and we could put the imaginary part in a trigonometric solution."

The king guessed the following solution:

$$x = e^{-1.5t}[A \sin (1.66t + B)]$$

"What does that function look like?" Recordis asked. "I never saw a solution like that before."

Igor drew a graph of the curve $x = e^{-t} \sin t\pi$ (Figure 15–1).

"That's right," the professor said. "The solution keeps going back and forth, as a ride on a spring should, but the amplitude of each swing is becoming less. The friction acts to damp out the sine wave."

"I could have told you that is what the motion of the ride would look like before we developed all this math," Builder said. "All you need is some physical intuition."

"We could call that kind of curve a *damped sine wave*," Recordis suggested. "But we better make sure that it is the right answer."

We differentiated our trial solution and put it back into Builder's equation. The first derivative required the product rule:

$$x = e^{-1.5t}[A \sin (1.66t + B)]$$

$$\frac{dx}{dt} = e^{-1.5t}A\,1.66 \cos (1.66t + B) - 1.5e^{-1.5t}A \sin (1.66t + B)$$

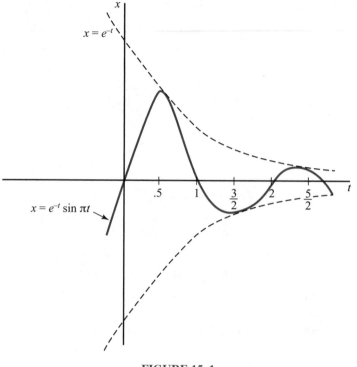

FIGURE 15–1.

We expressed this answer in slightly simpler form:

$$\frac{dx}{dt} = (Ae^{-1.5t})(1.66 \cos \theta - 1.5 \sin \theta)$$

(To save us some writing, we made the definition $\theta = 1.66t + B$.)

$$\frac{d^2x}{dt^2} = (Ae^{-1.5t})(-1.66^2 \sin \theta - 1.5 \times 1.66 \cos \theta) - (1.5Ae^{-1.5t})(1.66 \cos \theta - 1.5 \sin \theta)$$

$$= (1.5^2 - 1.66^2)Ae^{-1.5t} \sin \theta - 2(1.5 \times 1.66)Ae^{-1.5t} \cos \theta$$

$$= -0.5Ae^{-1.5t} \sin \theta - 4.98Ae^{-1.5t} \cos \theta$$

We recognized our expression for x:

$$x = e^{-1.5t} A \sin \theta$$

so we could rewrite the expression for the second derivative:

$$\frac{d^2x}{dt^2} = -0.5x - 4.98Ae^{-1.5t} \cos \theta$$

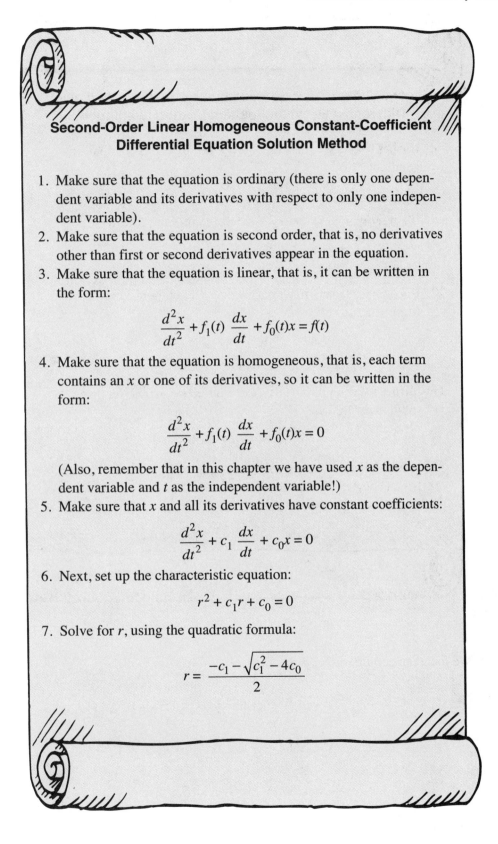

Second-Order Linear Homogeneous Constant-Coefficient Differential Equation Solution Method

1. Make sure that the equation is ordinary (there is only one dependent variable and its derivatives with respect to only one independent variable).
2. Make sure that the equation is second order, that is, no derivatives other than first or second derivatives appear in the equation.
3. Make sure that the equation is linear, that is, it can be written in the form:

$$\frac{d^2x}{dt^2} + f_1(t)\,\frac{dx}{dt} + f_0(t)x = f(t)$$

4. Make sure that the equation is homogeneous, that is, each term contains an x or one of its derivatives, so it can be written in the form:

$$\frac{d^2x}{dt^2} + f_1(t)\,\frac{dx}{dt} + f_0(t)x = 0$$

(Also, remember that in this chapter we have used x as the dependent variable and t as the independent variable!)

5. Make sure that x and all its derivatives have constant coefficients:

$$\frac{d^2x}{dt^2} + c_1\,\frac{dx}{dt} + c_0x = 0$$

6. Next, set up the characteristic equation:

$$r^2 + c_1 r + c_0 = 0$$

7. Solve for r, using the quadratic formula:

$$r = \frac{-c_1 - \sqrt{c_1^2 - 4c_0}}{2}$$

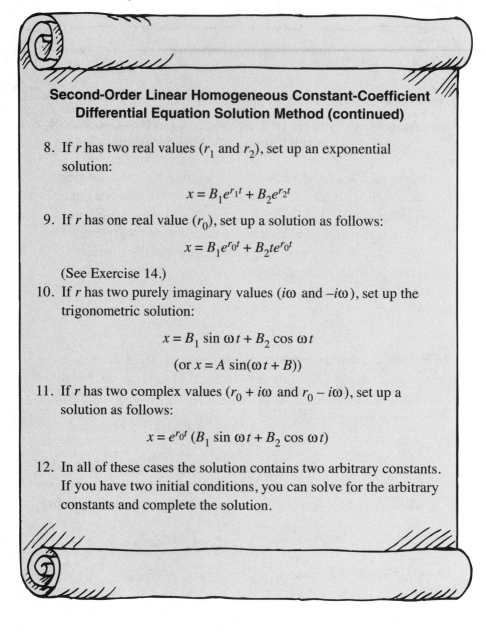

Second-Order Linear Homogeneous Constant-Coefficient Differential Equation Solution Method (continued)

8. If r has two real values (r_1 and r_2), set up an exponential solution:

$$x = B_1 e^{r_1 t} + B_2 e^{r_2 t}$$

9. If r has one real value (r_0), set up a solution as follows:

$$x = B_1 e^{r_0 t} + B_2 t e^{r_0 t}$$

(See Exercise 14.)

10. If r has two purely imaginary values ($i\omega$ and $-i\omega$), set up the trigonometric solution:

$$x = B_1 \sin \omega t + B_2 \cos \omega t$$

$$(\text{or } x = A \sin(\omega t + B))$$

11. If r has two complex values ($r_0 + i\omega$ and $r_0 - i\omega$), set up a solution as follows:

$$x = e^{r_0 t} (B_1 \sin \omega t + B_2 \cos \omega t)$$

12. In all of these cases the solution contains two arbitrary constants. If you have two initial conditions, you can solve for the arbitrary constants and complete the solution.

We found an expression for 3 dx/dt:

$$\frac{dx}{dt} = (A e^{-1.5t})(1.66^2 \cos \theta - 1.5 \sin \theta)$$

$$= 1.66 A e^{-1.5t} \cos \theta - 1.5x$$

$$3 \frac{dx}{dt} = 4.98 A e^{-1.5t} \cos \theta - 4.5x$$

Now that we know d^2x/dt^2, dx/dt, and x, we could put all the parts together in Builder's equation and see if they fit.

$$\frac{d^2x}{dt^2} + 3\frac{dx}{dt} + 5x = -0.5x - 4.98Ae^{-1.5t}\cos\theta + 4.98Ae^{-1.5t}\cos\theta - 4.5x + 5x$$

$$= -0.5x - 4.5x + 5x$$

$$= 0$$

"It does work!" Trigonometeris said, "We do need an exponential solution and a trigonometric solution for that type of problem!" We found that we could solve for the arbitrary constants A and B by using the initial conditions that Builder had given us. We wrote down a general procedure for solving second-order linear homogeneous differential equations with constant coefficients. (See page 297.)

"Solving differential equations is pretty complicated, but it may not be as bad as I thought it would be," Recordis said.

"I see that we have the solution," Builder added, "but that solution does not leave us with a very good ride. The children won't want a ride that just swings back and forth for a little bit before it damps out and stops. I'll put a driving motor on the ride so that it won't stop so quickly."

Before we could protest, Builder had designed a driving motor that would push the spring ride with a force equal to $F_{\text{driving}} = D\sin\Omega t$, where Ω is the driving angular frequency. (Ω is the upper case form of the Greek letter omega.)

"No!" Recordis cried. "Now we won't have a homogeneous equation anymore! That driving force term does not have any x's in it."

Builder gave us the equation for the driven ride:

$$\frac{d^2x}{dt^2} + b\frac{dx}{dt} + \omega^2 x = D\sin\Omega t$$

Again we used T to stand for the differential operator:

$$T = \frac{d^2}{dt^2} + b\frac{d}{dt} + \omega^2$$

so we could write the equation as:

$$Tx = D\sin\Omega t$$

"That's a second-order linear differential equation with constant coefficients, but it's not homogeneous," the professor objected. "How are we going to solve that?"

"The solution still must involve a sine function," Trigonometeris said. "The ride will still go back and forth, even with the driving force."

Trigonometeris guessed that the frequency of the ride should be close to the frequency of the driving force, so we guessed the following solution:

$$x = C \sin (\Omega t + B)$$

$$\frac{dx}{dt} = \Omega C \cos (\Omega t + B)$$

$$\frac{d^2 x}{dt^2} = -\Omega^2 C \sin (\Omega t + B)$$

We put these expressions back into Builder's equation:

$$Tx = -\Omega^2 C \sin (\Omega t + B) + b\Omega C \cos (\Omega t + B) + \omega^2 C \sin (\Omega t + B)$$

$$= (\omega^2 - \Omega^2)C \sin (\Omega t + B) + b\Omega C \cos (\Omega t + B)$$

Using the formula for the sine and cosine of the sum of two angles gave us:

$$Tx = (\omega^2 - \Omega^2)C(\sin \Omega t \cos B + \cos \Omega t \sin B) + b\Omega C(\cos \Omega t \cos B - \sin \Omega t \sin B)$$

We simplified this expression a bit:

$$Tx = [C \cos B(\omega^2 - \Omega^2) - b\Omega C \sin B] \sin \Omega t + [C \sin B(\omega^2 - \Omega^2) + b\Omega C \cos B] \cos \Omega t$$

Now we had the left-hand side of Builder's equation for the damped-driven ride. The next step was to set our equation for the left-hand side equal to the right-hand side of the equation:

$$Tx = D \sin \Omega t$$

(right-hand side) $= D \sin \Omega t$

(left-hand side) $= [C \cos B(\omega^2 - \Omega^2) - b\Omega C \sin B] \sin \Omega t + [C \sin B(\omega^2 - \Omega^2)$
$+ b\Omega C \cos B] \cos \Omega t$

If our alleged solution really worked, we wanted it to be true for all values of t. That meant that we could set the coefficient of $\sin \Omega t$ on the left equal to the coefficient of $\sin \Omega t$ on the right:

(1) $D = C \cos B(\omega^2 - \Omega^2) - b\Omega C \sin B$

coefficients of $\cos \Omega t$:

(2) $0 = C \sin B(\omega^2 - \Omega^2) + b\Omega C \cos B$

We could find the conditions that B and C must satisfy for these two equations to be correct. From Equation (2) above:

$$0 = (\omega^2 - \Omega^2) \sin B + b\Omega \cos B$$

$$(\Omega^2 - \omega^2) \sin B = b\Omega \cos B$$

$$\tan B = \frac{b\Omega}{\Omega^2 - \omega^2}$$

$$B = \arctan\left(\frac{b\Omega}{\Omega^2 - \omega^2}\right)$$

From Equation (1) above, we can see that C must satisfy this equation:

$$C = \frac{D}{\cos B(\omega^2 - \Omega^2) - b\Omega \sin B}$$

We found expressions for $\sin B$ and $\cos B$ (which we can do because we know $\tan B$):

$$\sin B = \frac{b\Omega}{\sqrt{b^2\Omega^2 + (\Omega^2 - \omega^2)^2}}$$

$$\cos B = \frac{\Omega^2 - \omega^2}{\sqrt{b^2\Omega^2 + (\Omega^2 - \omega^2)^2}}$$

Now we substituted these expressions into the equation for C:

$$C = \frac{D}{\dfrac{\Omega^2 - \omega^2}{\sqrt{b^2\Omega^2 + (\Omega^2 - \omega^2)^2}}(\omega^2 - \Omega^2) - b\Omega \dfrac{b\Omega}{\sqrt{b^2\Omega^2 + (\Omega^2 - \omega^2)^2}}}$$

$$= -\frac{\dfrac{D}{(\Omega^2 - \omega^2)^2 + b^2\Omega^2}}{\sqrt{b^2\Omega^2 + (\Omega^2 - \omega^2)^2}}$$

We were all exhausted at this point, but Recordis suddenly exclaimed, "We can simplify that!"

$$C = -\frac{D}{\sqrt{b^2\Omega^2 + (\Omega^2 - \omega^2)^2}}$$

Recordis put together the final answer:
The solution to the differential equation

$$\frac{d^2x}{dt^2} + b\,\frac{dx}{dt} + \omega^2 x = D \sin \Omega t$$

is given by the formula:

$$x = C \sin (\Omega t + B)$$

with B and C found from these formulas:

$$B = \arctan \left(\frac{b\Omega}{\Omega^2 - \omega^2} \right)$$

$$C = -\frac{D}{\sqrt{b^2\Omega^2 + (\Omega^2 - \omega^2)^2}}$$

"We've come up with an answer," Recordis said with relief. "Now all we have to do is put in the numbers. And this time we were able to solve for all the constants, rather than . . . " He suddenly stopped.

"Something is wrong!" the professor said. "We don't have any arbitrary constants left! We'll never be able to match the initial conditions now. We know that, whenever we have a second-order differential equation, the solution must contain two arbitrary constants."

We were stumped. "That solution doesn't have any place left to put any constants," Recordis complained.

We thought for a long time. "We'll have to add something to that function," I suggested.

$$x = x_1 + x_2$$
$$x_1 = C \sin (\Omega t + B)$$
$$x_2 = ?$$

"Whatever x_2 is, it must be some function that has two arbitrary constants in it."

"But that will wreck our solution to the equation," Recordis protested. "You can't just add any old thing to the solution and still expect to have a solution."

"Maybe we can find some special x_2 that will be all right," the king said hopefully. "Maybe there is one x_2 such that $x = x_1 + x_2$ will still be a solution."

We put this alleged solution back into the equation (written in operator form):

$$\left(\frac{d^2}{dt^2} + b\frac{d}{dt} + \omega^2\right)(x_1 + x_2) = D \sin \Omega t$$

$$T(x_1 + x_2) = D \sin \Omega t$$

"We know that

$$T = \frac{d^2}{dt^2} + b\frac{d}{dt} + \omega^2$$

is a linear operator," the professor said. "We can rewrite our equation like this."

$$Tx_1 + Tx_2 = D \sin \Omega t$$

"We know that $Tx_1 = D \sin \Omega t$, since that is the right answer for the equation that we just found."

$$D \sin \Omega t + Tx_2 = D \sin \Omega t$$

$$Tx_2 = 0$$

"That puts a restriction on x_2," the professor said. "We can safely add x_2 to our original solution without wrecking it if the following equality holds."

$$\left(\frac{d^2}{dt^2} + b\frac{d}{dt} + \omega^2\right)x_2 = 0$$

"Does anybody see any clues in that equation that will allow us to deterinine what x_2 is?"

We stared hard at the last equation. "We already solved that equation!" the king exclaimed suddenly. "That's just the homogeneous equation we had before Builder added the driving machine!"

"That means that x_2 is just the homogeneous solution," Recordis said in surprise.

$$x_2 = e^{-1.5t} A_1 \sin (1.66t + A_2)$$

"It even has two arbitrary constants, just as it is supposed to." Recordis wrote out the whole solution:

If you are given this differential equation:

$$\frac{d^2x}{dt^2} + b\frac{dx}{dt} + \omega^2 x = D \sin \Omega t$$

The complete solution for x is given by this formula:

$$x = C \sin (\Omega t + B) + e^{r_0 t} A_1 \sin (r_1 t + A_2)$$

The values for C, B, r_0, and r_1 are determined as follows:

$$C = -\frac{D}{\sqrt{b^2\Omega^2 + (\Omega^2 - \omega^2)^2}}$$

$$B = \arctan\left(\frac{b\Omega}{\Omega^2 - \omega^2}\right)$$

$$r_0 = \frac{-b}{2}$$

$$r_0 = \frac{\sqrt{4\omega^2 - b^2}}{2}$$

A_1 and A_2 are arbitrary constants whose values can be found if you know the initial conditions.

(Note that this solution holds when $4\omega^2 > b^2$. See Exercises 17 and 18.)

"What does that solution look like?" Recordis asked. "It looks as though it will be very complicated."

"It will be very complicated at first," Builder said thoughtfully. "But after a fair amount of time has passed, it looks as if we will be able to completely ignore the second term, since it will all damp out."

"That's right," the professor agreed. "As t becomes large, $e^{-1.5t}$ will go to zero. Then the only part of the solution we will have to worry about will be the first part.

"We could say that the x_2 part, which comes from the homogeneous equation solution, is a *transient* part of the solution. If we wait long enough, it will go away."

"The x_1 part will be a permanent part of the solution," Recordis said. "As t becomes large, the ride will keep oscillating. And that part doesn't even depend on the initial conditions."

"That makes sense," Builder concurred. "At first, the solution should depend a lot on the initial conditions. As time goes by, though, the effects of the initial conditions will be less important, until ultimately it will be only the driving force that makes a difference."

We drew two graphs of an example with $b = 0.75, D = 10, \Omega = 2\pi/5$, and $\omega = 2\pi$. Figure 15–2 shows the result with initial conditions $A_1 = 3$ and $A_2 = 2$. In Figure 15–3 everything is the same except $A_1 = 20$. The larger value of A, means that the ride starts out farther from the middle (which occurs at $x = 0$). This causes the ride to swing back and forth violently at first, but after a while the friction slows this motion down. In both Figures 15–2 and 15–3, the ride eventually settles down so that it oscillates smoothly with the frequency of the driving motor (Ω). If you erased the parts of these two curves before $t = 20$, they would be practically indistinguishable, showing that, after a certain time has passed, the difference in the initial conditions doesn't matter.

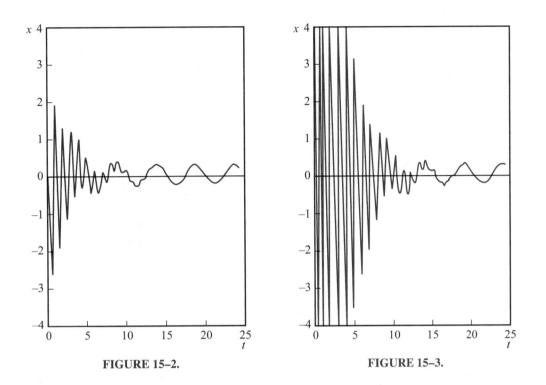

FIGURE 15–2.   **FIGURE 15–3.**

After Builder had worked on his drawings for a few moments, he noticed something else. "It is a good thing that the friction is there," he said. "Suppose there was no friction ($b = 0$). Then the permanent solution would be as follows."

$$x = -\frac{D}{\Omega^2 - \omega^2}\sin\Omega t$$

"I am trying to figure out the driving frequency that would give the ride the maximum amplitude. It looks as though the maximum amplitude would occur if the driving frequency Ω equals ω. But if that happens, the amplitude of the ride will be infinity, which means that it will shoot all over space. The children would probably think that was a fun ride, but I don't think it would be very safe."

"I remember what ω is," the king said. "That was the angular frequency of the ride before you added the driving motor. We could call that the *natural frequency* of the ride. It looks as though the most amplitude occurs if the frequency of the driving motor is close to the natural frequency of the ride." We decided to call this occurrence *resonance*.

"Fortunately, the friction is there," Builder noted. "That means that the ride will not go off to infinity, even if we use the resonant driving frequency."

"I've had enough of differential equations," Recordis said. "It's time to go back to work on the party."

We wrote down a general procedure for solving linear differential equations:

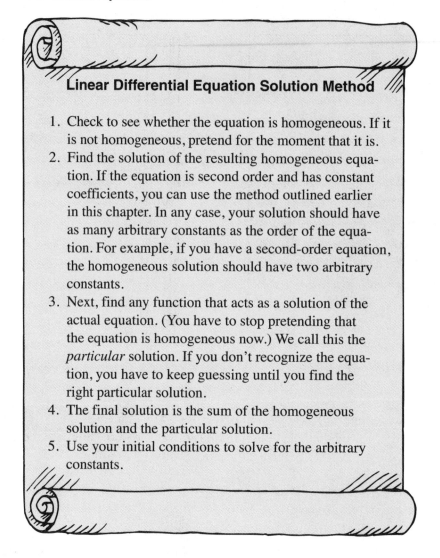

Linear Differential Equation Solution Method

1. Check to see whether the equation is homogeneous. If it is not homogeneous, pretend for the moment that it is.
2. Find the solution of the resulting homogeneous equation. If the equation is second order and has constant coefficients, you can use the method outlined earlier in this chapter. In any case, your solution should have as many arbitrary constants as the order of the equation. For example, if you have a second-order equation, the homogeneous solution should have two arbitrary constants.
3. Next, find any function that acts as a solution of the actual equation. (You have to stop pretending that the equation is homogeneous now.) We call this the *particular* solution. If you don't recognize the equation, you have to keep guessing until you find the right particular solution.
4. The final solution is the sum of the homogeneous solution and the particular solution.
5. Use your initial conditions to solve for the arbitrary constants.

"We'd better not advertise that we can solve differential equations yet," the professor said, "or someone may come to us with an equation that is too complicated for us to solve."

Nevertheless, we realized that our investigation of differential equations had made it possible for us to solve a completely new type of problem.

(Note: We learned more about solving differential equations in Chapter 17.)

NOTE TO CHAPTER 15

The phenomenon of resonance is very important in physics. A radio station is tuned to the correct station by adjusting the resonance frequency of an electrical oscillator circuit. A severe windstorm caused oscillations at the resonance frequency of the Tacoma Narrows Suspension Bridge in Washington State in 1940, causing the collapse of the bridge.

Exercises

Which of these differential equations are linear? Which are homogeneous?

1. $\dfrac{d^2y}{dx^2}\, y^2 + 2\, \dfrac{dy}{dx} = 0$

2. $\dfrac{dy}{dx} = f(x)$

3. $3\left(\dfrac{dy}{dx}\right)^2 + 4y = 5x$

4. $\dfrac{dy}{dx} = f(y)$

5. $x\, \dfrac{dy}{dx} - 3x^2 + 2y = 4\, \dfrac{dy}{dx}$

6. $\dfrac{d^2x}{dt^2}\, \dfrac{dx}{dt} = 4t$

7. $\dfrac{d^2y/dx^2}{[1 + (dy/dx)^2]^{1.5}} = K$

8. For
$$T = \dfrac{d^n}{dt^n} + f_{n-1}(t)\, \dfrac{d^{n-1}}{dt^{n-1}} + \cdots + f_2(t)\, \dfrac{d^2}{dt^2} + f_1(t)\, \dfrac{d}{dt} + f_0(t)$$

show that $T(ax_1 + bx_2) = aTx_1 + bTx_2$ (a and b are constants; x_1 and x_2 are functions of t).

Solve these differential equations. Then solve for the two arbitrary constants, using the initial conditions that are given:

9. $\dfrac{d^2x}{dt^2} + \dfrac{dx}{dt} - 6x = 0$; $t = 0, x = 21, dx/dt = -3$

10. $\dfrac{d^2x}{dt^2} + \dfrac{dx}{dt} + x = 0$, $t = 0$; $x = 4/3, dx/dt = (\sqrt{3} - 2)/3$

11. $\dfrac{d^2x}{dt^2} + 9x = 0; t = 0, x = 20, dx/dt = 480$

12. $\dfrac{d^2x}{dt^2} + 2\dfrac{dx}{dt} + x = 0; t = 0, x = -4, dx/dt = -3$

13. $\dfrac{d^2x}{dt^2} + 4\dfrac{dx}{dt} = 0; t = 0, x = 20, dx/dt = 12$

14. Step 9 of the method for solving a homogeneous equation says that, if the characteristic equation has one real root (r), the solution will be $x = Ae^{rt} + Bte^{rt}$. Differentiate this solution to show that it is the correct one.

15. Find the motion of Builder's ride if $b = 3$, $\omega^2 = 5$, $\Omega = 2$, and $D = -4$.

16. Solve for the arbitrary constants in the solution to Exercise 15 by using these initial conditions: when $t = 0$, $x = -0.6484$ and $dx/dt = 1.884$.

17. Consider an undriven ride with greater friction. Let $\omega = 2$, $b = 4$. What does the motion of the ride look like?

18. Consider another undriven ride with an even greater force of friction: $b = 5$ and $\omega = 2$. What does the motion of the ride look like?

19. Suppose a flexible string is hung between two posts, and let $y(x)$ represent the shape of the curve (as in Chapter 14). In physics it can be found that y must satisfy this differential equation:

$$a\frac{d^2y}{dx^2} = \left[1 + \left(\frac{dy}{dx}\right)^2\right]^{1/2}$$

Verify that the function $y = 1/2a(e^{x/a} + e^{-x/a})$ satisfies the differential equation.

20. If an electrical oscillator circuit contains an inductor with inductance L, a capacitor with capacitance C, and a resistor with resistance R, then the voltage V at time t is given by the solution to this differential equation:

$$LC\frac{d^2V}{dt^2} + RC\frac{dV}{dt} + V = 0$$

Determine the formula for V. Assume $(RC)^2 < 4LC$.

Partial Derivatives and Vectors

The king decided that we should print programs showing the events scheduled for the party. We hired the publisher of *Carmorra Magazine* to do the printing. When we met with her to place the order, she asked for our help with what seemed to be a routine problem.

"I am publishing a new book, *Deluxe Travel Guide to Carmorra*," she explained. "I have found that the quantity of books that I will sell is given by the formula $Q = 420 - 3x$, where x represents the price of the book (in cents). Can you tell me what price I should charge in order to earn the most revenue?"

"Sure," Recordis said. "We know revenue (R) is equal to price times quantity, so we can write the revenue as a function of the quantity: $R = 420x - 3x^2$. Then we take the derivative: $dR/dx = 420 - 6x$; set the derivative equal to zero: $420 - 6x = 0$; and we find that $x = 70$."

The publisher was very grateful, and she left to work on the programs. However, when she returned with our order the next week she was very upset. "I have discovered that the situation is ever so much more complicated than I had thought," she complained. "It turns out that the number of deluxe travel guides that I sell depends not only on the price of the deluxe travel guides, but also on the price of my standard travel guides. If the price of the standard guide is lower, people will buy fewer of the deluxe guides, everything else being equal. To make things even worse, the price of the deluxe guides also affects the number of standard guides that I sell." She gave us these formulas:

$$Q_1 = 100 - 3x + 5y \qquad \text{(quantity of deluxe guides sold)}$$
$$Q_2 = 40 + 3x - 8y \qquad \text{(quantity of standard guides sold)}$$
$$x = \text{price of deluxe guides}$$
$$y = \text{price of standard guides}$$

"We can write a formula for your total revenue," Recordis offered helpfully.

$$R = xQ_1 + yQ_2$$
$$= 100x - 3x^2 + 5xy + 40y + 3xy - 8y^2$$
$$= -3x^2 - 8y^2 + 100x + 40y + 8xy$$

"Now in order to find the maximum total revenue, you just have to tell me which is the independent variable, and then we can take the derivative."

Recordis had to explain to the publisher what an independent variable was. "You get to choose the value of the independent variable. Once that value is chosen, then the function rule determines the value for the dependent variable (in this case R)."

"But I can choose the values for both x and y!" the publisher protested. "By your definition they are both independent variables."

"Help!" Recordis screamed, calling the rest of us to the Main Conference Room. "We're in real trouble now! There's no way we can solve an optimal value problem when we have more than one independent variable! We can only take a derivative when we know which is the independent variable."

When he had calmed down, Recordis carefully explained the publisher's problem.

The professor had an idea. "All we need to do is write a formula that expresses y in terms of x, then substitute that formula in the place of y, and then we have turned the problem into a standard one-variable problem." Just like our fence problems (see page 106).

"Haven't you been listening to what I've been saying?" Recordis cried. "We can't express y in terms of x because they are both independent. The publisher is free to choose whatever she wants for either x or y."

The professor suddenly realized Recordis was right. "This is more complicated than anything we have done before. We have never before had a function with multiple independent variables."

We worried about this problem for a long time.

"If only we could pretend that y was constant," Recordis moaned. "Then we could easily calculate dR/dx:

$$dR/dx = -6x + 100 + 8y \qquad \text{(assuming } y \text{ is constant)}$$

"Or if only we could pretend that x was constant. Then we could calculate dR/dy:

$$dR/dy = -16y + 40 + 8x \qquad \text{(assuming } x \text{ is constant)}$$

"I think you have a good idea!" the astonished professor said. "When we have a function with several independent variables, there is no reason we can't calculate the derivative with respect to one of the independent variables while pretending that the other independent variables are all held constant. It won't be the derivative of the total function, but we could call it a *partial derivative*."

The king agreed with the professor's plan, but Recordis had three anguished complaints. "I don't want to learn a new procedure to find a new kind of derivative. And even if we do calculate a partial derivative, what does it mean? And, most importantly (at least from my point of view), what symbol are we going to use to represent it?"

We decided that the symbol for partial derivative should look almost like the symbol for a regular derivative, so we decided that we would use a curly d (∂) in the place of a regular d to represent a partial derivative. The king issued a proclamation:

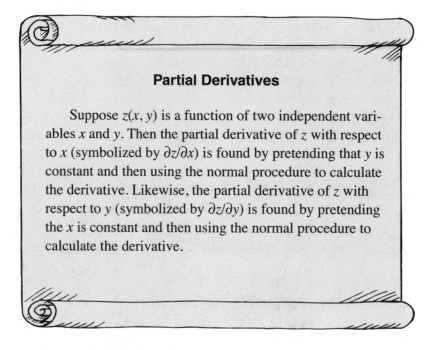

Partial Derivatives

Suppose $z(x, y)$ is a function of two independent variables x and y. Then the partial derivative of z with respect to x (symbolized by $\partial z/\partial x$) is found by pretending that y is constant and then using the normal procedure to calculate the derivative. Likewise, the partial derivative of z with respect to y (symbolized by $\partial z/\partial y$) is found by pretending the x is constant and then using the normal procedure to calculate the derivative.

Recordis brightened considerably when he realized that we did not need to learn a new procedure for calculating partial derivatives. "I see!" he exclaimed. "We already know how to calculate partial derivatives, since the procedure is the same trusty procedure we have used to calculate ordinary derivatives."

For our example we had:

$$R = -3x^2 - 8y^2 + 100x + 40y + 8xy$$

$$\partial R/\partial x = -6x + 100 + 8y$$

$$\partial R/\partial y = -16y + 40 + 8x$$

"We still need to figure out what a partial derivative means," the king said.

The professor had an idea. "An ordinary derivative represents the rate of change of the dependent variable as the independent variable changes. So, a partial derivative should represent the rate of change of the dependent variable with respect to that independent variable. For example, if $\partial R/\partial y$ is positive, that must mean that R will increase if the value of y is increased, and if $\partial R/\partial y$ is negative then R will increase if y is decreased."

"But we don't want R to increase if y is increased or decreased," Recordis said. "If that is the case, then we can't be at the optimum. If we were at the optimum, then it would not be possible to increase the value of R, no matter what we did."

The professor's eyes widened. "That's it!" she exclaimed. "We cannot be at the optimum if either $\partial R/\partial y$ is positive or $\partial R/\partial y$ is negative. Therefore, at the optimum, we must have $\partial R/\partial y$ equal to zero! That is just the same as we found in the one-variable case—the optimum occurs where the derivative is equal to zero."

"The same reasoning works for $\partial R/\partial x$," the king, realized. "Therefore, if we are at the optimum we must have both $\partial R/\partial x$ and $\partial R/\partial y$ equal to zero."

We set both of the partial derivatives equal to zero:

$$-6x + 100 + 8y = 0$$

$$-16y + 40 + 8x = 0$$

The result was a two-equation system with two variables, which we were able to solve to find that $x = 60$ and $y = 32.5$. We were able to convince the publisher that the optimum value of revenue would occur if the deluxe travel guides were sold at a price of 60 and the standard travel guides were sold at a price of 32.5. (See Exercise 1.)

The concept of partial derivatives still bothered Recordis. "I am much more comfortable with a strange concept if we can draw a picture to represent it," he said. "I wish we could draw a picture that could represent a function with two independent variables."

The king realized that this would be very difficult. "We are able to draw a graph that represents a function of one independent variable, since paper has two dimensions. We can use one dimension to represent the independent variable and one dimension to represent the dependent variable. Matters are much more complicated when there are two independent variables. We would need one dimension for each of the independent variables, but then we would still need another dimension to represent the dependent variable."

At that moment Builder came in to ask for help. He was building a domed pavilion for the party. "The dome will be a hemisphere of radius r. I need to set up several temporary posts to support the domed roof during the construction process. I know the location of each post, but I need to know how tall to make the post."

"In circumstances such as this we should draw a coordinate system," the professor said. "We can put the origin at the center of the floor of the dome, with the x axis pointing east and the y axis pointing north. We can use z to represent the height of the post at a particular location."

"I know how tall the post at the exact center of the dome needs to be," Builder said. "That post will have a height equal to r."

"In our notation, we can say if $x = 0$ and $y = 0$, then $z = r$," the professor said.

"We need a formula that will allow us to calculate the height of the post for any given values of x and y," the king said. "Then Builder could plug the values for x and y into the formula, and then he would know how tall the post should be at that location."

According to the definition of a sphere, all points on the dome must be the same distance from the center of the dome (the point [0,0].) Therefore, we could write:

$$\sqrt{x^2 + y^2 + z^2} = r$$

"We can solve for z in terms of x and y," Recordis said helpfully.

$$z(x,y) = \sqrt{r^2 - x^2 - y^2}$$

(provided that $x^2 + y^2 \leq r^2$)

"That means that z is a function of the two independent variables x and y!" the king said. "Now I see how we can make a graph of a function of two independent variables. We need to make the graph in three dimensions. In this case the two horizontal dimensions represent the two independent variables x and y and the vertical dimension represents the dependent variable z." (See Figure 16–1.)

(Note: In Excel you can create a surface graph that creates the illusion that it is three dimensional.)

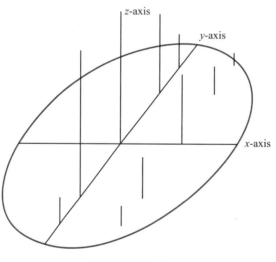

FIGURE 16–1.

Builder left to use this formula to calculate the heights of the support posts. The professor thought we should calculate the two partial derivatives of z. "Remember that in this problem r really is a constant, since the radius does not change. Even though y is a variable, we will pretend that it is constant while we calculate $\partial z/\partial x$."

$$\partial z/\partial x = 1/2(r^2 - x^2 - y^2)^{-1/2}\,(-2x)$$

$$\frac{\partial z}{\partial x} = \frac{-x}{\sqrt{r^2 - x^2 - y^2}}$$

We found a similar expression for $\partial z/\partial y$:

$$\frac{\partial z}{\partial y} = \frac{-y}{\sqrt{r^2 - x^2 - y^2}}$$

"Now I can picture what the partial derivative represents!" Recordis exclaimed. "Suppose I took a giant knife and cut a gash through the dome along the values where $y = 2$. (See Figure 16–2.) Then we can look at a cross section of the dome. If we substitute $y = 2$ in the formula

$$z(x,y) = \sqrt{r^2 - x^2 - y^2}$$

we can find the formula for the cross section of the dome.

$$z(x,2) = \sqrt{r^2 - x^2 - 4}$$

"This cross section represents a function of one variable (x), which in this case happens to be a portion of a circle. If we substitute $y = 2$ into the formula for the partial derivative

$$\frac{\partial z}{\partial x} = \frac{-x}{\sqrt{r^2 - x^2 - y^2}}$$

we find

$$\frac{\partial z}{\partial x} = \frac{-x}{\sqrt{r^2 - x^2 - 4}}$$

"Therefore, if you choose a fixed value for y, then the partial derivative $\partial z/\partial x$ represents the slope of the cross section of the dome for that value of y."

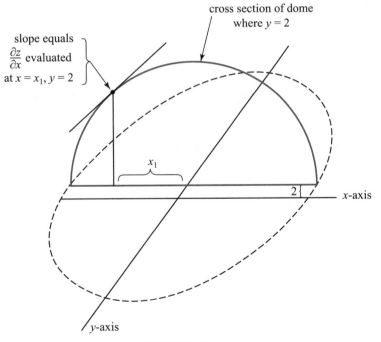

FIGURE 16–2.

Trigonometeris wanted to contribute something, so he suggested that we could construct a vector whose x component was $\partial z/\partial x$, and whose y component was $\partial z/\partial y$. (A *vector* is a quantity for which it is necessary to specify both the magnitude [length] and the direction. We had learned several interesting properties of vectors when we had studied trigonometry.) They asked me to suggest a name, so I suggested the term *gradient* for a vector consisting of the partial derivatives of a function. We symbolized the gradient of a function z by writing grad z, or ∇z:

$$\text{grad } z = \left(\frac{\partial z}{\partial x}, \frac{\partial z}{\partial y} \right)$$

Note that a vector can be described by listing its components, surrounded by parentheses and separated by commas. In our specific case we had:

$$\text{grad } z = \left(\frac{-x}{\sqrt{r^2 - x^2 - y^2}}, \frac{-y}{\sqrt{r^2 - x^2 - y^2}} \right)$$

Trigonometeris investigated the matter further to find out in which direction the vector pointed. He found that, in this case, the gradient vector always pointed to the origin (the center of the dome). This was true for all values of x and y. (See Figure 16–3.)

"If you are standing at any point on the dome, then the direction where the dome is increasing most steeply is the direction that faces toward the center of the dome," Trigonometeris noticed. "Therefore, the gradient vector points in the direction where the rate of increase in the height of the dome is the greatest. If you are trying to climb up to the top of the dome as quickly as possible, then you should always climb in the direction where the gradient vector points."

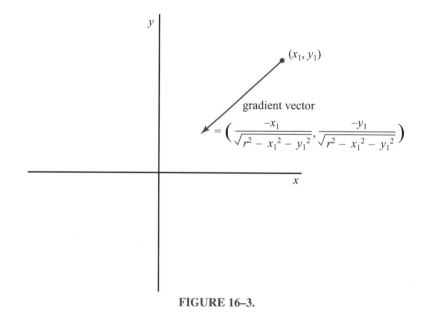

FIGURE 16–3.

In general, we found that the gradient vector of a function of several variables points in the direction where that function is increasing most rapidly. "Unless, of course, each component of the gradient vector is zero," the professor said. "In that case all of the partial derivatives are zero. Then we are already at the maximum point, and the function will not increase in any direction."

Builder quickly completed the construction of the dome and then he removed the temporary posts. We went inside the dome to watch him test the temperature control and ventilation systems.

"How do you tell what the temperature is at any point in the dome!" Recordis asked.

"I have specially trained bugs," Builder said. He released a few large purple bugs that began slowly flying around the dome. "These bugs are trained to always fly in the direction where the increase in the temperature is the greatest. By watching the flight of the bugs I can see how the temperature changes at different points in the dome."

The professor suddenly had an idea. "We can identify any point in the dome by specifying three coordinates: x, y, and z, where x is the distance east of the center of the dome, y is the distance north of the center of the dome, and z is the height above the center of the dome. Suppose T represents the temperature of the air at a specific point in the dome with coordinates (x,y,z). Then we can express T as a function of three variables."

$$\text{temperature} = T(x,y,z)$$

"No!" Recordis screamed. "You will be very sorry if you start talking about functions with three independent variables! I can see how we can use three-dimensional space to make a graph of a two-independent-variable function, even though that is much harder than using two-dimensional paper to draw a graph of a one-independent-variable function. It will be absolutely impossible to make a graph of a function of three independent variables! You would need to use a four-dimensional space, and you and I know there is no such thing."

"It is true that we cannot ever draw a graph of a function with three independent variables," the professor conceded. "Nevertheless, whether you like it or not, the temperature is a function of the three independent variables x, y, and z."

"We could find the gradient, though," the king said. "In this case the gradient will be a vector with three components."

$$\text{grad } T = \left(\frac{\partial T}{\partial x}, \frac{\partial T}{\partial y}, \frac{\partial T}{\partial z} \right)$$

"And I bet the gradient will still point in the direction of the greatest increase of the function," Trigonometeris said. "We know that the purple bugs always fly in the direction of the greatest increase in the temperature, so that means that they are able to calculate the gradient of the temperature function and then fly in the direction where the gradient vector points."

After finishing our work in the dome, Builder described three more rides for the children. First, Pal would push a sleigh on frictionless Ice Skating Lake.

"How much work will Pal have to do when he pushes the sleigh?" the king asked.

"It depends on how hard he pushes," Builder said. "I call the amount of push the *force*. If he pushes twice as hard, then he has to do twice as much work."

"It also makes a difference how far he pushes the sleigh," the king observed. "Sometimes he gives the sleigh a little tap. Other times he extends his long arms and pushes the sleigh for quite a distance before he lets go. If he pushes the sleigh for twice the distance, he has to work twice as hard."

We set up an equation:

$$\text{work } (W) = \text{force } (F) \text{ times distance moved } (\Delta x)$$

$$W = F \, \Delta x$$

"The force might not be constant," the professor worried. "It might be a function of position: $F = F(x)$."

"Then we should use an integral," the king declared.

$$\text{work} = W = \int F(x) \, dx$$

After doing some testing, it was clear that when Pal did more work to push the sleigh, the sleigh moved faster. Once he let go, the sleigh moved with constant velocity in a straight line along the perfectly frictionless lake. We looked for a relation between the work and the velocity.

"Remember that force = mass times acceleration, where acceleration = dv/dt, or d^2x/dt^2," Builder reminded us.

We inserted the formula $m \, dv/dt$ in place of F into the integral for work:

$$W = \int F \, dx = \int m \frac{dv}{dt} \, dx$$

Replace dx with $dx/dt \, dt$, and pull the mass m outside the integral since it is a constant:

$$W = m \int \frac{dv}{dt} \frac{dx}{dt} \, dt$$

Since $dx/dt = v$:

$$W = m \int \frac{dv}{dt} \, v \, dt$$

$$W = m \int v \, dv$$

From the power rule:

$$W = \frac{1}{2} \, mv^2$$

(The arbitrary integration constant will be zero if the velocity is zero before Pal starts pushing, since the work (W) is zero before he starts.)

Recordis had an idea. "When Pal does work, he uses up energy. But I don't think that energy disappears—instead, it gets transferred to the sleigh. Therefore, the formula $mv^2/2$ gives the amount of energy of the sleigh. We'll say that Builder does work *on* the sleigh when he pushes it; that work gets transferred into the motion energy of the sleigh."

Builder designed another ride where Pal slowly lifted a basket straight up, letting the children have a view from high in the sky before he let the basket fall onto a giant pillow.

"Make sure the pillow is very soft," the king ordered.

"Why doesn't the basket continue moving straight up when Pal stops pushing it?" the professor wondered.

"If that isn't obvious!" Recordis exclaimed. "Gravity pulls it down."

"But where does the energy go?" the professor asked. "The work that Pal does is still $W = \int F \, dx$."

"I see one important difference," the king said. "When he pushes the sleigh on the ice, no force is pushing back against him, so all of the work he does gets transferred into the motion energy of the sleigh. When he lifts the basket up, he has to fight against the force of gravity that pulls down. So, the work is not transferred into the motion of the basket; instead, somehow the energy is stored up as gravitational energy. After Pal lets the basket go, it quickly gains speed as gravity pulls it down. That means it is using up its stored energy and converting it into motion energy."

We called the stored energy the *gravitational potential energy*, and we called the motion energy *kinetic energy*. Using E for total energy, U for potential energy, and KE for kinetic energy, we wrote the rule stating that the work done on an object equals its total energy (assuming the object had zero energy to start with):

$$W = \int F \, dx = U + KE = U + \frac{1}{2}mv^2$$

We had established the formula for kinetic energy: $KE = mv^2/2$, but we did not yet know a formula for the potential energy.

If v is zero when Pal finishes doing work on the object (as it is when he lifts the basket and holds it in the air), all of the work will have been converted into potential energy:

$$W = \int F \, dx = U$$

"If potential energy is the integral of force, it means that force is the derivative of potential energy, with respect to x," the professor said with surprise. After thinking another moment, she added, "The force of gravity pulls down, which is opposite to the upward-pointing force of Pal lifting the basket. Therefore, the force of gravity must be the negative of the derivative of the gravitional potential energy."

$$F = -\frac{dU}{dx}$$

"I have one little concern," Recordis said. "If potential energy is defined as an integral, there must be an arbitrary constant. That means we have to arbitrarily choose some point and say that the potential energy of that point is zero."

"Not a problem," Builder said confidently. "I only care about the difference in potential energy between two locations, so you can define the zero location for potential energy to be wherever it is convenient for you."

For the third ride, Builder attached the sleigh to a long rope that swiveled around a post stuck in the ice. This time, when Pal pushed the sleigh, it moved in a circle about the post.

"So far we have considered motion in only one dimension," the professor realized. "In reality we need to take into account motion in three dimensions, so we will need to use a vector with three components: x, y, and z."

"We need a short notation for the vector (x,y,z)," Recordis suggested. "We also need another letter to represent the vector that's not x, y, or z; how about r?"

"We'll write **r** in boldface when it represents a vector," the king decided.

$$\mathbf{r} = (x,y,z)$$

(When you write the name of a vector by hand, you typically put an arrow over the name to indicate that it is a vector.)

The professor started to systematically make a list of things you can do to vectors. (See page 321.)

"Now to find the formulas for the motion of the circular ride," the professor said. "We'll put the origin of the coordinate system at the post at the center of the circle. We'll orient the xy plane to coincide with the plane of the ice. Since the ride always stays on the ice, this means that the z component of the vector will always stay zero. Now we need formulas for the x and y components."

"Since the motion is a circle, we have:

$$x^2 + y^2 = R^2$$

"where R is the radius of the circle," Recordis contributed.

"That's a start, but it's not enough," Builder said. "In order to understand the motion, we need to know x and y as a function of time. We need a vector something like this:"

$$\mathbf{r}(t) = (x(t), y(t), z(t))$$

For the moment we were stymied, not knowing what formulas of time would produce circular motion.

"I can see the x coordinate will oscillate back and forth as the ride moves from one side of the circle to the other," Trigonometeris whispered to me. "The answer is right on the tip of my tongue. Just give me a minute."

"If we did know the components of the vector, we could find the derivative of the vector by finding the derivative of each component," the professor tried to cheer us up. (See page 322.)

Vectors

Addition

The sum of two vectors $\mathbf{a} = (a_x, a_y, a_z)$ and $\mathbf{b} = (b_x, b_y, b_z)$ is found by adding the corresponding components:

$$\mathbf{a} + \mathbf{b} = (a_x + b_x, a_y + b_y, a_z + b_z)$$

Multiplication by a Scalar

A vector can be multiplied by a number n by multiplying each component by n:

$$n\mathbf{a} = (na_x, na_y, na_z)$$

(Since this operation changes the scale of a vector, we decided numbers could also be called *scalars*.)

See Figure 16–4.

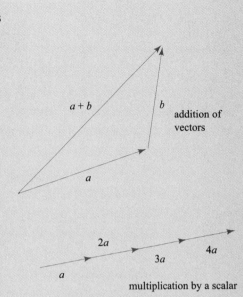

addition of vectors

multiplication by a scalar

FIGURE 16–4.

Length

The length of the vector $\mathbf{a} = (a_x, a_y, a_z)$ can be found from the Pythagorean theorem:

$$\|\mathbf{a}\| = \sqrt{a_x^2 + a_y^2 + a_z^2}$$

We used $\|\mathbf{r}\|$ as a notation for the length of the vector $\mathbf{r}$. (The length is also called the *magnitude* or *norm* of the vector.)

Dot Product

The formula for length suggests it will be convenient to define an operation of multiplying the corresponding components of two vectors, and then adding up the products. This is called the *dot product* of the two vectors. (Recordis liked the dot notation because a dot was easy to write.)

The dot product of two vectors $\mathbf{a} = (a_x, a_y, a_z)$ and $\mathbf{b} = (b_x, b_y, b_z)$ is:

$$\mathbf{a} \cdot \mathbf{b} = a_x b_x + a_y b_y + a_z b_z$$

Note that the dot product of two vectors is a number (or scalar), not a vector. It is also sometimes called the *scalar product*.

The length of a vector $\mathbf{a}$ is

$$\|\mathbf{a}\| = \sqrt{\mathbf{a} \cdot \mathbf{a}}$$

If $\mathbf{a}$ and $\mathbf{b}$ are perpendicular, then $\mathbf{a} \cdot \mathbf{b}$ is zero. In general:

$$\mathbf{a} \cdot \mathbf{b} = \|\mathbf{a}\| \times \|\mathbf{b}\| \times \cos \theta$$

where θ is the angle between vectors $\mathbf{a}$ and $\mathbf{b}$. See Exercise 36.

(This box discussed vectors in three dimensions. The analogous definitions apply for other dimensions.)

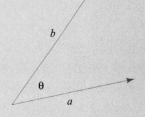

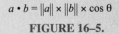
$a \cdot b = \|a\| \times \|b\| \times \cos \theta$

FIGURE 16–5.

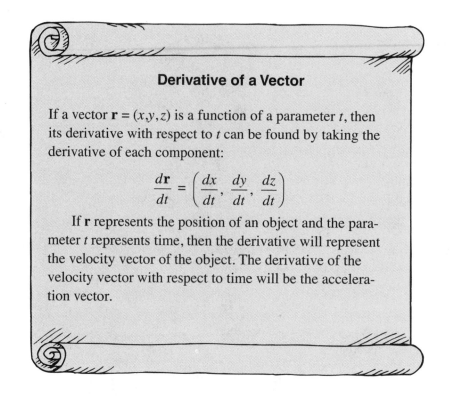

Derivative of a Vector

If a vector $\mathbf{r} = (x, y, z)$ is a function of a parameter t, then its derivative with respect to t can be found by taking the derivative of each component:

$$\frac{d\mathbf{r}}{dt} = \left(\frac{dx}{dt}, \frac{dy}{dt}, \frac{dz}{dt} \right)$$

If $\mathbf{r}$ represents the position of an object and the parameter t represents time, then the derivative will represent the velocity vector of the object. The derivative of the velocity vector with respect to time will be the acceleration vector.

Recordis breathed a huge sigh of relief when he realized that the same trusty methods we used to find the derivative of a one-variable function could also be used to find the derivatives of the components of a vector. "Let's work this out for the simpler case of the first sleigh ride, which moves in a straight line," he suggested. "If the sleigh starts at position (x_0, y_0, z_0) and moves in a straight line, its position vector is":

$$\mathbf{r} = (x_0 + v_x t,\ y_0 + v_y t,\ z_0 + v_z t)$$

The velocity vector is:

$$\frac{d\mathbf{r}}{dt} = \left(\frac{dx}{dt}, \frac{dy}{dt}, \frac{dz}{dt} \right) = (v_x, v_y, v_z)$$

We found the magnitude of the velocity vector:

$$\|\mathbf{v}\| = v = \sqrt{\mathbf{v} \cdot \mathbf{v}} = \sqrt{v_x^2 + v_y^2 + v_z^2}$$

We decided to call the magnitude of the velocity vector the *speed*, and write it as v. (In common usage, the terms velocity and speed are sometimes used interchangeably, but technically velocity refers to the vector $\mathbf{v}$, which has magnitude and direction, while speed refers to the magnitude of the velocity vector, which is a scalar.)

We can take the second derivative of the position, with respect to time, and the result will be the acceleration vector:

$$\mathbf{a} = \frac{d^2\mathbf{r}}{dt^2} = \left(\frac{d^2x}{dt^2}, \frac{d^2y}{dt^2}, \frac{d^2z}{dt^2} \right)$$

For the straight-line sleigh ride the velocity components are constants, meaning their derivatives are zero, so the acceleration of the ride is zero:

$$\mathbf{a} = \frac{d\mathbf{v}}{dt} = (0, 0, 0)$$

In one dimension, we have $F = ma$, so in three dimensions we have a force vector $\mathbf{F}$ (with components F_x, F_y, and F_z) that equals mass times the acceleration vector:

$$\mathbf{F} = (F_x, F_y, F_z) = m\mathbf{a} = \left(m\frac{d^2x}{dt^2}, m\frac{d^2y}{dt^2}, m\frac{d^2z}{dt^2} \right)$$

"Any progress on finding the formula for motion of the circular ride?" Builder was getting impatient.

"I've figured it out," Trigonometeris said. "The position vector of an object moving in a circle will be:

$$\mathbf{r} = (R \cos (\omega t), R \sin (\omega t), 0)$$

"where R is the radius of the circle, and ω is a constant I call angular velocity, which determines how fast the sleigh moves around the circle." Recordis was about to object to the use of trigonometry, but fortunately he remembered he had decided to be extra nice to Trigonometeris after the secant integral incident.

"The magnitude of this alleged position vector had better be constant," the professor warned. "Since the origin is at the center of the circle, the magnitude of the position vector represents the distance from the sleigh to the origin. If that distance isn't constant, we don't have circular motion."

We did the calculation (recall that $\sin^2 \omega t + \cos^2 \omega t = 1$):

$$\|\mathbf{r}\| = \sqrt{\mathbf{r} \cdot \mathbf{r}} = \sqrt{R^2 \cos^2(\omega t) + R^2 \sin^2(\omega t) + 0^2} = R$$

Trigonometeris was right; the motion described by the vector $\mathbf{r}$ was circular.

"We can find the velocity vector by finding the derivatives of the components," the king said.

$$\mathbf{v} = \frac{d\mathbf{x}}{dt} = \left(\frac{dx}{dt}, \frac{dy}{dt}, \frac{dz}{dt} \right) = (-R\omega \sin (\omega t), R\omega \cos (\omega t), 0)$$

We found the speed v (which is the magnitude of the velocity vector):

$$\|\mathbf{v}\| = \sqrt{\mathbf{v} \cdot \mathbf{v}} = \sqrt{R^2\omega^2 \sin^2(\omega t) + R^2\omega^2 \cos^2(\omega t)}$$

$$= R\omega$$

We noticed something interesting when we found the dot product $\mathbf{r} \cdot \mathbf{v}$:

$$\mathbf{r} \cdot \mathbf{v} = (R \cos (\omega t), R \sin (\omega t), 0) \cdot (-R\omega \sin (t), + R\omega \cos (t), 0)$$

$$= -R^2\omega \cos (\omega t) \sin (\omega t) + R^2\omega \sin (\omega t) \cos (\omega t) + 0^2 = 0$$

"Since the dot product is zero, it means the two vectors are perpendicular," the professor noted.

Next, we found the acceleration vector **a** by finding the derivative of the velocity:

$$\mathbf{a} = \frac{dv}{dt} = \left(\frac{dv_x}{dt}, \frac{dv_y}{dt}, \frac{dv_z}{dt} \right) = (-R\omega^2 \cos{(\omega t)}, -R\omega^2 \sin{(\omega t)}, 0)$$

"Look what happens if we pull the ω^2 in front of the vector," Recordis said.

$$\mathbf{a} = \frac{dv}{dt} = -\omega^2 (R \cos{(\omega t)}, -R \sin{(\omega t)}, 0) = -\omega^2 \mathbf{r}$$

"The position vector **r** points from the center of the circle to the sleigh. The acceleration vector points exactly in the opposite direction: from the sleigh to the center." See Figure 16–6.

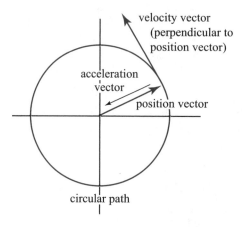

FIGURE 16–6.

"Now I could use some help with the motion of a particle in three dimensions under the force of gravity," Builder reminded us.

We worked on revising our formulas for energy to account for three dimensions. The formula for kinetic energy became

$$KE = \frac{1}{2} m\mathbf{v} \cdot \mathbf{v}$$

In one dimension, force was the negative of the derivative of the potential energy function with respect to x, so in three dimensions we should take the gradient of the potential energy function:

$$\mathbf{F} = -\mathbf{grad}\ U = \left(-\frac{\partial U}{\partial x}, -\frac{\partial U}{\partial y}, -\frac{\partial U}{\partial z} \right)$$

We investigated to see whether the total energy of an object moving in a gravitational field would be constant, assuming no external force (like Pal) is pushing or pulling on the object. We started with the equation for total energy E:

$$E = U + KE$$

where U is the potential energy, which is a function of the position vector; and KE is the kinetic energy, which is $1/2 m\mathbf{v} \cdot \mathbf{v}$ where m is the mass of the object, and $\mathbf{v}$ is its velocity vector. The rate of change in energy with time can be found from the derivative:

$$\frac{dE}{dt} = \frac{dU}{dt} + \frac{d(KE)}{dt}$$

U is a function of the position vector $\mathbf{r}$, but then the position vector $\mathbf{r}$ is a function of time:

$$\text{potential energy} = U(\mathbf{r}(t))$$

"We've had functions inside of other functions before!" Recordis remembered the chain rule. If we had $U(r(t))$, where r was an ordinary scalar function of t, then the chain rule says that:

$$\frac{dU}{dt} = \frac{dU}{dr}\frac{dr}{dt}$$

We figured that the same basic idea should work in three dimensions, only we would use the gradient (**grad** U) instead of dU/dx, and we would have to use the dot product instead of the simple multiplication:

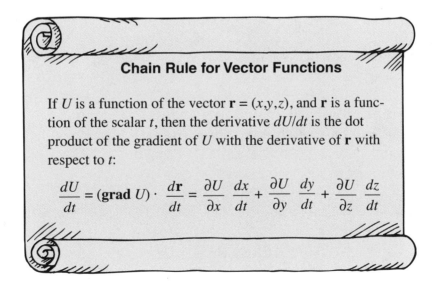

Chain Rule for Vector Functions

If U is a function of the vector $\mathbf{r} = (x,y,z)$, and $\mathbf{r}$ is a function of the scalar t, then the derivative dU/dt is the dot product of the gradient of U with the derivative of $\mathbf{r}$ with respect to t:

$$\frac{dU}{dt} = (\mathbf{grad}\ U) \cdot \frac{d\mathbf{r}}{dt} = \frac{\partial U}{\partial x}\frac{dx}{dt} + \frac{\partial U}{\partial y}\frac{dy}{dt} + \frac{\partial U}{\partial z}\frac{dz}{dt}$$

"Be sure to notice which ones are partial derivatives and which are ordinary derivatives," Recordis reminded us. Since the force is the negative of the gradient of the potential energy, we had:

$$\frac{dU}{dt} = -\mathbf{F} \cdot \frac{d\mathbf{r}}{dt} = -\mathbf{F} \cdot \mathbf{v}$$

"Now to find the derivative with respect to time of the kinetic energy *KE*." The professor led us through the process.

$$KE = \frac{1}{2} m\mathbf{v} \cdot \mathbf{v}$$

$$\frac{d(KE)}{dt} = \frac{m}{2} \frac{d}{dt}(v_x^2 + v_y^2 + v_y^2)$$

$$\frac{d(KE)}{dt} = \frac{m}{2}\left(2v_x \frac{dv_x}{dt} + 2v_y \frac{dv_y}{dt} + 2v_z \frac{dv_z}{dt} \right)$$

"Cancel the 2's, and replace the derivatives of the velocity components with the acceleration components."

$$\frac{d(KE)}{dt} = m(v_x a_x + v_y a_y + v_z a_z)$$

"We can write that as a dot product of the velocity vector (v_x, v_y, v_z) and the acceleration vector (a_x, a_y, a_z):"

$$\frac{d(KE)}{dt} = m\mathbf{v} \cdot \mathbf{a}$$

We moved the *m* next to **a** (see Exercise 34).

$$\frac{d(KE)}{dt} = \mathbf{v} \cdot (m\mathbf{a})$$

"And *m***a** is mass times acceleration, which is the force vector:"

$$\frac{d(KE)}{dt} = \mathbf{v} \cdot \mathbf{F}$$

"We previously found the derivative of the potential energy with respect to time,"

$$\frac{dU}{dt} = -\mathbf{F} \cdot \mathbf{v}$$

"and we can see that dU/dt and $d(KE)/dt$ are negatives of each other. This means that,"

$$\frac{dE}{dt} = \frac{dU}{dt} + \frac{d(KE)}{dt} = 0$$

"If the derivative with respect to time is zero, it means that the total energy is a constant," the king declared.

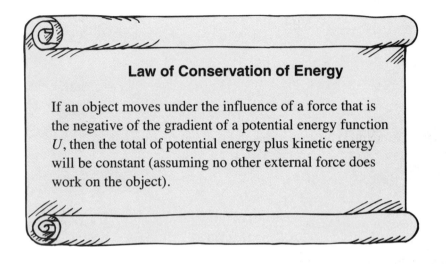

Law of Conservation of Energy

If an object moves under the influence of a force that is the negative of the gradient of a potential energy function U, then the total of potential energy plus kinetic energy will be constant (assuming no other external force does work on the object).

Builder quickly performed some measurements to determine that the potential energy of the basket being lifted in the air was mgz, where z is the height of the basket above the ground, m is the mass of the basket, and g is a constant (which equals 9.8 meters per second per second when converted from the Carmorra measurement system to the metric system).

We found the gradient of the potential energy to determine the force vector for gravity:

$$U = mgz$$

$$\mathbf{F} = -\mathbf{grad}\ U = \left(-\frac{\partial U}{\partial x},\ -\frac{\partial U}{\partial y},\ -\frac{\partial U}{\partial z} \right) = (0, 0, -mg)$$

We found the acceleration vector $\mathbf{a} = (1/m)\mathbf{F}$:

$$\mathbf{a} = (0, 0, -g)$$

"The negative sign for the z component indicates that the force of gravity pulls things down," the professor said.

Then we set up the differential equations for the components of the velocity:

$$\frac{d\mathbf{v}}{dt} = \mathbf{a}$$

$$\frac{dv_x}{dt} = 0$$

$$\frac{dv_y}{dt} = 0$$

$$\frac{dv_z}{dt} = -g$$

"These differential equations are a piece of cake," Recordis said.

$$v_x = v_{x0}$$

$$v_y = v_{y0}$$

$$v_z = -gt + v_{z0}$$

where the three constants of integration (v_{x0}, v_{y0}, v_{z0}) represent the components of the initial velocity vector at time $t = 0$.

Next, we set up the differential equations for the components of the position vector $\mathbf{r}$:

$$\frac{d\mathbf{r}}{dt} = \mathbf{v}$$

$$\frac{dx}{dt} = v_{x0}$$

$$\frac{dy}{dt} = v_{y0}$$

$$\frac{dz}{dt} = -gt + v_{z0}$$

These differential equations were also easy to solve:

$$x = v_{x0}t + x_0$$

$$y = v_{y0}t + y_0$$

$$z = -\frac{1}{2}gt^2 + v_{z0}t + z_0$$

where the arbitrary integration constants (x_0, y_0, z_0) represent the components of the vector giving the initial position of the basket at time 0.

"v_{x0} and v_{y0} both happen to be zero, since Pal drops the basket straight down with no initial velocity in the x or y direction," Builder said. "Also, v_{z0} is zero, since he is holding the basket motionless at $t = 0$ (right before he drops it). For convenience, put the origin of the coordinate system directly under the basket, so x_0 and y_0 are both zero. And z_0 is the height of the basket above the ground at the moment Pal releases the basket. Let's call this height h."

With this information, we wrote the acceleration, velocity, and position vectors for the basket:

$$\mathbf{a} = (0, 0, -g)$$

$$\mathbf{v} = (0, 0, -gt)$$

$$\mathbf{r} = \left(0,\ 0,\ h - \frac{1}{2}gt^2\right)$$

We also found:

kinetic energy $= 1/2m\mathbf{v} \cdot \mathbf{v} = 1/2mg^2t^2$

potential energy $= mgz = mgh - 1/2mg^2t^2$

total energy $= 1/2mg^2t^2 + mgh - 1/2mg^2t^2 = mgh$, which is constant

"Now that we know about gravity, we can determine the motion of the moon," Trigonometeris said excitedly. However, our formula for the gravitational potential energy mgz did not work for the moon. Builder performed some more experiments and told us that the formula mgz only works if an object is close to the surface of Earth. In general, the potential energy function of a mass m moving under the influence of gravity of Earth is:

$$U(x, y, z) = -\frac{GMm}{\sqrt{x^2 + y^2 + z^2}}$$

M is the mass of Earth; the constant G determines the strength of the gravitational force. This formula assumes: (a) the origin of the coordinate system is at the center of the Earth; (b) you are on or above the surface of Earth, which is usually always the case; (c) Earth is a perfect sphere (not exactly true, but not far from the truth); (d) other objects are so far away that their gravity can be ignored (obviously this assumption is not realistic if you start traveling very far away in space); (e) potential energy is arbitrarily assumed to be zero at an infinite distance away from Earth. This means that the value of the potential energy is actually negative, but Builder assured us that would be no problem. "It's only the difference in potential energy between two points that has physical significance," he insisted. "You can define the potential energy to be zero wherever it is convenient for you."

We defined $r = \|\mathbf{r}\| = \sqrt{x^2 + y^2 + z^2}$ to be the distance from the object to the center of Earth, so the potential could be written:

$$U = \frac{-GMm}{r}$$

To determine the force vector, we found the gradient:

$$\frac{\partial U}{\partial x} = -\frac{-1}{2} GMm(x^2 + y^2 + z^2)^{-3/2}(2x) = \frac{GMmx}{r^3}$$

$$\frac{\partial U}{\partial y} = -\frac{-1}{2} GMm(x^2 + y^2 + z^2)^{-3/2}(2y) = \frac{GMmy}{r^3}$$

$$\frac{\partial U}{\partial z} = -\frac{-1}{2} GMm(x^2 + y^2 + z^2)^{-3/2}(2z) = \frac{GMmz}{r^3}$$

Therefore, the force vector became:

$$\mathbf{F} = -\mathbf{grad}\ U = \left(\frac{-GMmx}{r^3}, \frac{-GMmy}{r^3}, \frac{-GMmz}{r^3} \right)$$

We could write the force $\mathbf{F}$ in terms of the position vector $\mathbf{r}' = (x, y, z)$:

$$\mathbf{F} = -\mathbf{grad}\ U = -\frac{GMm}{r^3}\mathbf{r}$$

"The minus sign indicates that the force of gravity pulls in the opposite direction of the vector $\mathbf{r}$, which makes sense since we know gravity will always pull the object back toward the origin," the professor noted.

"Sometimes I get confused because a vector represents both direction and magnitude," Recordis said. "Let's define a new vector $\hat{\mathbf{r}}$ (r hat) that points in the same direction as $\mathbf{r}$ but has length 1. Then:

$$\hat{\mathbf{r}} = \left(\frac{x}{r}, \frac{y}{r}, \frac{z}{r} \right)$$

Or, we could say that $\mathbf{r} = r\hat{\mathbf{r}}$. Then the force of gravity becomes:

$$\mathbf{F} = -\mathbf{grad}\ U = -\frac{GMm}{r^3} r\hat{\mathbf{r}} = -\frac{GMm}{r^2} \hat{\mathbf{r}}$$

"I like that version better, since it makes clear that the magnitude of the force is proportional to $1/r^2$. In other words, if you double the distance between two objects, the force of gravity becomes one-fourth as strong," Builder said.

We were all exhausted by that point, but we felt we had accomplished a tremendous amount that day. Trigonometeris promised to work on more details of the moon's motion under the influence of Earth's gravity (see Exercise 46).

Only four days remained until the party. "There is one thing that worries me," the king said.

"What's that?" I asked.

"We haven't heard from the gremlin lately, and it is unlike him to be quiet this long."

NOTE TO CHAPTER 16

The measured value of the gravity constant is

$$G = 6.67 \times 10^{-11} \ \frac{\text{meters}^3}{\text{kilograms} \times \text{seconds}^2}.$$

The mass of Earth is $M = 5.97 \times 10^{24}$ kilograms; the radius of Earth is $R = 6,378,000$ meters. An object of mass m on the surface of Earth will feel a gravitational force with magnitude

$$\|\mathbf{F}\| = \frac{GMm}{R^2}$$

The magnitude of its acceleration is known as g:

$$g = \frac{\|\mathbf{F}\|}{m} = \frac{GM}{R^2} = \frac{6.67 \times 10^{-11} \ \frac{\text{meters}^3}{\text{kilograms} \times \text{seconds}^2} \times 5.97 \times 10^{24} \ \text{kilograms}}{(6,378,000 \ \text{meters})^2}$$

$g = 9.8$ meters per second per second, which is also written as $9.8 \ \frac{\text{meters}}{\text{second}^2}$ (read as 9.8 meters per second squared). Note that small g is used for the acceleration of gravity on the surface of Earth; capital G is used for the general gravitational constant.

Exercises

1. Calculate the value of the revenue for the publisher for each of the following pairs of values for x and y:

   $(x = 60, y = 32.5)$; $(x = 60.1, y = 32.5)$; $(x = 59.9, y = 32.5)$;
   $(x = 60, y = 32.6)$; $(x = 60, y = 32.4)$.

Find $\partial z/\partial x$ and $\partial z/\partial y$ for each of the following functions. (Treat all letters other than x, y, and z as constants.)

2. $z = x + y$

3. $z = x^2 + y^2$

4. $z = xy$

5. $z = x^2y^2$

6. $z = x^ay^b$

7. $z = \sqrt{x^2 + y^2}$

8. $z = e^{ax + by}$

9. $z = ax^2 + bxy + cy^2 + dx + ey + f$

10. $z = (x - a)^n + (y - b)^m$

11. $z = \sin(ax + by)$

For Exercises 12 to 16, find the point where both partial derivatives are equal to zero. These points may be either maximum points, minimum points, or neither. As in the one-variable case, it is necessary to check the second derivative to determine which is the case. Unfortunately, the conditions involving the second derivative are more complicated when there are multiple independent variables.

Once you have found the point where $\partial z/\partial x$ and $\partial z/\partial y$ are both zero, perform the following test, using this notation for the three second-order derivatives:

$$z_{xx} = \frac{\partial^2 z}{\partial x^2} = \frac{\partial}{\partial x}\frac{\partial z}{\partial x}$$

$$z_{yy} = \frac{\partial^2 z}{\partial y^2} = \frac{\partial}{\partial y}\frac{\partial z}{\partial y}$$

$$z_{xy} = \frac{\partial^2 z}{\partial x \partial y} = \frac{\partial}{\partial x}\frac{\partial z}{\partial y}$$

If $z_{xx}z_{yy} > (z_{xy})^2$, there is a local maximum or minimum. To tell the difference, if z_{xx} and z_{yy} are positive, you have a local minimum. This means that a cross section of the surface will be concave upward. If z_{xx} and z_{yy} are negative, you have a local maximum.

If $(z_{xy})^2 > z_{xx}z_{yy}$, the point is neither a maximum nor a minimum (it is something called a *saddle point*, which means you are at a minimum if you cut a cross section of the surface in one direction, but you are at a maximum if you cut a cross section in another direction).

If $(z_{xy})^2 = z_{xx}z_{yy}$, you cannot tell from this test whether you have a maximum, minimum, or saddle point.

12. $z = 3x^2 + xy + 2y^2 + 9x - 7y + 10$

13. $z = x^2 + 10xy + y^2 - x - 5y - 8$

14. $z = -x^2 - 2xy + 3y^2 + 8x - 4y - 17$

15. $z = -0.5x^2 + 0.2xy - 1.4y^2 + 1.8x - 1.2y + 0.75$

16. $z = x^3y + 2x - y$

17. A production function with two inputs (K and L) is given by the formula

$$Q = (L^{-2} + K^{-2})^{-1/2}$$

where Q is the quantity of output produced. The profit is given by this formula:

$$Z(L,K) = PQ - wL - rK$$

where P, w, and r are constants. Determine the profit maximizing ratio between K and L.

18. Consider this function of three variables:

$$f(x,y,z) = (x - 1)^2 + (y - 3)^2 + (z - 2)^2$$

Determine the coordinates of the point where all components of the gradient vector are zero.

For Exercises 19 to 27, find $\mathbf{a} + \mathbf{b}$ and $\mathbf{a} \cdot \mathbf{b}$ for the two vectors given.

19. $\mathbf{a} = (2, 6)$; $\mathbf{b} = (8, 3)$

20. $\mathbf{a} = (4, 3)$; $\mathbf{b} = (7, 1)$

21. $\mathbf{a} = (8, 5)$; $\mathbf{b} = (3, 9)$

22. $\mathbf{a} = (3, 7)$; $\mathbf{b} = (2, 4)$

23. $\mathbf{a} = (20, 12)$; $\mathbf{b} = (-20, -12)$

24. $\mathbf{a} = (60, 10)$; $\mathbf{b} = (-59, -10)$

25. $\mathbf{a} = (4, 0)$; $\mathbf{b} = (0, 6)$

26. $\mathbf{a} = (1, 1)$; $\mathbf{b} = (-1, 1)$

27. $\mathbf{a} = (0.5000, 0.8660)$; $\mathbf{b} = (0.8660, 0.5000)$

For Exercises 28 to 32, find the dot product of the two vectors **a** and **b**, and find the angle between the two vectors:

28. $(3, 4)$ and $(4, -3)$

29. $(18, 25)$ and $(54, 75)$

30. $\left(\dfrac{\sqrt{3}}{2}, \dfrac{1}{2}\right)$ and $\left(\dfrac{1}{\sqrt{2}}, \dfrac{1}{\sqrt{2}}\right)$

31. $\left(\dfrac{\sqrt{3}}{2}, \dfrac{1}{2}\right)$ and $\left(\dfrac{1}{2}, \dfrac{\sqrt{3}}{2}\right)$

32. $\left(\dfrac{\sqrt{3}}{2}, \dfrac{1}{2}\right)$ and $(0, 1)$

For two-dimensional vectors, verify these properties:

33. $\mathbf{a} \cdot \mathbf{b} = \mathbf{b} \cdot \mathbf{a}$

34. $n\mathbf{a} \cdot \mathbf{b} = \mathbf{a} \cdot (n\mathbf{b})$

35. $(\mathbf{a} + \mathbf{b}) \cdot \mathbf{c} = \mathbf{a} \cdot \mathbf{c} + \mathbf{b} \cdot \mathbf{c}$

36. Verify that $\mathbf{a} \cdot \mathbf{b} = \|\mathbf{a}\| \times \|\mathbf{b}\| \cos \theta$.

37. At what time will the basket ride hit the ground? How fast will the basket be traveling when it hits the ground? Recall that the position vector of the basket is $(0, 0, h - 1/2gt^2)$.

38. Determine the value of t that minimizes the distance between the line $(a + tu, b + tv, c + tw)$ and the point (h, j, k). (All letters except t represent constants.)

39. The equation $ax + by + cz = a^2 + b^2 + c^2$ defines a plane.
   (a) Solve the equation for z in terms of x and y. (Treat $a, b,$ and c as constants.)
   (b) The distance from the origin to a point in this plane is $\sqrt{x^2 + y^2 + z^2}$. Determine formulas for $x, y,$ and z that minimize the distance.

40. An xy coordinate system is drawn on a frictionless table. The table is then rotated by an angle θ and tilted by an angle ϕ. According to the original coordinate system, a marble on the table will be subject to a gravitational potential energy given by the formula:

$$U = -gx \sin \theta \, \sin \phi + gy \cos \theta \, \sin \phi$$

All letters except x and y are constants. Determine the force vector.

41. A marble rolling around a curved frictionless bowl faces a potential energy function $U = x^2 + y^2$. What is the force acting on the marble?

42. When a force **F** causes an object to move in three dimensions, the work performed is given by this integral:

$$W = \int_0^{t_2} \mathbf{F} \cdot \mathbf{v} \, dt$$

What is the work if the force is $(0, 0, -g)$ and:
    (a) $\mathbf{v} = (0, 0, v_{up})$
    (b) $\mathbf{v} = (v_x, v_y, 0)$
    (c) $\mathbf{v} = (v \cos \theta \, \cos \phi, v \sin \theta \, \cos \phi, v \sin \phi)$

43. If the temperature T at point (x, y, z) is $T = (x^2 + y^2)^a z^b$, find the rate of change of temperature with time (dT/dt) for an object whose position vector is $\mathbf{r} = (x_0 + v_x t, y_0 + v_y t, z_0 + v_z t)$.

44. If a probe is launched into space with a velocity greater than or equal to the *escape velocity*, it will not return to Earth. Find the value of the escape velocity at the surface of Earth, which is the value of v that makes the initial energy (E_0) zero:

$$E_0 = -\frac{GMm}{r_0} + \frac{1}{2}mv^2$$

M is the mass of Earth; r_0 is the radius of Earth; m is the mass of the probe.

45. The gravitational potential energy of an object at height h above the surface of Earth is $-GMm/(R + h)$, where R is the radius of Earth. What is the difference between the potential at this point and a point on the surface of Earth (which has potential $-GMm/R$). Assume R is much, much bigger than h.

46. For this exercise, assume that the moon moves in a circular orbit about the Earth. (The actual path is an ellipse; see the next exercise.) The position vector is:

$$\mathbf{r} = (R \cos (\omega t), R \sin (\omega t), 0)$$

    (a) Find the velocity vector **v** and the speed $v = \|\mathbf{v}\|$.
    (b) Find the acceleration vector **a**. Also find $\|\mathbf{a}\|$.
    (c) The potential energy is $U = -GMm/r$. Find the force vector. (M is the mass of Earth; m is the mass of the moon.)
    (d) Use the equation $\mathbf{F} = m\mathbf{a}$ to find a formula for ω in terms of G, M, and R. Then, use the result from (a) to find a formula for v in terms of G, M, and R.
    (e) Find a formula for the amount of time it takes to complete one orbit, in terms of ω.
    (f) Find the potential energy, kinetic energy, and total energy of the moon.

47. Let r represent the distance of a planet from the sun at time t, and let θ represent the angle formed by the x axis and the line connecting the planet to the sun. The speed of the planet about its orbit is given by:

$$v = \sqrt{r^2\theta'^2 + r'^2}$$

The prime symbol ($'$) represents the derivative of a quantity with respect to time. The kinetic energy (KE) of any object is given by $KE = 1/2mv^2$, where m is the mass of the object, so the kinetic energy of the planet is:

$$\frac{1}{2}m\,(r^2\theta'^2 + r'^2)$$

The potential energy (U) of the planet is $U = -k/r$, where k is a constant. The Lagrangian expression (L) for an object is:

$$L(r,\theta,r',\theta') = KE - U$$

L is a function of four variables: r, θ, r', and θ'. Once the Lagrangian for an object is known, then the motion of that object is described by a set of differential equations known as Euler's equations:

$$\frac{\partial L}{\partial\theta} = \frac{d}{dt}\frac{\partial L}{\partial\theta'}$$

$$\frac{\partial L}{\partial r} = \frac{d}{dt}\frac{\partial L}{\partial r'}$$

(a) Determine the specific form of the first Euler equation for the example of the planetary motion.

(b) The total energy (E) of the planet is equal to the sum of the kinetic energy plus the potential energy:

$$E = KE + U = \frac{1}{2}m\,(r^2\theta'^2 + r'^2) - k/r$$

The angular momentum (A) of the planet is given by:

$$A = mr^2\theta'$$

The angular momentum remains constant. Rewrite the formula for the total energy in terms of A, m, r, r', and k.

(c) Rewrite the formula from part (b) so that r' is expressed in terms of the other quantities in the formula.

(d) Rewrite the formula from part (c) so that it expresses θ as this integral:

$$\theta = \int \frac{1/r^2}{\sqrt{\dfrac{2mE}{A^2} + \dfrac{2mk}{A^2 r} - \dfrac{1}{r^2}}}\, dr$$

(e) Perform the integration from part (d) in order to arrive at a formula for $\cos\theta$ in terms of r and the constants m, E, A, and k. Hint: Make the substitution $v = 1/r$, and then use the result from Exercise 12–40.

(f) Simplify the formula from part (e) by making these substitutions:

$$B = A^2/mk \qquad e = \sqrt{1 + 2A^2E/mk^2}$$

Can you recognize the resulting formula as the equation of a curve written in polar coordinates?

(In this problem, it has been assumed that the sun is much more massive than the planet, so it can be treated as stationary. Also, the fact that each planet is affected by the gravity of the other planets has been ignored.)

48. Suppose you are trying to design a ramp so that a ball rolling down the ramp will reach the bottom as quickly as possible. The function giving the optimal shape must solve Euler's equation:

$$\frac{\partial L}{\partial y} = \frac{d}{dx}\frac{\partial L}{\partial y'}$$

where L is defined as follows:

$$L(x,y,y') = \sqrt{\frac{1 + y'^2}{-2gx}}$$

Solve for y. Hint: Use the result from Exercise 12–41.

49. (a) Spaceship 1 is orbiting Earth along a circular orbit with radius $r_1 = 6{,}540{,}000$ meters (measured from the center of the Earth). Calculate its velocity using the formula from Exercise 46(d). [$G = 6{:}67 \times 10^{-11}$ (gravitational constant); $M = 5.97 \times 10^{24}$ (mass of Earth in kilograms)].

(b) Spaceship 2 orbits the Earth along a circular orbit with radius $r_2 = 7{,}180{,}000$ meters. Calcluate its velocity.

(c) For an object traveling along an elliptical orbit around the Earth, the distance to the center of the Earth at its closest approach is known as the *perigee*, and the distance at the highest point of the orbit is the *apogee*. In order to rendezvous with spaceship 2, spaceship 1 will have to fire its rocket to increase its speed to put it along an elliptical orbit with perigee r_1, apogee r_2, and

eccentricity $= e = \frac{r_2 - r_1}{r_2 + r_1}$. When an object moves along an elliptical orbit, its velocity at a point where it is at distance r from the center of the Earth is

$$v = \sqrt{GM\left(\frac{2}{r} - \frac{1}{a}\right)}$$

where a (the semi-major axis of the ellipse) is the average of the apogee and perigee ($a = \frac{r_1 + r_2}{2}$). Determine the velocity that this elliptical orbit has at its perigee $r_1 = a(1 - e)$.

(d) By how much does spaceship 1 need to increase its velocity from part (a) to part (c)?

(e) After spaceship 1 has entered the elliptical orbit from part (c), it will eventually reach its apogee at a distance r_2. Calculate its velocity at that point.

(f) When it reaches its apogee at r_2, spaceship 1 needs to fire its rockets to increase its speed to that of the circular orbit of spaceship 2 (from part (b)). Calculate the increase in velocity needed.

50. Determine the gravitational potential energy from a spherical planet with radius R and density ρ (both constants). You are located at the origin, and the planet is centered at the point $(D, 0, 0)$. You will need to evaluate a triple integral:

$$PE_{\text{sphere}} = -\int_{x=D-R}^{x=D+R} \int_{r=0}^{r=y} \int_{\theta=0}^{\theta=2\pi} Gm\rho u^{-1} r \, d\theta \, dr \, dx$$

As with a double integral, you start with the inner integral first.

Step 1: Evaluate the inner integral to find the potential energy for a ring of radius r, centered at the point $(x, 0, 0)$ (see Figure 16–7). Let $u = \sqrt{x^2 + r^2}$ be the distance from your location to a point on the ring. G is the gravitational constant, and m is your mass, which is constant. While evaluating the inner integral treat u as a constant,

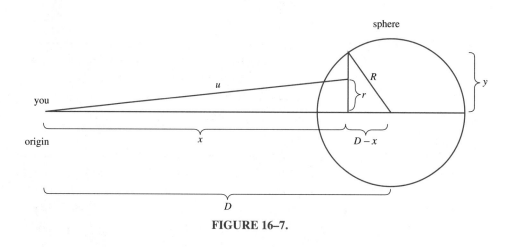

FIGURE 16–7.

$$PE_{\text{ring}} = \int_{\theta=0}^{\theta=2\pi} Gm\rho u^{-1} r \, d\theta$$

Step 2: Find the potential for a plate of radius y centered at $(x, 0, 0)$ by evaluating the second integral.

$$PE_{\text{plate}} = \int_{r=0}^{r=y} Gm\rho u^{-1} r(2\pi) \, dr$$

Step 3: Find the potential energy for the sphere by evaluating this integral:

$$PE_{\text{sphere}} = \int_{x=D-R}^{x=D+R} 2\pi Gm\rho(\sqrt{x^2 + r^2} - x) \, dx$$

Numerical Methods, Taylor Series, and Limits

A piercing scream echoed throughout the palace the next morning. We ran to Recordis' room and found him staring at his problem closet. "An intruder has been rummaging around in here!"

"Look for clues!" the king ordered.

We searched the nearby area. "The culprit must have been in a hurry and dropped this scrap of paper," Trigonometeris called to us in a few minutes. He showed us a torn shred of paper with the words "Secret Plan Draft 1."

"That's the gremlin's handwriting!" the king cried.

"The gremlin must be developing a secret plan to disrupt the party, and he thought there was information in Recordis' problem closet that would help him," the professor stated the obvious.

We quickly went to work to improve security for the party. "In case of trouble, we'll have the children gather in the elliptical rose garden," the king decided.

"Then we will need a fence around the garden to keep them safe," Recordis said.

"I'll build the fence," Builder said. "You tell me how long the fence needs to be."

We set to work to find the distance around an ellipse with semimajor axis a and semi-minor axis b, given by the equation:

$$\frac{x^2}{a^2} + \frac{y^2}{b^2} = 1$$

We solved this equation for y:

$$\frac{y^2}{b^2} = 1 - \frac{x^2}{a^2}$$

$$y^2 = b^2\left(1 - \frac{x^2}{a^2}\right)$$

$$y = b\sqrt{1 - \frac{x^2}{a^2}}$$

Then we found the derivative dy/dx:

$$\frac{dy}{dx} = b\left(\frac{1}{2}\right)\left(1 - \frac{x^2}{a^2}\right)^{-1/2}\left(\frac{-2x}{a^2}\right)$$

$$\frac{dy}{dx} = \frac{-xb}{a^2}\left(1 - \frac{x^2}{a^2}\right)^{-1/2}$$

We used our arc-length integral for the length of the arc from $x = 0$ to $x = a$. Then we realized that the total circumference of the ellipse (S) would be four times the length of this arc. (See Figure 17–1.)

$$S = 4\int_0^a \sqrt{1 + \left(\frac{dy}{dx}\right)^2}\ dx$$

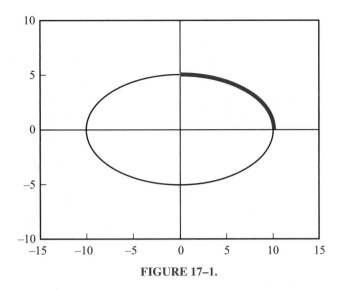

FIGURE 17–1.

We found $(dy/dx)^2$:

$$\left(\frac{dy}{dx}\right)^2 = \frac{x^2 b^2}{a^4}\left(1 - \frac{x^2}{a^2}\right)^{-1}$$

Next, we inserted this formula into the integral for S:

$$S = 4\int_0^a \sqrt{1 + \frac{x^2 b^2}{a^4}\left(1 - \frac{x^2}{a^2}\right)^{-1}}\; dx$$

We worked on simplifying this expression:

$$S = 4\int_0^a \sqrt{1 + \frac{x^2 b^2}{a^4}\left(1 - \frac{x^2}{a^2}\right)^{-1}}\; dx$$

$$S = 4\int_0^a \sqrt{1 + \frac{x^2 b^2}{a^4}\left(\frac{a^2}{a^2 - x^2}\right)}\; dx$$

$$S = 4\int_0^a \sqrt{1 + \frac{x^2 b^2}{a^2(a^2 - x^2)}}\; dx$$

$$S = 4\int_0^a \sqrt{1 + \frac{x^2 b^2}{a^4 - a^2 x^2}}\; dx$$

$$S = 4\int_0^a \sqrt{\frac{a^4 - a^2 x^2 + x^2 b^2}{a^4 - a^2 x^2}}\; dx$$

$$S = 4\int_0^a \sqrt{\frac{a^4 + (b^2 - a^2)x^2}{a^4 - a^2 x^2}}\; dx$$

"It sure would be nice if $a = b$," Recordis said wistfully. "Then the integral would become:"

$$S = 4 \int_0^a \sqrt{\frac{a^4}{a^4 - a^2 x^2}} \, dx = 4 \int_0^a \sqrt{\frac{a^2}{a^2 - x^2}} \, dx = 4a \int_0^a \frac{1}{\sqrt{a^2 - x^2}} \, dx$$

"I think I've seen that integral before," the professor said with a sense of deja vu.

"That's just the integral for the circumference of a circle, only this time we're calling the radius a!" Trigonometeris exclaimed. (See page 268.) "And we know that an ellipse with $a = b$ is the same as a circle."

"That is good to know, but we must get back to our actual problem: finding the circumference of an ellipse when a does not equal b."

$$S = 4 \int_0^a \sqrt{\frac{a^4 + (b^2 - a^2)x^2}{a^4 - a^2 x^2}} \, dx$$

We recalled that the eccentricity (call it k) of an ellipse measured the shape of the ellipse, ranging from $k = 0$ for a circle up to k close to 1 for a very flat ellipse. Since

$$k^2 = \frac{a^2 - b^2}{a^2}$$

we could divide both the top and the bottom of the fraction by a^2 and then rewrite the integral:

$$S = 4 \int_0^a \sqrt{\frac{a^2 + \left(\frac{b^2 - a^2}{a^2}\right)x^2}{a^2 - x^2}} \, dx$$

$$S = 4 \int_0^a \sqrt{\frac{a^2 - \left(\frac{a^2 - b^2}{a^2}\right)x^2}{a^2 - x^2}} \, dx$$

$$S = 4 \int_0^a \sqrt{\frac{a^2 - k^2 x^2}{a^2 - x^2}} \, dx$$

For the elliptical garden, $a = 10$ and $b = 5$, so $k^2 = \dfrac{10^2 - 5^2}{10^2} = 3/4$. We tried the substitution $x = au$, $u = x/a$, $dx = a \, du$.

$$S = 4 \int_0^1 \sqrt{\frac{a^2 - k^2 a^2 u^2}{a^2 - a^2 u^2}} \, a \, du$$

Factor out a^2:

$$S = 4 \int_0^1 \sqrt{\frac{a^2(1 - k^2 u^2)}{a^2(1 - u^2)}} \, a \, du$$

$$S = 4a \int_0^1 \sqrt{\frac{1 - k^2 u^2}{1 - u^2}} \, du$$

"Try the substitution $u = \sin\theta$, $du = \cos\theta\, d\theta$," Trigonometeris suggested.

$$S = 4a \int_0^1 \sqrt{\frac{1-k^2u^2}{1-u^2}}\, du = 4a \int_{\arcsin 0}^{\arcsin 1} \sqrt{\frac{1-k^2\sin^2\theta}{1-\sin^2\theta}} \cos\theta\, d\theta$$

$$= 4a \int_0^{\pi/2} \sqrt{\frac{1-k^2\sin^2\theta}{\cos^2\theta}} \cos\theta\, d\theta$$

$$= 4a \int_0^{\pi/2} \frac{\sqrt{1-k^2\sin^2\theta}}{\cos\theta} \cos\theta\, d\theta$$

$$= 4a \int_0^{\pi/2} \sqrt{1-k^2\sin^2\theta}\, d\theta$$

"We'll call this an *elliptic integral*, since it determines the circumference of an ellipse," the professor said.

"As if giving it a name will help us solve it," Recordis snorted.

We worked for hours trying to solve this integral. None of our usual methods worked. Our panic gradually increased as we wondered what the drastic consequences would be if the gremlin performed his diabolical secret plan before we had built a safe fenced-in enclosure for the children.

Soon we had to set aside that integral when another urgent problem arose. "We need to decide what size chairs to have at the party," Builder declared. "We need to know the heights of the children that will be there."

"We'll assume that the children at the party will be a representative sample of the children in the kingdom," the king decided. "We'll visit the Royal Census office and find that information."

We put a sign saying, "We'll be back at 2 P.M." on the Main Conference Room door and walked across town to the census office. Unfortunately, that office had a sign on the door saying, "We'll be back at 2 P.M."

"Now what!" Recordis said. "We said we'd be back at the castle at 2 P.M.!"

With growing frustration we walked back to the castle. To our surprise, we met the Royal Census Taker walking back to her office, grumbling that the people she had been trying to meet had been away.

"Just the person we're looking for!" the king said.

"Just the people I'm looking for!" the census taker said upon seeing us.

"We need you to tell us what proportion of the children in the kingdom have heights in a particular range," the king said.

"I know the general idea," the census taker said. "Suppose the population height has an average, or mean value, of μ (mu), and suppose it has a standard deviation of σ (sigma). (The standard deviation is a measure of how far spread out the values in the population are.) Then the fraction of people with heights between two values a and b will be the area under this curve:

$$f(x) = \frac{1}{\sqrt{2\pi}} e^{-x^2/2}$$

between the values $(a - \mu)/\sigma$ and $(b - \mu)/\sigma$."

"What does that curve look like?" Recordis asked. The census taker made a quick sketch (Figure 17–2).

"It looks like a bell!" Recordis exclaimed. "We've never seen a curve like that before."

"The bell-shaped curve is called the *normal distribution*," the census taker explained. "I was coming to ask you to find the area under the curve for me; I've heard that this is the kind of problem you can solve. For example, if the mean height is $\mu = 48$ and the standard deviation is $\sigma = 4$, then the fraction of children with heights between 48 and 52 will be the area under the curve between $(48 - 48)/4 = 0$ and $(52 - 48)/4 = 1$.

"We can set up the area as an integral," the professor demonstrated.

$$\text{area} = \int_0^1 \frac{1}{\sqrt{2\pi}} e^{-x^2/2} \, dx$$

As we worked on that integral, though, the grim reality set in: This integral also defied our usual means of solution.

"What are the odds of getting two integrals that we can't solve in one day!" Recordis moaned.

"If we can't find an exact answer, we'll have to find an approximate answer," the king said finally as we became increasingly desperate.

"Is there any shape that we could use as an approximation to the area under a curve like that?" the professor asked.

Recordis shuffled through his notes. "When we first tried to find the area under a curve, we set up the area as a series of n rectangular strips:

$$\int_a^b f(x) \, dx \approx \sum_{i=1}^{n} f(x_i) \, \Delta x$$

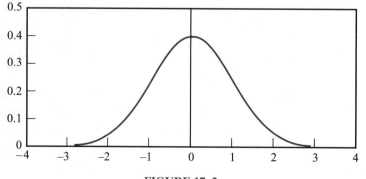

FIGURE 17–2.

"where $\Delta x = (b - a)/n$, $x_1 = a$, $x_2 = a + \Delta x$, $x_3 = a + 2\,\Delta x$, etc. To get the exact value for the integral, we had to take the limit as Δx goes to zero and n goes to infinity. However, you may recall that this method didn't work for finding the area."

"It will work for finding an approximation to the area," the king declared. "Even if we can't actually use an infinite number of rectangular strips, if n is large enough we still should get a reasonable approximation. This method won't be as good as finding a formula for the integral, but we can use it in case it is impossible to find a formula."

Igor illustrated how a group of five rectangles could approximate the area under the curve (Figure 17–3).

"There is a lot of area under the curve not included in the rectangles," the professor worried. "Notice that the right side of each rectangle has the same height as the curve at that point. Maybe we would be better off by making the *left* side of each rectangle have the same height as the curve." (See Figure 17–4.)

"But now there is a lot of area included in the rectangles that is not under the curve!" Recordis protested.

"I think the best approach will be to make the height of each rectangle be the same as the height of the curve at the midpoint of the rectangle," the king suggested. See Figure 17–5.

"For example, for the rectangle between $x = 0.6$ and $x = 0.8$, we will make the height of the rectangle equal to $f(0.7)$, since $0.7 = (0.6 + 0.8)/2$." For each rectangle there is a little roughly triangular-shaped piece on the left that is below the curve but not included in the rectangle, but then there is another triangular-shaped piece on the right that is part of the rectangle but not below the curve. If the curve happened to be a straight line, then both of these pieces would be triangles with equal areas, and the area of one triangle would cancel out the area of the other triangle and therefore the area under the curve would be the same as the area of

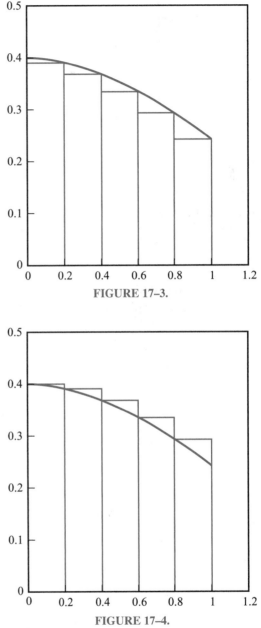

FIGURE 17–3.

FIGURE 17–4.

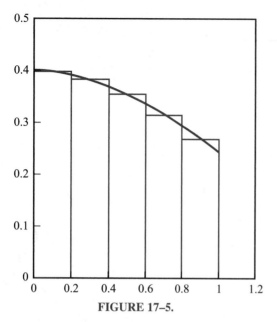

FIGURE 17–5.

the rectangle. Since we are dealing with a curve, rather than a straight line, the little pieces will not exactly be triangles, but we can still hope that the areas of the two pseudo-triangles will be roughly the same, and the area of the rectangle will be a good approximation for the area of the curve.

"How many rectangles will we have to use?" Recordis gulped, thinking of all the work involved.

"I can help with the calculations," Builder said. "I've been working with Igor to reprogram his screen to show what I call a *spreadsheet*. It's a table of numbers that you can do calculations with."

He demonstrated. The spreadsheet screen showed a table of numbers, with columns labeled A, B, C, etc., and rows labeled 1, 2, 3, etc. In column A, he put the x values $0.1, 0.3, 0.5, 0.7,$ and 0.9 (these were the x values at the midpoints of the rectangles in Figure 17–5).

	A
1	x
2	0.1000
3	0.3000
4	0.5000
5	0.7000
6	0.9000

In cell B2 he entered the formula for $f(x)$, the height of the curve. Builder translated the mathematical formula

$$f(x) = \frac{1}{\sqrt{2\pi}}e^{-x^2/2}$$

into this form for use in the spreadsheet:

```
=EXP(-A2^2/2)/SQRT(2*3.14159)
```

"I use A2 instead of x because the A2 is the cell address of the cell containing the x value," Builder explained.

"We need to have that formula in the other cells in column B," the king reminded him. Builder showed how he could copy the formula from cell B2 to cells B3 to B11.

The result was:

	A	B
1	x	$f(x)$
2	0.1000	0.3970
3	0.3000	0.3814
4	0.5000	0.3521
5	0.7000	0.3123
6	0.9000	0.2661

Next, we were going to create a column for the area of each rectangle, but then Recordis realized one trick that would save some calculations. Since the area (A) comes from the formula:

$$A = f(x_1)\,\Delta x + f(x_2)\,\Delta x + f(x_3)\,\Delta x + \cdots + f(x_n)\,\Delta x$$

we could factor out the Δx:

$$A = [f(x_1) + f(x_2) + f(x_3) + \cdots + f(x_n)]\,\Delta x$$

This way, we could add up all of the f values, and then multiply by Δx at the end (saving us from multiplying by Δx for each rectangle).

In cell B7, Builder entered the formula for the sum of that column:

	A	B
1	x	$f(x)$
2	0.1000	0.3970
3	0.3000	0.3814
4	0.5000	0.3521
5	0.7000	0.3123
6	0.9000	0.2661
7	Sum =	1.7087

Then we multiplied the sum by Δx:

area of rectangles = approximate area under curve = $1.7087 \times 0.2 = 0.3417$

"We should try it again with more rectangles," the professor insisted. Fortunately, Igor's spreadsheet did the calculations for us. We tried 20 rectangles, each with width $(b - a)/20 = (1 - 0)/20 = 0.05$:

	A	B
1	*x*	*f*(*x*)
2	0.0250	0.3988
3	0.0750	0.3978
4	0.1250	0.3958
5	0.1750	0.3929
6	0.2250	0.3890
7	0.2750	0.3841
8	0.3250	0.3784
9	0.3750	0.3719
10	0.4250	0.3645
11	0.4750	0.3564
12	0.5250	0.3476
13	0.5750	0.3382
14	0.6250	0.3282
15	0.6750	0.3177
16	0.7250	0.3067
17	0.7750	0.2955
18	0.8250	0.2839
19	0.8750	0.2721
20	0.9250	0.2601
21	0.9750	0.2480
22	Sum =	6.8274

$$\text{area} = 6.8274\,\Delta x = 6.8274 \times 0.05 = 0.3414$$

"I could sleep easier if we tried it again with more rectangles," the professor said, "since we know the approximation will be more accurate if we include more rectangles."

For 100 rectangles, we found the area was 0.341346, and for 1,000 rectangles we found 0.341345. The professor was satisified that the best four-decimal-place approximation of the true area under the curve was 0.3413.

We called this the method of *numerical integration.*

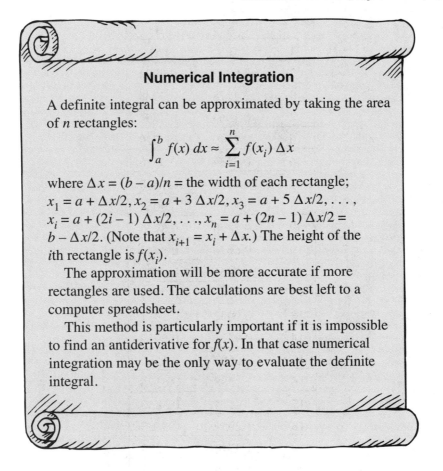

Numerical Integration

A definite integral can be approximated by taking the area of n rectangles:

$$\int_a^b f(x)\,dx \approx \sum_{i=1}^{n} f(x_i)\,\Delta x$$

where $\Delta x = (b - a)/n$ = the width of each rectangle; $x_1 = a + \Delta x/2$, $x_2 = a + 3\,\Delta x/2$, $x_3 = a + 5\,\Delta x/2$, . . . , $x_i = a + (2i - 1)\,\Delta x/2$, . . . , $x_n = a + (2n - 1)\,\Delta x/2 = b - \Delta x/2$. (Note that $x_{i+1} = x_i + \Delta x$.) The height of the ith rectangle is $f(x_i)$.

The approximation will be more accurate if more rectangles are used. The calculations are best left to a computer spreadsheet.

This method is particularly important if it is impossible to find an antiderivative for $f(x)$. In that case numerical integration may be the only way to evaluate the definite integral.

Now we could use numerical integration to calculate the circumference S for the elliptical rose garden:

$$S = 4a \int_0^{\pi/2} \sqrt{1 - k^2 \sin^2 \theta}\ d\theta$$

(with k = eccentricity = $\sqrt{\frac{3}{4}} = 0.8660$). We used twenty rectangles of width $\Delta x = (\pi/2 - 0)/20 = 0.07854$. The midpoint x value for the first rectangle was $0.07854/2 = 0.03927$. We created the table:

	A	B
1	*x*	*f(x)*
2	0.0393	0.9994
3	0.1178	0.9948
4	0.1963	0.9856
5	0.2749	0.9720
6	0.3534	0.9540
7	0.4320	0.9320
8	0.5105	0.9061
9	0.5890	0.8766
10	0.6676	0.8441
11	0.7461	0.8090
12	0.8247	0.7717
13	0.9032	0.7331
14	0.9817	0.6939
15	1.0603	0.6550
16	1.1388	0.6176
17	1.2174	0.5830
18	1.2959	0.5525
19	1.3744	0.5278
20	1.4530	0.5103
21	1.5315	0.5012
22	Sum =	15.4196

The value of the integral was $15.4196 \times \Delta x = 1.2111$. To find the complete circumference, we had to multiply by the semimajor axis $a = 10$, and then multiply by 4 since we had found the length of arc for only one-quarter of the complete circumference. The result was:

circumference of ellipse with $a = 10$ and eccentricity= $\sqrt{3}/2$:

$$S = 4 \times 10 \times 15.4196\,\Delta x = 4 \times 10 \times 15.4196 \times 0.0785 = 48.4422$$

"That seems about right," the professor said. "If $a = 1$, instead of 10, the circumference would be 4.8442. We know the circumference of an ellipse with $a = 1$ must be between 4 and 2π. If the eccentricity approached 1, then the circumference would approach 4, since the ellipse would degenerate into two line segments each of length 2. On the other hand, if the eccentricity approached zero, then the circumference would approach 2π (about 6.28), since then we would have a circle with radius 1."

While we were working on extending the numerical integration spreadsheet so it would include more rectangles, Builder had said, "It sure would help a lot if you could give me a formula for e^x."

"I do have an approximation method I use sometimes," Recordis said sheepishly. "Sometimes I approximate a curve by pretending it really is a straight line."

Suppose $f(x)$ represents some curve, and suppose I happen to know the value of $f(x_0)$, but I need to know the value of $f(x_0 + h)$. I draw a line through the point $(x_0, f(x_0))$, which has the same slope as the curve at that point. The slope, of course, is given by the derivative:

$$\text{slope} = f'(x_0)$$

"Then, since

$$f'(x_0) \approx \frac{f(x_0 + h) - f(x_0)}{h}$$

"I can write this."

$$f(x_0 + h) \approx f(x_0) + f'(x_0)h$$

See Figure 17–6.

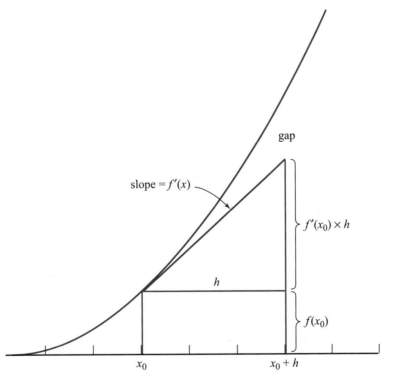

FIGURE 17–6.

"I don't think that will give a very good approximation," the professor said scornfully. "Look at that giant gap between the curve and the tangent line where x is equal to $x_0 + h$."

"It will be an OK approximation if h is very very small!" Recordis cried defensively.

"So all you can do will be to find the value of f at points that are very very close to points where you already know the value of the function. What good will that do when we need to find the value of f at other points?" the professor responded.

"I see some potential in this method," the king said. "Except, instead, of approximating the curve by a straight line, approximate it by a curve, like a parabola."

Suppose we have this parabola:

$$f(x) = a_0 + a_1 x + a_2 x^2$$

"If you're going to approximate an arbitrary curve f like this (when f isn't really a parabola), you're going to need a method to determine the value of a_0, a_1, and a_2," the professor pointed out.

"Good point."

We worked out $f(x_0 + h)$ for the parabola:

$$f(x_0 + h) = a_0 + a_1(x_0 + h) + a_2(x_0 + h)^2$$

$$f(x_0 + h) = a_0 + a_1 x_0 + a_1 h + a_2 x_0^2 + 2a_2 x_0 h + a_2 h^2$$

Recordis thought he saw a pattern. We reordered the terms, and then the pattern was apparent:

$$f(x_0 + h) = (a_0 + a_1 x_0 + a_2 x_0^2) + a_1 h + 2a_2 x_0 h + a_2 h^2$$

"Since $a_0 + a_1 x_0 + a_2 x_0^2$ equals $f(x_0)$, we have:"

$$f(x_0 + h) = f(x_0) + a_1 h + 2a_2 x_0 h + a_2 h^2$$

"I see something else. Write it as:"

$$f(x_0 + h) = f(x_0) + (a_1 + 2a_2 x_0)h + a_2 h^2$$

"Then, since $f(x) = a_0 + a_1 x + a_2 x^2$, we have the derivative $f'(x_0) = a_1 + 2a_2 x_0$. Therefore, we can write:"

$$f(x_0 + h) = f(x_0) + f'(x_0)h + a_2 h^2$$

"If we left out the last term, the result would be the same as my good old approximation method," Recordis said.

All of a sudden we noticed that the second derivative of $f(x)$ was $2a_2$ (for any value of x), so we could rewrite our formula as:

$$f(x_0 + h) = f(x_0) + f'(x_0)h + \frac{f''(x_0)}{2}h^2$$

"It looks like we can approximate a curve in terms of its derivatives," the king said.

"This only works for a second-degree polynomial," Recordis said. "I wonder if something like this works for a third-degree polynomial."

We tried an example:

$$f(x) = a_0 + a_1 x + a_2 x^2 + a_3 x^3$$

Then:

$$f(x_0 + h) = a_0 + (a_1 x_0 + a_1 h) + (a_2 x_0^2 + 2a_2 x_0 h + a_2 h^2) + (a_3 x_0^3 + 3a_3 h x_0^2 + 3a_3 x_0 h^2 + a_3 h^3)$$

Collecting the terms with like powers of h:

$$f(x_0 + h) = (a_0 + a_1 x_0 + a_2 x_0^2 + a_3 x_0^3) + (a_1 h + 2a_2 x_0 h + 3a_3 x_0^2 h) + (a_2 h^2 + 3a_3 x_0 h^2) + a_3 h^3$$

We recognized that the term in the first set of parentheses was $f(x_0)$:

$$f(x_0 + h) = f(x_0) + h[a_1 + 2a_2 x_0 + 3a_3 x_0^2] + \frac{h^2}{2}[2a_2 + 6a_3 x_0] + \frac{h^3}{6}[6a_3]$$

The first three derivatives of f are:

$$f'(x) = a_1 + 2a_2 x + 3a_3 x^2$$

$$f''(x) = 2a_2 + 6a_3 x$$

$$f'''(x) = 6a_3$$

Therefore:

$$f(x_0 + h) = f(x_0) + hf'(x_0) + \frac{h^2 f''(x_0)}{2!} + \frac{h^3 f'''(x_0)}{3!}$$

The exclamation mark symbolizes the *factorial* function, defined as:

$$0! = 1$$
$$1! = 1$$
$$2! = 2 \times 1 = 2$$
$$3! = 3 \times 2 \times 1 = 6$$
$$4! = 4 \times 3 \times 2 \times 1 = 24$$
$$5! = 5 \times 4 \times 3 \times 2 \times 1 = 120$$

and so on.

We guessed that we would get a more accurate approximation if we took more and more derivatives, following the same pattern:

$$f(x+h) = f(x) + hf'(x) + \frac{h^2 f''(x)}{2!} + \frac{h^3 f'''(x)}{3!} + \frac{h^4 f''''(x)}{4!} + \cdots + \frac{h^i f^{[i]}(x)}{i!} + \cdots$$

where $f'''(x)$ is the third derivative of $f(x)$, $f''''(x)$ is the fourth derivative, $f^{[i]}$ is the ith derivative, and so on.

"Can we use this method to find e^h?" Builder asked.

Using the formula:

$$e^{x0+h} = e^{x0} + hf'(x_0) + \frac{h^2 f''(x_0)}{2!} + \frac{h^3 f'''(x_0)}{3!} + \frac{h^4 f''''(x_0)}{4!} + \cdots$$

"Now what was the derivative of e^x?" Recordis started shuffling through a giant stack of papers.

"That's the indestructable function whose derivative is equal to itself!" the professor reminded him.

$$f(x) = e^x$$
$$f'(x) = e^x$$
$$f''(x) = e^x$$
$$f'''(x) = e^x$$
$$f''''(x) = e^x$$

and so on.

"That will make things simple," Recordis agreed. "Now we need to choose our value for x_0. It has to be an easy value so that we know the value of e^{x0}. Hmmm . . . the easiest value would be $x_0 = 0$, because $e^0 = 1$." We tried that in the formula:

$$e^{x0+h} = e^{x0} + hf'(x_0) + \frac{h^2 f''(x_0)}{2!} + \frac{h^3 f'''(x_0)}{3!} + \frac{h^4 f''''(x_0)}{4!} + \cdots$$

$$e^{0+h} = e^0 + he^0 + \frac{h^2 e^0}{2!} + \frac{h^3 e^0}{3!} + \frac{h^4 e^0}{4!} + \cdots$$

This simplified dramatically, since $e^0 = 1$:

$$e^h = 1 + h + \frac{h^2}{2!} + \frac{h^3}{3!} + \frac{h^4}{4!} + \cdots$$

We decided it would be easier to change the h to x and write the formula as:

$$e^x = 1 + x + \frac{x^2}{2!} + \frac{x^3}{3!} + \frac{x^4}{4!} + \cdots$$

For an example, try $x = 1$, so

$$e^1 = 1 + 1 + \frac{1^2}{2!} + \frac{1^3}{3!} + \frac{1^4}{4!} + \cdots$$

$$e = 2 + \frac{1}{2!} + \frac{1}{3!} + \frac{1}{4!} + \cdots$$

This provides another method to track down the value of e. We worked out the first few terms in the series:

$$e \approx 2 + \frac{1}{2!} + \frac{1}{3!} + \frac{1}{4!} + \frac{1}{5!} + \frac{1}{6!} + \frac{1}{7!}$$

$$e \approx 2 + \frac{1}{2} + \frac{1}{6} + \frac{1}{24} + \frac{1}{120} + \frac{1}{720} + \frac{1}{5,040}$$

$$e \approx 2 + 0.5 + 0.16667 + 0.04167 + 0.00833 + 0.00139 + 0.00020 = 2.71825$$

"That's quite suitable as an approximate value of e," the king declared. "We could get a more accurate value by taking more terms in the series, if we needed to." (The actual five-decimal-place approximation of e is 2.71828.)

(In our land this type of series is known as a *Taylor series*.) The professor did some more investigation and prepared a summary.

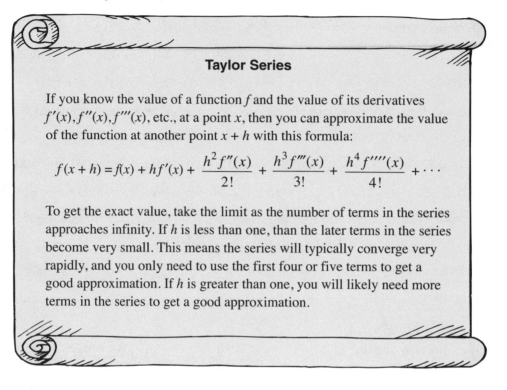

Taylor Series

If you know the value of a function f and the value of its derivatives $f'(x), f''(x), f'''(x)$, etc., at a point x, then you can approximate the value of the function at another point $x + h$ with this formula:

$$f(x + h) = f(x) + hf'(x) + \frac{h^2 f''(x)}{2!} + \frac{h^3 f'''(x)}{3!} + \frac{h^4 f''''(x)}{4!} + \cdots$$

To get the exact value, take the limit as the number of terms in the series approaches infinity. If h is less than one, than the later terms in the series become very small. This means the series will typically converge very rapidly, and you only need to use the first four or five terms to get a good approximation. If h is greater than one, you will likely need more terms in the series to get a good approximation.

Trigonometeris was excited by the method because he realized it provided a way to approximate the values of the trigonometric functions. Since $\sin \theta = \sin (0 + \theta)$, we can form the Taylor series:

$$\sin \theta = \sin 0 + \theta \cos 0 - \frac{\theta^2 \sin 0}{2!} - \frac{\theta^3 \cos 0}{3!} + \frac{\theta^4 \sin 0}{4!} + \cdots$$

(using the fact that $d \sin \theta / d\theta = \cos \theta$, and $d \cos \theta / d\theta = -\sin \theta$). ($\theta$ is in radians.)

Since $\sin 0 = 0$ and $\cos 0 = 1$, we have

$$\sin \theta = \theta - \frac{\theta^3}{3!} + \frac{\theta^5}{5!} - \frac{\theta^7}{7!} + \frac{\theta^9}{9!} - \cdots$$

For the cosine function we found:

$$\cos \theta = 1 - \frac{\theta^2}{2!} + \frac{\theta^4}{4!} - \frac{\theta^6}{6!} + \cdots$$

"It will also help if we could find numeric solutions for differential equations," the king declared. "There is no guarantee we will always be able to find a formula for the solution, so we need some other way to find the solution." He suggested we try some examples, starting with $dy/dx = -x/y$.

"We could easily plug in particular values of x and y and then determine dy/dx at that point," Recordis suggested. "Since the derivative gives the slope of the curve at that point, we would at least know what direction the curve is going at that point."

"If we knew the direction the curve was going at a whole bunch of points, it would give us a good feel for where the curve was going," the professor realized.

"Except for one problem," the king reminded us. "There will be a whole bunch of curves that satisfy the differential equation. We would need an initial condition to solve for the arbitrary constant and thereby determine which curve actually applies to our situation."

"Even so, it would be useful to see how dy/dx varies at a bunch of different points," the professor said. "In fact, I have a brilliant idea about how to do this. At each point (x,y) we will draw a little line segment with the same slope as dy/dx at that point."

Igor illustrated this concept for $dy/dx = -x/y$. For example, at $x = 3$, $y = 3$, we drew a line segment with slope -1. At $x = -2$, $y = 1$, we drew a line segment with slope 2. See Figure 17–7.

"I'm getting dizzy watching the way those line segments whirl around in circles like that," Recordis complained.

"I think you're right—the diagram illustrates that the solution to our differential equation $dy/dx = -x/y$ will be a circle," the king realized.

We called this type of diagram a *slope field*, since it was like we were looking down on a big field with a bunch of markers representing different slopes. (See the web page for this book for a spreadsheet that will generate slope fields for you.)

We created some other examples. For $dy/dx = xy$, we could see the solution would be some kind of a U-shaped curve (Figure 17–8).

For $dy/dx = -xy$, Recordis said, "It must be some kind of upside-down U-shaped curve; perhaps our new friend: the bell curve." (See Figure 17–9.)

(See Exercises 13 to 15 for more about these solutions.)

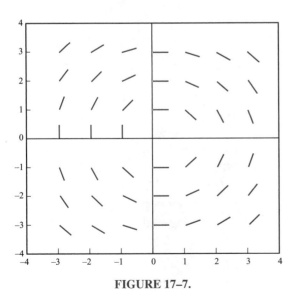

FIGURE 17–7.

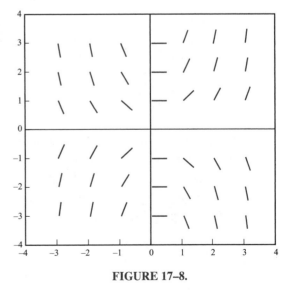

FIGURE 17–8.

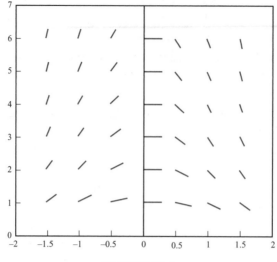

FIGURE 17–9.

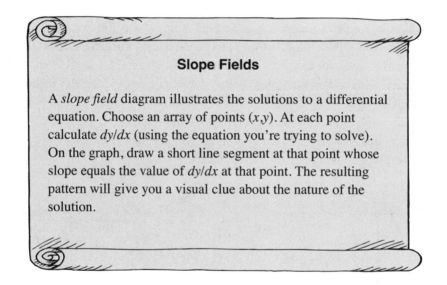

Slope Fields

A *slope field* diagram illustrates the solutions to a differential equation. Choose an array of points (x,y). At each point calculate dy/dx (using the equation you're trying to solve). On the graph, draw a short line segment at that point whose slope equals the value of dy/dx at that point. The resulting pattern will give you a visual clue about the nature of the solution.

The Royal Astronomer came to us to show his plans for a space probe. "Pal will throw the probe straight up with great velocity. As we know, gravity will cause the probe to slow down, until it eventually comes to a stop and then will fall back to Earth. Could you tell me what its height will be at any particular time? (Also note that this probe is highly unusual in that it will not be affected by the resistance of the atmosphere.)"

"It's a good thing we have our results from yesterday," the king said.

$$\text{total energy} = \text{potential energy} + \text{kinetic energy}$$

$$\text{total energy} = -\frac{GMm}{r} + \frac{mv^2}{2}$$

where G is the gravity constant; M is the mass of Earth; m is the mass of the probe; r is the distance from the probe to the center of Earth; and v is the velocity of the probe.

"Initially, the distance from the probe to the center of Earth will be r_0, the radius of Earth," the astronomer said. "And the initial velocity v_0 depends on how hard Pal throws it.

$$\text{initial energy} = -\frac{GMm}{r_0} + \frac{mv_0^2}{2}$$

"The total energy will always equal the initial energy. After Pal lets the probe go, it has no more work done on it, so it does not acquire any more energy."

$$-\frac{GMm}{r} + \frac{mv^2}{2} = -\frac{GMm}{r_0} + \frac{mv_0^2}{2}$$

"We know that $v = dr/dt$, since velocity is the derivative of position with respect to time," Recordis contributed. We rearranged the equation:

$$\frac{mv^2}{2} = \frac{GMm}{r} - \frac{GMm}{r_0} + \frac{mv_0^2}{2}$$

Multiply both sides by $2/m$:

$$v^2 = \frac{2GM}{r} - \frac{2GM}{r_0} + v_0^2$$

$$v = \frac{dr}{dt} = \sqrt{\frac{2GM}{r} - \frac{2GM}{r_0} + v_0^2} = \sqrt{\frac{2GM}{r} + E_0}$$

where E_0 is the initial energy.

Recordis moaned, "My worst nightmare! A differential equation with a square root sign."

"We do know the initial condition: $r = r_0$ when $t = 0$. We'll work on a numerical method for finding the other points," the king declared.

"How?" Recordis asked.

"Recordis already solved this problem," the professor said.

"I did?" Recordis asked. "I mean, of course I did." He waited for the professor to remind him how he had solved it.

"We use the first two terms of the Taylor series approximation," the professor said.

$$f(x_0 + h) \approx f(x_0) + f'(x_0)h$$

"As long as we know the value of the function and its derivative at one point, we can approximate the value of the function at another nearby point. We just have to make sure that h is small so the approximation will be reasonable. Then, once we know the value at the new point, we can find the value at another nearby point. Eventually we'll know the whole curve. This method will require a lot of calculations, but alas I fear that will always be our fate when we try numerical methods."

The astronomer gave us numerical values (for the convenience of the reader, these have been translated into metric units): gravity constant = $G =$ $6.67 \times 10^{-11} \frac{\text{meters}^3}{\text{kilograms} \times \text{seconds}^2}$; mass of Earth = $M = 5.97 \times 10^{24}$ kilograms; radius of Earth = $r_0 = 6,378,000$ meters; initial velocity of projectile = $v_0 = 8,000$ meters per second.

We used a time step of 10 seconds. Since the initial velocity is 8,000, the probe will move $8,000 \times 10 = 80,000$ meters in that time interval, giving a new value of $r = 6,378,000 + 80,000 = 6,458,000$. Using this new value of r, we had to calculate the new value of velocity:

$$v = \frac{dr}{dt} = \sqrt{\frac{2GM}{r} - \frac{2GM}{r_0} + v_0^2} = 7,902.7$$

With the new velocity value we calculated the new r value, and then we kept repeating the procedure. Here is a table of the first few results:

t	r	$v = dr/dt$
0.00	6,378,000	8,000.0000
10.00	6,458,000	7,902.6674
20.00	6,537,027	7,807.7097
30.00	6,615,104	7,715.0140
40.00	6,692,254	7,624.4745
50.00	6,768,499	7,535.9928

The final table was much larger. We made a graph showing the probe go up and then go down. (See Figure 17–10.)

In our land this method is known as *Euler's method*.

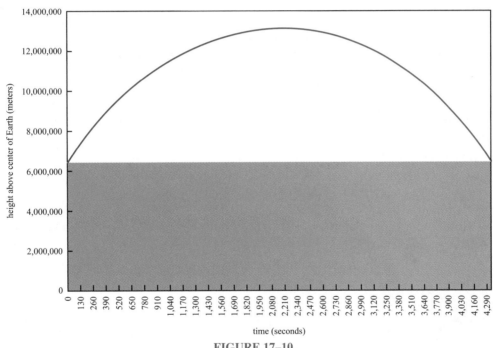

FIGURE 17–10.

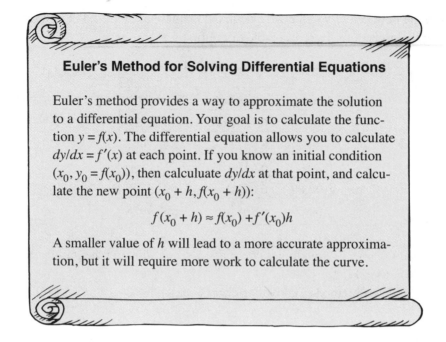

Euler's Method for Solving Differential Equations

Euler's method provides a way to approximate the solution to a differential equation. Your goal is to calculate the function $y = f(x)$. The differential equation allows you to calculate $dy/dx = f'(x)$ at each point. If you know an initial condition $(x_0, y_0 = f(x_0))$, then calculate dy/dx at that point, and calculate the new point $(x_0 + h, f(x_0 + h))$:

$$f(x_0 + h) \approx f(x_0) + f'(x_0)h$$

A smaller value of h will lead to a more accurate approximation, but it will require more work to calculate the curve.

The next morning we were again woken by a terrible scream, but this time it came from the professor. "I've been having a terrible nightmare. How do we know that the values of limits really have the values that we have calculated?"

She explained her concern. "Suppose we have been trying to find the limit of a function $f(x)$ as x approaches a particular value a. What we've done is let x approach very close to a, and then observe that $f(x)$ appears to converge to a particular value, call it b:

$$\lim_{x \to a} f(x) = b \quad \text{(allegedly)}$$

"How do we know that the value of f doesn't veer off suddenly, and not be equal to b, as x approaches a?"

We decided that we would informally describe limits with this statement: "If x is close to a, then $f(x)$ is close to b." This statement means: "The limit of $f(x)$, as x approaches a, is equal to b."

"How close must it be for us to say it is close?" the king wondered. We tried a more formal statement:

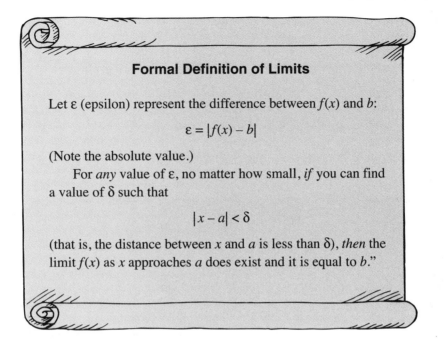

Formal Definition of Limits

Let ε (epsilon) represent the difference between $f(x)$ and b:

$$\varepsilon = |f(x) - b|$$

(Note the absolute value.)

For *any* value of ε, no matter how small, *if* you can find a value of δ such that

$$|x - a| < \delta$$

(that is, the distance between x and a is less than δ), *then* the limit $f(x)$ as x approaches a does exist and it is equal to b."

"I think an example would help," Recordis suggested confidently, trying to conceal the fact he was rather mystified by the whole discussion.

The professor helped us work out the following: If $f(a)$ is a continuous curve, and $f(a)$ exists and is equal to b, then the limit is trivial. For example, if $f(x) = 8 + 5x$, then $f(2) = 18$, and also $\lim_{x \to 2} f(x) = 18$. Then, if $|f(x) - 18| < \varepsilon$, we can find that

$$-\varepsilon < f(x) - 18 < \varepsilon$$

and therefore $f(x)$ must be between $18 - \varepsilon$ and $18 + \varepsilon$. If $f(x) = 18 - \varepsilon$, then we can find x:

$$f(x) = 8 + 5x$$

$$18 - \varepsilon = 8 + 5x$$

Subtract 8 from both sides:

$$10 - \varepsilon = 5x$$

Divide by 5:

$$2 - \frac{\varepsilon}{5} = x$$

By a similar method, we can find that if $f(x) = 18 + \varepsilon$, then

$$2 + \frac{\varepsilon}{5} = x$$

Therefore:

$$\text{if } 18 - \varepsilon < f(x) < 18 + \varepsilon, \text{ then } 2 - \frac{\varepsilon}{5} < x < 2 + \frac{\varepsilon}{5}$$

This can be rewritten as:

$$\text{if } |f(x) - 18| < \varepsilon, \text{ then } |x - 2| < \frac{\varepsilon}{5}$$

Therefore, if we let $\delta = \varepsilon/5$, we have satisfied the condition that δ exists, and we have $|x - 2| < \delta$.

Matters are more interesting if $f(x)$ doesn't exist. For example, if

$$f(x) = \frac{5x^2 - 2x - 16}{x - 2}$$

then $f(2)$ doesn't exist because it would lead to 0/0.

We used the spreadsheet to create a table of values:

x	$5x^2 - 2x - 16$	$x - 2$	$f(x)$
0	−16	−2	8
1	−13	−1	13
2	0	0	ERROR: DIVIDE BY 0
3	23	1	23
4	56	2	28
5	99	3	33
6	152	4	38
7	215	5	43
8	288	6	48

We drew the graph of the function, and Recordis said happily: "It's a line."

"No, it's a line with a hole in it, since $f(x)$ is undefined for $x = 2$," the professor said.

See Figure 17–11. The little circle in the graph represents the hole in the function.

"If it has a hole in it, then you can't draw it without lifting your pen off the paper," Recordis said in puzzlement.

We decided to say that a curve would be called *continuous* if we could draw it without lifting the pencil from the paper. That is, the curve must not have any holes in it. (The formal definition will be given later.) If the graph of $f(x)$ is continuous, and $f(a)$ exists and is equal to b, then it seemed clear that the limit $f(x)$ as x approaches a would be equal to b.

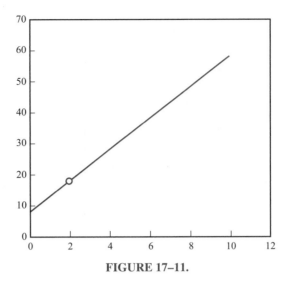

FIGURE 17–11.

However, if the curve is discontinuous at the point $x = a$, as in our example $f(x) = \frac{5x^2 - 2x - 16}{x - 2}$, then it is more complicated to determine the limit. We found we could factor the numerator:

$$f(x) = \frac{(5x + 8)(x - 2)}{x - 2}$$

As long as we don't let x equal 2, we can see that $f(x)$ will equal $5x + 8$. In this form, it is clear that as x approaches 2, then $f(x)$ approaches 18.

We developed a formal definition of continuous. The graph of $f(x)$ is a continous curve at a point $x = a$ if three conditions are met:

(1) $f(a)$ exists
(2) the limit of $f(x)$ as x approaches a exists
(3) the limit of $f(x)$ as x approches a is equal to $f(a)$.

Almost all of the curves we have worked with in this book have been continuous. Some examples of discontinuous curves are $f(x) = 1/x$ (discontinuous at $x = 0$), $f(x) = (8x + x^2)/x$ (discontinuous at $x = 0$ because $f(0)$ does not exist), $f(x) = (3x + 4)/(x - 5)$ (discontinuous at $x = 5$), and $f(x) = \tan x$ (discontinuous at $x = \pi/2$ and other values).

We continued work on the party emergency plan. "Once the children are safe inside the fenced elliptical rose garden, we can have Pal carry our valuables to safety."

"What if the gremlin chases after Pal?"

"We'll be safe," Recordis said confidently. "Let a represent Pal's speed, and b represent the gremlin's speed. Then, Pal's position will be at, and the gremlin's position will be bt, since they both start at position zero when t is zero. Since $a > b$, Pal will always be ahead." See Figure 17–12.

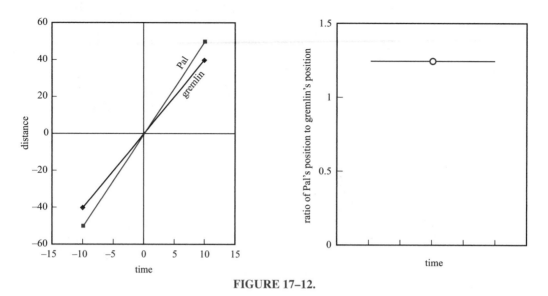

FIGURE 17–12.

"In fact, we can say confidently that the ratio of Pal's position to the gremlin's position is always $at/bt = a/b$."

"Correct, except for one minor complication: the ratio is undefined if $t = 0$." (See the right half of Figure 17–12.)

"The limit of at/bt at $t = 0$ is still a/b," Recordis insisted.

"You don't know that!" the professor said.

"You can't take the ratio at/bt when $t = 0$."

"I can take the ratio of their slopes," Recordis said. "The slope of Pal's position line is a, and the slope of the gremlin's position line is b, and their ratio is a/b."

"The ratio of the slopes of the lines is not the same as the ratio of the positions!" the professor exclaimed. "Unless maybe they are," she said with sudden thoughtfulness. "Perhaps you have discovered an important new property. If two lines ($y_1 = ax$) and ($y_2 = bx$) both cross at the origin, then the ratio of their y values ($y_1/y_2 = ax/bx = a/b$) is the same as the ratio of their slopes a/b. However, the ratio y_1/y_2 is undefined if $x = 0$, but we can still say that:

$$\lim_{x \to 0} \frac{ax}{bx} = \frac{a}{b}$$

"I wonder if this works for curves," Recordis wondered.

We investigated the case of where $y(x)$ was the ratio of two other functions of x:

$$y(x) = \frac{f(x)}{g(x)}$$

We were investigating a point $x = a$ where $f(a)$ and $g(a)$ were both zero, so $y(a)$ does not exist. However, we would like to see if the limit $f(a)/g(a)$ would be the ratio of their slopes, or derivatives, just as it worked for lines:

$$\lim_{x \to a} \frac{f(x)}{g(x)} = ? \lim_{x \to a} \frac{f'(x)}{g'(x)}$$

We knew that the ratio of the derivatives at $x = a$ would be approximately:

$$\frac{f'(a)}{g'(a)} \approx \frac{\dfrac{f(a + \Delta x) - f(a)}{\Delta x}}{\dfrac{g(a + \Delta x) - g(a)}{\Delta x}}$$

(In order to get the exact value we would have to take the limit Δx goes to zero.)

"We can cancel out Δx," Recordis noticed.

$$\frac{f'(a)}{g'(a)} \approx \frac{f(a + \Delta x) - f(a)}{g(a + \Delta x) - g(a)}$$

"We're investigating a point where $f(a)$ and $g(a)$ are both zero," the king reminded us. That simplified the formula:

$$\frac{f'(a)}{g'(a)} \approx \frac{f(a + \Delta x)}{g(a + \Delta x)}$$

We took the limit Δx goes to zero on both sides:

$$\lim_{\Delta x \to 0} \frac{f'(a)}{g'(a)} = \lim_{\Delta x \to 0} \frac{f(a + \Delta x)}{g(a + \Delta x)}$$

"It looks like the right-hand side becomes $f(a)/g(a)$ when you take the limit," the professor said.

$$\lim_{\Delta x \to 0} \frac{f'(a)}{g'(a)} = \lim_{\Delta x \to 0} \frac{f(a)}{g(a)}$$

We reversed the equation to put it in the order that we normally use it, and we gave a name for the rule (in our land it is known as L'Hôpital's rule):

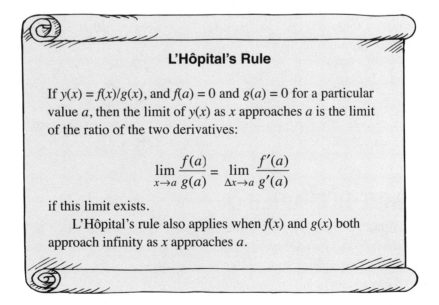

L'Hôpital's Rule

If $y(x) = f(x)/g(x)$, and $f(a) = 0$ and $g(a) = 0$ for a particular value a, then the limit of $y(x)$ as x approaches a is the limit of the ratio of the two derivatives:

$$\lim_{x \to a} \frac{f(a)}{g(a)} = \lim_{\Delta x \to a} \frac{f'(a)}{g'(a)}$$

if this limit exists.

L'Hôpital's rule also applies when $f(x)$ and $g(x)$ both approach infinity as x approaches a.

That night the Royal Banker asked us to help him calculate the monthly payment for a home loan, from the formula:

$$M = rP \frac{(1 + r)^n}{(1 + r)^n - 1}$$

where r is the monthly interest rate, n is the number of months until the loan is repaid, P is the principal amount of the loan (the amount borrowed), and M is the monthly payment. Everything went fine until the banker said, "In some special charity cases we will make loans where the interest rate is zero." We put $r = 0$ into the formula, and disaster struck:

$$C = 0 \times P \frac{(1 + 0)^n}{(1 + 0)^n - 1} = \frac{0}{0}$$

"Zero over zero is undefined!" the professor complained.

"Our new rule might help," Recordis suggested. We let $f(r) = rP(1 + r)^n$, and $g(r) = (1 + r)^n - 1$. Then,

$$C = \frac{f(r)}{g(r)}$$

Since $f(0) = 0$ and $g(0) = 0$, L'Hôpital's rule told us that

$$\lim_{r \to 0} rP \frac{(1 + r)^n}{(1 + r)^n - 1} = \lim_{x \to 0} \frac{f(r)}{g(r)} = \lim_{x \to 0} \frac{f'(r)}{g'(r)}$$

We found the derivatives:

$$f'(r) = rP[n(1 + r)^{n-1}] + (1 + r)^n P$$

$$g'(r) = n(1 + r)^{n-1}$$

In the limit r approaches zero, we could see $f'(r)$ approached P, and $g'(r)$ approached n. Therefore:

$$\lim_{r \to 0} rP \frac{(1 + r)^n}{(1 + r)^n - 1} = \frac{P}{n}$$

"That makes perfect sense," the banker said. "If the interest rate is zero, the monthly payment should simply be the total amount borrowed divided by the number of months."

NOTE TO CHAPTER 17

Although in theory your numerical integration calculations will be more accurate as you take more rectangles, in practice you will run into a problem if the rectangles are too narrow because there are limits to how accurately a computer can do arithmethic.

Exercises

For Exercises 1 to 9, use a computer spreadsheet to perform a numerical integration to find:

1. the area of a circle with radius 1

2. the area under one arch of the curve $y = \sin x$

3. the length of one arch of the curve $y = \sin x$

4. the length of the curve $y = e^x$ from $x = 1$ to $x = 10$

5. the circumference of an ellipse with semimajor axis 1 and eccentricity 0.5.

6. the area under the curve $y = \dfrac{1}{\sqrt{2\pi}} e^{-x^2/2}$ from -1.96 to 1.96.

7. the area under the curve $y = \dfrac{1}{\sqrt{2\pi}} e^{-x^2/2}$ from -10 to 10.

8. the surface area of a sphere with radius 1

9. the surface area of the ellipsoid formed by rotating the ellipse $x^2/25 + y^2/16 = 1$ about the x-axis

For Exercises 10 to 12, find slope fields for the solutions for the differential equations.

10. $\dfrac{dy}{dx} = y$

11. $\dfrac{dy}{dx} = \sqrt{1 - y^2}$

12. $\dfrac{dy}{dx} = x + y$

For Exercises 13 to 15, find solutions to the differential equations. Compare the solutions with the slope fields given in the chapter.

13. $\dfrac{dy}{dx} = -x/y$

14. $\dfrac{dy}{dx} = xy$

15. $\dfrac{dy}{dx} = -xy$

16. Find the Taylor series for: $y = \ln(1 + x)$

17. (a) Multiply $(1 - z)(1 + z + z^2 + z^3 + z^4 + \cdots + z^n)$
    (b) From the result of part (a), we can find:

$$1 + z + z^2 + z^3 + z^4 + \cdots + z^n = \frac{1 - z^{n+1}}{1 - z}$$

Let n approach infinity, assume $|z| < 1$, and let $z = -x^2$, and integrate both sides of the equation.

18. An infinite series may or may not add up to a finite sum. If the sum is finite, the series is said to converge to that value. One way to test whether or not a series converges is the *ratio test*. The series $a_0 + a_1 + a_2 + a_3 + \cdots$ converges if $\lim_{i \to \infty} |a_{i+1}/a_i| < 1$. Use the ratio test to see whether this series converge:

$$1 + x + \frac{x^2}{2!} + \frac{x^3}{3!} + \frac{x^4}{4!} + \cdots$$

For Exercises 19 to 32, use L'Hôpital's rule to evaluate these limits:

19. $\lim\limits_{x \to 0} \dfrac{\sin x}{x}$

20. $\lim\limits_{x \to 0} \ln(1 + x)^{1/x}$

21. $\lim\limits_{x \to \infty} \dfrac{10x + 4}{19x + 7}$

22. $\lim\limits_{x \to \infty} \dfrac{x^n}{e^x}$

23. $\lim\limits_{x \to 8} \dfrac{x^3 - 19x^2 + 118x - 240}{x^2 - 11x + 24}$

24. $\lim\limits_{x \to 7} \dfrac{x^3 - 19x^2 + 111x - 189}{x^2 - 8x + 7}$

25. $\lim\limits_{x \to 3} \dfrac{x^3 - 16x^2 + 79x - 120}{x^2 - 12x + 27}$

26. $\lim\limits_{x \to 2} \dfrac{x^3 - 12x^2 + 41x - 42}{x^2 - 6x + 8}$

27. $\lim\limits_{x \to -9} \dfrac{x^3 - 23x^2 + 48x + 2{,}160}{x^2 + 14x + 45}$

28. $\lim\limits_{x \to 9} \dfrac{x^3 + x^2 - 66x - 216}{x^2 - 6x - 27}$

29. $\lim\limits_{x \to -10} \dfrac{x^3 - 2x^2 - 184x - 640}{x^2 - 8x - 180}$

30. $\lim\limits_{x \to 24} \dfrac{x^3 - 26x^2 - 51x + 2{,}376}{x^2 - 18x - 144}$

31. $\lim\limits_{x \to 14} \dfrac{x^3 - 44x^2 + 321x + 1{,}386}{x^2 - 22x + 112}$

32. $\lim\limits_{x \to 5} \dfrac{x^5 - 51x^4 + 951x^3 - 8{,}201x^2 + 33{,}060x - 50{,}400}{x^4 - 20x^3 + 146x^2 - 460x + 525}$

Comprehensive Test
of Calculus Problems

Two days before the party, when everyone was in the Main Conference Room, there was a knock at the door, and we opened it to find, to our surprise, the gremlin. He was dressed in the same evil-looking cape that he always wore, but this time he had not come flying in the window and he had toned down the cackle in his laugh. He was carrying a large scroll, which he tossed on the table.

"Take that!" he commanded.

"Thank you," Recordis said. "We appreciate this very much. What is it?"

"Fool!" He slyly slithered over to the trembling professor. "So you are intending to write a book, are you? Before you can do that, you must solve these. These are all problems—calculus problems. If you can do them all, then you will have something to write about. But I think you will not be able to." He turned to the king. "There is no gimmick this time. I will not engulf the kingdom in fire or anything else if (I mean *when*) you fail. I shall simply proclaim to the people at the party that you, who have spent so much time on this calculus, cannot even solve calculus problems. The people of Carmorra themselves will turn against you, and I shall be proclaimed king!"

"We've defeated you every other time so far," Recordis said. "Your win-loss record is 0–17."

"Ah! But I realize where I have been going wrong," the gremlin said. "I have only fought you one at a time on theoretical grounds. Theory, unfortunately, is your strong point. Now you must apply everything you have done. That is where I think you will fail."

He reached inside his cloak and pulled out a giant conical-shaped hourglass. "The sand in this glass has been set to run for 48 hours. Once it is empty, I shall return and witness your defeat. I hope you appreciate the trouble I had rounding up these problems. It has been fun knowing you!" With a sudden burst of his horrible laugh he swept out the window.

"We surely got rid of him!" Recordis said.

"What do you mean, 'we got rid of him'?" the professor exclaimed. "We still have to do all these problems before he comes back!"

"Since you're writing the book, we'll let you do the problems for us," Recordis said, packing his notebooks and heading for the door.

"Stop it!" the king commanded. "We have no time for any of this arguing. We all helped get us into this, and we're all going to help get us out."

There was a long silence while we contemplated the scroll on the table. Pal began to cry again.

"We may as well get started," the professor said. Recordis began reading through the 45 problems that were our final test:

1. Find the volume of the top half of the hourglass that is even now measuring the time until your doom. (The hourglass is a cone: the radius of the base is $r = 15$, and the height is $h = 30$.)

2. The sand falls through the opening in the hourglass at the rate of u cubic units per unit time. What is the rate of change in the height of the sand in the cone when $x = 5$? ($u = 0.04$.) (See Figure 18–1.)

3. You have 48 hours to solve 45 problems. I predict that the rate at which you solve problems will be given by:

$$\frac{dn}{dt} = 181e^{-2t} - 1$$

where n is the number of problems you have solved and t is the time measured in hours ($t = 0$ at the time you start working on this test). If you follow my prediction, how many problems will you have finished by the time $t = 48$?

Find the derivative of the following. functions:

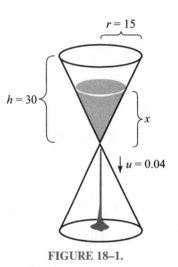

$r = 15$

$h = 30$

x

$u = 0.04$

FIGURE 18–1.

4. $f(x) = x^3 - 3x^2 + 4x + 10$

5. $f(x) = x^{96} + 10x^{43} - 3x^{20} + 11x$

6. $f(x) = x^2 + x + 1 + 1/x + 1/x^2 + 1/x^3$

7. $f(x) = \sqrt{4 - x^3}$

8. $f(x) = (10 + x^4)/(3 - x^2)$

9. $f(x) = [3 + (x + 4)^3]^2$

10. $f(x) = x^3 \sqrt{3 + x^2}$

11. $f(x) = \sqrt{(x + 5)(x - 3)/(x + 2)(x - 1)}$

12. $f(x) = \sin^2 x + \cos^2 x$

13. $f(x) = x^2 e^x \sin(x) \ln x$

14. $f(x) = e^{\ln x}$

15. When you are planning your escape from the kingdom, you will want to reach Nowhere Island as quickly as possible. Nowhere Island is $a = 10$ km directly offshore from River Mouth. When you arrive at Shore Rock, you will be $b = 12$ km away from River Mouth. Assume that you are carrying your boat with you. You can row $v = 3$ km/hr, and you can walk $u = 5$ km/hr. What course should you follow to arrive at Nowhere Island as quickly as possible? (See Figure 18–2.)

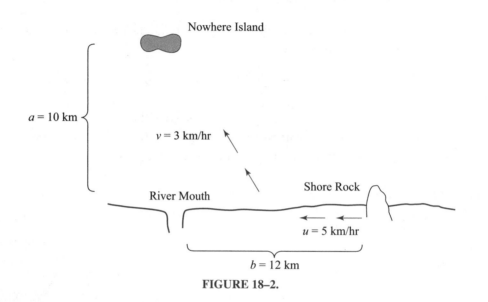

Nowhere Island

$a = 10$ km

$v = 3$ km/hr

River Mouth

Shore Rock

$u = 5$ km/hr

$b = 12$ km

FIGURE 18–2.

Find dy/dx for the following relations (r, a, and b are constants):

16. $x^2 + y^2 = r^2$

17. $x^2/a^2 + y^2/b^2 = 2$

18. $(x + y)^{1/2} = (x - y)^{-1/2}$

19. $(x - 3)y^2 = x^2 + 4$

20. If Rutherford starts running on a block of ice where his speed is given by $v = 3 - t^2$, how far will he have traveled from the time when $t = 0$ until he stops?

21. If Pal jumps on a muddy hillside where his speed changes at the rate $dv/dt = -2v$, with $v = 5$ when $t = 0$, how long will it take until his speed is equal to 0.01, and how far will he have traveled by the time this happens?

Find the values of the following integrals:

22. $\int (4x^3 - 2x + 5)\, dx$

23. $\int x \sqrt{x^2 - 1}\ dx$

24. $\int \dfrac{1}{1 + x^2}\, dx$

25. $\int 4x^2 \sin^2 x\, dx$

Find the equation for y, given the following conditions:

26. $dy/dx = 5x + 4;$ $\qquad\qquad\qquad$ $y = 3$ when $x = 2$

27. $d^2y/dt^2 = 5;$ $\qquad\qquad\qquad$ $dy/dt = 3$ and $y = 2$ when $t = 0$

28. $d^2y/dt^2 = 2t;$ $\qquad\qquad\qquad$ $dy/dt = 0$ and $y = 0$ when $t = 0$

29. Pal went for a ride on a spring where he was acted upon by the following forces: $F_{spring} = -500x$; $F_{friction} = -10\, dx/dt$. The mass of Pal is 1,000. At time $t = 0$, his speed was zero ($dx/dt = 0$) and his position was 20 ($x = 20$). Set up the differential equation that represents his motion, and solve it to find x as a function of t.

30. Suppose that the *work* required to move Irving Electron from point a to point b against a force F is given by:

$$W = \int_a^b F(x)\, dx$$

What is the work if $a = 100$, $b = 10$, and $F(x) = -kx^{-2}$?

Find the values of these definite integrals:

31. $\int_4^5 \dfrac{x-1}{(x-2)(x-3)}\, dx$

32. $\int_0^{10} 4x^2\, dx$

33. $\int_0^{\pi/6} \tan 2x\, dx$

34. $\int_0^{10} (e^x - x^3)\, dx$

35. $\int_1^2 x^4 \sin(x^5)\, dx$

36. $\int_0^{\pi/2} \sin x \cos x \, dx$

37. $\int_2^3 x^2 \sin x \, dx$

38. Find the area between the curves $y_1 = \sin x$ and $y_2 = x^2 - \pi x$.

39. When I take over the kingdom, I will want a new concert hall designed with a triangular stage. (See Figure 18–3.) Find the point where it will balance (the center of mass).

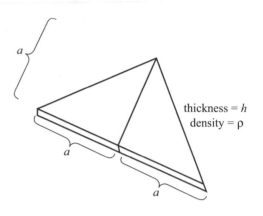

thickness = h
density = ρ

FIGURE 18–3.

40. If a football is formed by rotating the sine curve $y = \sin x$ from zero to π about the x axis, find the volume of the football.

41. Sketch a graph of the curve $y = e^{-x^2}$. Make a list of places with horizontal tangents and points of inflection.

42. Solve the differential equation $d^2y/dx^2 + 2\,dy/dx + 2y = 10$.

43. The two towers of the Swirling Pass Suspension Bridge are 100 units apart. The roadway is 40 units below the top of the towers. The main cable of the bridge hangs freely between the two towers, with its lowest point just touching the roadway. What is the length of the main cable? (Equation of main cable: $y = 18(e^{x/36} + e^{-x/36}) - 36$.)

44. Find the area between the curves $y_1 = x^2 - 5$ and $y_2 = x - 5$.

45. Find the derivative of the function $y = |x|$ (y is the absolute value of x, defined by $|x| = x$ for $x \geq 0$ and $|x| = -x$ for $x < 0$) at the point where $x = 0$.

"Some of these are easy, and some are hard," Recordis said. "Which should we do first?"

"We'll split up," the king told him. We broke into teams and set to work on the problems. The next two days passed in a blur. The only things we saw were problems. I made no effort to keep track of the steps we took chronologically, but I did make a record of the solutions.

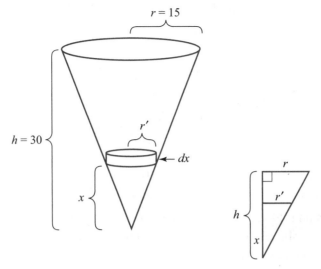

FIGURE 18–4.

1. We divided the hourglass into little cylindrical disks (Figure 18–4). The volume of each cylinder is given by $dV = \pi r'^2 \, dx$, and r' can be found using similar triangles: $r'/x = r/h$. So $r' = rx/h$. Then we integrated from $x = 0$ to $x = h$:

$$\int dv = V = \int_0^h \frac{\pi r^2 x^2 \, dx}{h^2} = \frac{\pi r^2}{h^2} \frac{1}{3} x^3 \Big|_0^h = \frac{\pi r^2 h}{3}$$

"That formula should be good for any cone," the professor noted. Recordis inserted the numbers: $V = 7{,}069$.

"That scoundrel!" Recordis cried. "He made the hourglass just small enough!"

2. The first thing the professor suggested was that we write down everything that we knew:

$$\frac{dv}{dx} = 0.04$$

(v is the volume of sand left in the top of the hourglass at time t).

$$\frac{r'}{r} = \frac{x}{h}$$

(from a similar triangle relationship).

Now we were stuck, until we realized that we needed the result from Problem 1 so that we knew the volume of a cone:

$$v = \frac{\pi r'^2 x}{3}$$

The professor noticed that we could substitute $r' = rx/h$:

$$v = \frac{\pi r^2 x^3}{3h^2}$$

Now we could determine dv/dx:

$$\frac{dv}{dx} = \frac{\pi r^2 x^2}{h^2}$$

The problem was to find dx/dt. We knew dv/dx and dv/dt. The king realized that we could use the chain rule:

$$\frac{dv/dt}{dv/dx} = \frac{dv}{dt}\frac{dx}{dv} = \frac{dx}{dt}$$

$$\frac{dx}{dt} = \frac{0.04h^2}{\pi r^2 x^2}$$

This expression told us the rate at which the height of the sand was changing for any value of the height x, but the gremlin asked us only for the answer when $x = 5$. We put these numbers into the equation:

$$\frac{dx}{dt} = 2 \times 10^{-3}$$

"That monster!" cried Recordis, who had figured out that the sand would indeed run dry after 48 hours.

3. "We'll integrate to find the answer," the king said.

$$\frac{dn}{dt} = 181e^{-2t} - 1$$

$$n = \int (181e^{-2t} - 1)\, dt$$

We used the sum rule and the multiplication rule:

$$n = 181 \int e^{-2t}\, dt - \int dt$$

$$= 181 \int e^{-2t}\, dt - t$$

Trigonometeris suggested that we try the substitution $u = -2t$; $dt = -\frac{1}{2}du$

$$n = 181 \int e^u \left(-\frac{1}{2}\right) du - t$$

We recognized e^u as the indestructible function whose derivative and integral both equal itself:

$$n = -90.5e^u - t + C$$

We made the reverse substitution, $u = -2t$:

$$n = -90.5e^{-2t} - t + C$$

Next, we solved for the arbitrary constant C by using the initial condition $n = 0$ when $t = 0$ (because, unfortunately, we had solved zero problems when the gremlin first gave us the test).

$$0 = -90.5e^0 - 0 + C$$

$$C = 90.5$$

The final equation for n became:

$$n = -90.5e^{-2t} - t + 90.5$$

We found the value of n when $t = 48$:

$$n = -90.5e^{-96} - 48 + 90.5$$

$$= (-90.5)(2 \times 10^{-42}) + 42.5$$

"If you put that in round numbers, 10^{-42} is practically zero," the professor noted. "That means $n \simeq 43$."

"That villain!" Recordis cried. "We need to do 45 problems! If he's right, we'll never get finished in time. I hope he's wrong."

4. Polynomials can be differentiated using the sum rule and the power rule for derivatives:

$$f'(x) = 3x^2 - 6x + 4$$

5. $f'(x) = 96x^{95} + 430x^{42} - 60x^{19} + 11$

6. $f'(x) = 2x + 1 + 0 - x^{-2} - 2x^{-3} - 3x^{-4}$

7. The professor wrote down the chain rule as it applied to a function raised to a power:

$$f(x) = u^n$$

$$f'(x) = nu^{n-1}\frac{du}{dx}$$

We wrote our function like this:

$$u = 4 - x^3$$

$$f(u) = u^{1/2}$$

$$f'(x) = \tfrac{1}{2}(4 - x^3)^{-1/2}(-3x^2) = \frac{-3}{2}(x^2)(4 - x^3)^{-1/2}$$

8. We wrote this function as a product and used the product rule:

$$f(x) = (x^4 + 10)(3 - x^2)^{-1}$$

$$f'(x) = (x^4 + 10)\,\frac{d}{dx}(3 - x^2)^{-1} + (3 - x^2)^{-1}\,\frac{d}{dx}(x^4 + 10)$$

$$= \frac{2x(x^4 + 10)}{(3-x^2)^2} + \frac{4x^3}{3-x^2}$$

9. Let $u = (x + 4)^3$. Also,

$$f(u) = (3 + u)^2$$

$$\frac{du}{dx} = 3(x + 4)^2$$

$$\frac{df}{du} = 2(3 + u)$$

Using the chain rule gives:

$$f'(x) = 6[3 + (x + 4)^3](x + 4)^2$$

10. "This calls for the product rule," Recordis said.

$$f'(x) = x^3\,\frac{d}{dx}(3 + x^2)^{1/2} + (3 + x^2)^{1/2}\,\frac{d}{dx}x^3$$

$$= \frac{x^4}{\sqrt{3 + x^2}} + 3x^2\sqrt{3 + x^2}$$

11. Recordis almost fainted until he remembered the method of logarithmic implicit differentiation.

$$y = \sqrt{\frac{(x + 5)(x - 3)}{(x + 2)(x - 1)}}$$

$$\ln y = \tfrac{1}{2}[\ln(x + 5) + \ln(x - 3) - \ln(x + 2) - \ln(x - 1)]$$

$$\frac{1}{y}\frac{dy}{dx} = \frac{1}{2}\frac{d}{dx}[\ln(x + 5) + \ln(x - 3) - \ln(x + 2) - \ln(x - 1)]$$

$$\frac{dy}{dx} = \frac{y}{2}\left[\frac{1}{x + 5} + \frac{1}{x - 3} - \frac{1}{x + 2} - \frac{1}{x - 1}\right]$$

$$= \frac{1}{2}\sqrt{\frac{(x + 5)(x - 3)}{(x + 2)(x - 1)}}\left[\frac{1}{x + 5} + \frac{1}{x - 3} - \frac{1}{x + 2} - \frac{1}{x - 1}\right]$$

12. Without thinking, we used the chain rule and the power rule:

$$f(x) = \sin^2 x + \cos^2 x$$

$$f'(x) = 2 \sin x \cos x - 2 \cos x \sin x$$

$$= 0$$

"Of course it should be zero!" Trigonometeris said, realizing the obvious. "We know that $\sin^2 x + \cos^2 x = 1$ for any value of x, so this function is really a constant function. It better have a derivative of zero."

13. "This is hopeless!" the king said.
"We need to use the product rule several times," the professor told him.

$$f'(x) = x^2 \frac{d}{dx} (e^x \sin x \ln x) + e^x \sin x \ln x \frac{d}{dx} (x^2)$$

"The last one is easy," Recordis said, "$(d/dx)x^2 = 2x$."
We used the product rule again on the first part:

$$\frac{d}{dx} (e^x \sin x \ln x) = e^x \frac{d}{dx} (\sin x \ln x) + (\sin x \ln x) \frac{d}{dx} e^x$$

"That last one is easy," Recordis said again, "$(d/dx)e^x = e^x$."

$$\frac{d}{dx} (\sin x \ln x) = \sin x \frac{d}{dx} \ln x + \ln x \frac{d}{dx} \sin x$$

"Those two are both easy!" Recordis said.

$$\frac{d}{dx} \ln x = \frac{1}{x}$$

$$\frac{d}{dx} \sin x = \cos x$$

Now all we had to do was put the pieces together:

$$f'(x) = xe^x \sin x + x^2 e^x \ln x \cos x + x^2 e^x \sin x \ln x + 2xe^x \sin x \ln x$$

14. $f(x) = e^{\ln x}$

"That's trivial!" Recordis said. "By definition, $e^{\ln x} = x$. That means $f(x) = x$, so $f'(x) = 1$."

15. "It's obvious what the approximate shape of our course would be," the king said, "although I do *not* intend to escape anywhere. We would travel straight along the shore for some distance (call it x) and then board our boat and sail in a straight line to the island." (Figure 18–5.)

"That means that all we need to do is figure out what x is," Recordis said.

"I know what the total time is," the professor said.

$$(\text{total time}) = T = \frac{(\text{distance on land})}{(\text{speed on land})} = \frac{(\text{distance on water})}{(\text{speed on water})}$$

$$T = \frac{x}{u} + \frac{y}{v}$$

"We can figure out y in terms of x," the professor noted.

$$a^2 + (b-x)^2 = y^2$$

$$T = \frac{x}{u} + \frac{\sqrt{a^2 + (b-x)^2}}{v}$$

"T is the function we want to minimize," the king said. "We now have T expressed as a function of x, so all we need to do is find the derivative dT/dx and set it equal to zero."

$$\frac{dT}{dx} = \frac{1}{u} + \frac{1}{v}\,\frac{1}{2}[a^2 + (b-x)^2]^{-1/2}(2)(b-x)(-1)$$

$$= \frac{1}{u} - \frac{b-x}{v\sqrt{a^2 + (b-x)^2}}$$

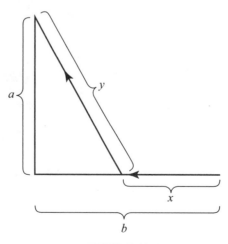

FIGURE 18–5.

Setting the derivative equal to zero to find the optimum x gave:

$$\frac{1}{u} = \frac{b-x}{v\sqrt{a^2 + (b-x)^2}}$$

$$\frac{1}{u^2} = \frac{(b-x)^2}{v^2[a^2 + (b-x)^2]}$$

$$v^2[a^2 + (b-x)^2] = u^2(b-x)^2$$

$$v^2 a^2 + v^2(b^2 - 2bx + x^2) = u^2(b^2 - 2bx + x^2)$$

$$v^2 a^2 + v^2 b^2 - 2bxv^2 + v^2 x^2 = u^2 b^2 - 2bxu^2 + u^2 x^2$$

$$x^2(v^2 - u^2) + x(2bu^2 - 2bv^2) + (v^2 a^2 + v^2 b^2 - u^2 b^2) = 0$$

"That looks pretty hopeless," the professor said.

"We know the numbers," Recordis pointed out. "It will help if we put those numbers in the expressions for the coefficients of x."

The gremlin had given us these numbers:

$$v = 3, \qquad u = 5, \qquad b = 12, \qquad a = 10$$

$$v^2 - u^2 = 9 - 25 = -16$$

$$2bu^2 - 2bv^2 = (24)(16) = 384$$

$$v^2 a^2 + v^2 b^2 - u^2 b^2 = 9 \times 100 + 9 \times 144 - 25 \times 144 = -1{,}404$$

We rewrote the equation for x with the numerical coefficients:

$$-16x^2 + 384x - 1{,}404 = 0$$

We used the quadratic formula to find x:

$$= \frac{-384 \pm \sqrt{384^2 - (4)(-16)(-1{,}404)}}{-32}$$

$$= \frac{-384 \pm 240}{-32}$$

$$= 19.5 \qquad \text{or} = 4.5$$

The king pointed out, "It's clear from the map that x can't be greater than 12."

"Then the answer must be $x = 4.5$," the professor said.

16. The professor decided to use the method of implicit differentiation, so we took d/dx and applied it to both sides of the equation:

$$\frac{d}{dx}(x^2 + y^2) = \frac{d}{dx}r^2$$

$$2x\frac{dx}{dx} + 2y\frac{dy}{dx} = 0$$

$$\frac{dy}{dx} = \frac{-x}{y}$$

17. We applied the same method to this problem. (Notice that this is the equation of an ellipse.)

$$\frac{d}{dx}\left(\frac{x^2}{a^2} + \frac{y^2}{b^2}\right) = 0$$

$$\frac{x}{a^2} + \frac{y}{b^2}\frac{dy}{dx} = 0$$

$$\frac{dy}{dx} = -\frac{b^2 x}{a^2 y}$$

18. First we simplified the expression:

$$(x + y)^{1/2}(x - y)^{1/2} = 1$$
$$(x^2 - xy + xy - y^2)^{1/2} = 1$$
$$(x^2 - y^2)^{1/2} = 1$$
$$x^2 - y^2 = 1$$

"There are two choices now," the professor said. "We can solve for y first, or we can use the implicit differentiation method." The king and Recordis disagreed about which method would be quicker, so they had a race.

Implicit Method (King)	Explicit Method (Recordis)
$\dfrac{d(x^2 - y^2)}{dx} = \dfrac{d}{dx}1$	$y^2 = x^2 - 1$
$2x - 2y\dfrac{dy}{dx} = 0$	$y = \sqrt{x^2 - 1}$
$2y\dfrac{dy}{dx} = 2x$	$\dfrac{dy}{dx} = \frac{1}{2}(x^2 - 1)^{-1/2}(2x)$
$\dfrac{dy}{dx} = \dfrac{x}{y}$	$\dfrac{dy}{dx} = \dfrac{x}{\sqrt{x^2 - 1}}$

"We got different answers!" Recordis exclaimed.

"No, you did not get different answers," the professor said. "If you make the substitution $y = \sqrt{x^2 - 1}$, you will see that you have exactly the same answer that the king has."

19. Using the implicit method again gave us:

$$\frac{d}{dx}(x-3)y^2 = \frac{d}{dx}(x^2 + 4)$$

$$(x-3)\frac{d}{dx}y^2 + y^2\frac{d}{dx}(x-3) = 2x$$

$$(x-3)2y\frac{dy}{dx} + y^2 = 2x$$

$$\frac{dy}{dx} = \frac{2x - y^2}{2y(x - 3)}$$

20. "This is a regular integration problem," the professor said. We let s stand for Rutherford's position:

$$\frac{ds}{dt} = 3 - t^2; \qquad s = 0 \text{ when } t = 0$$

$$s = \int (3 - t^2)\, dt$$

$$= 3t - \frac{t^3}{3} + C$$

Solving for C, we had:

$$0 = C$$

$$s = 3t - \frac{t^3}{3}$$

Now we needed to find out what t equaled when Rutherford came to a stop. That meant we had to find t when $ds/dt = 0$:

$$0 = 3 - t^2_{\text{stop}}$$

$$t_{\text{stop}} = \sqrt{3}$$

We put that result back into the equation for s:

$$s = 3\sqrt{3} - \frac{3\sqrt{3}}{3} = 2\sqrt{3} = 3.46$$

21. We set up an integral:

$$\frac{dv}{v} = -2\,dt$$

$$\int v^{-1}\,dv = -2t$$

$$\ln v = -2t + C$$

We used the initial condition:

$$\ln 5 = C$$

$$\ln v = -2t + \ln 5$$

The professor suggested we take both sides of the last equation and raise e to that power:

$$e^{\ln v} = e^{-2t}\,e^{\ln 5}$$

$$v = 5e^{-2t}$$

Now we had to solve for the value of t that made $v = 0.01$:

$$0.01 = 5e^{-2t}$$

$$t = \tfrac{1}{2}\ln 500$$

$$= 3.1$$

"Now we have to integrate the equation $ds/dt = 5e^{-2t}$ to find the position," Recordis said.

$$s = \frac{-5}{2}e^{-2t} + C$$

Using the initial condition $s = 0$ when $t = 0$ gave us.

$$C = \frac{5}{2}$$

$$s = \frac{-5}{2}e^{-2t} + \frac{5}{2}$$

"Now we just about have it," the king said. "All we need to do is put the value $t = 3.1$ into the equation for s, and then we will know how far Pal travels before his speed equals 0.01."

$$s = \frac{-5}{2}e^{-6.2} + \frac{5}{2}$$

$$= 2.49$$

22. By this time Recordis was an expert at integrating polynomials (since they are his favorite kind of function):

$$x^4 - x^2 + 5x + C$$

23. The professor suggested the substitution:

$$u = x^2 - 1$$

$$\frac{du}{dx} = 2x$$

$$\int x\sqrt{x^2 - 1}\ dx = \tfrac{1}{2}\int u^{1/2}\ du = \tfrac{1}{3}u^{3/2} + C$$

$$= \tfrac{1}{3}(x^2 - 1)^{3/2} + C$$

24. "We already did this one!" Recordis said gleefully. "We can just look it up in our table."

$$\int \frac{1}{1 + x^2}\ dx = \arctan x + C$$

25. We used Trigonometeris' identity for $\sin^2 x$ to simplify matters a little bit:

$$I = \int 4x^2 \sin^2 x\ dx = \int 4x^2(\tfrac{1}{2})(1 - \cos 2x)\ dx$$

$$= 2\int x^2\ dx - 2\int x^2 \cos 2x\ dx$$

$$= \tfrac{2}{3}x^3 - 2\int x^2 \cos 2x\ dx$$

The professor suggested the substitution $z = 2x$:

$$I = \tfrac{2}{3}x^3 - \tfrac{1}{4}\int z^2 \cos z\ dz$$

"We'll have to use integration by parts," the professor said. We made a hazardous guess at the parts and hoped it would work:

Let $u = z^2$. Also,

$dv = \cos z \, dz$. Then:

$du = 2z \, dz$

$v = \sin z$

Using the formula for integration by parts gave us:

$$I_2 = \int z^2 \cos z \, dz$$

$$= uv - \int v \, du$$

$$= z^2 \sin z - \int \sin z \, (2z) \, dz$$

"The new integral is simpler," the professor said encouragingly. "Maybe if we try integration by parts once more, we'll be able to get it."

$$I_3 = \int z \sin z \, dz$$

Let $u = z$. Also,

$dv = \sin z \, dz$. Then:

$du = dz$

$v = -\cos z$

$$I_3 = uv - \int v \, du$$

$$= -z \, \cos z + \int \cos z \, dz$$

"That integral is easy," Recordis said.

$$I_3 = -z \cos z + \sin z + C$$

Now all we had to do was put all the other subintegrals back together and make the reverse substitution, $z = 2x$:

$$\int 4x^2 \sin^2 x \, dx = \frac{2}{3}x^3 - x^2 \sin 2x - x \cos 2x + \frac{1}{2}\sin 2x + C$$

26. $dy/dx = 5x + 4$. Also,

$$y = \int (5x + 4) \, dx$$

$$= \frac{5}{2}x^2 + 4x + C$$

$$C = -15$$

$$y = \frac{5}{2}x^2 + 4x - 15$$

27. "We should be able to integrate this equation twice," the professor said. We let $v = dy/dt$, so

$$\frac{dv}{dt} = 5$$

$$v = 5t + C$$

"We have an initial condition for dy/dt," Recordis noted. "That's lucky," the professor said.

$$v = \frac{dy}{dt} = 5t + 3$$

Integrating again:

$$y = \frac{5}{2}t^2 + 3t + C$$

Using the other initial condition, we got:

$$C = 2$$

The final answer became:

$$y = \frac{5}{2}t^2 + 3t + 2$$

28. We did this problem in the same way as the preceding one:

$$v = t^2 + 0$$

$$y = \frac{1}{3}t^3$$

29. Two forces were acting on Pal, so the total force was

$$F = F_{spring} + F_{friction} = -500x - 10\,\frac{dx}{dt}$$

We used the equation that Builder gave us:

$$F = m\,\frac{d^2x}{dt^2}$$

$$-500x - 10\,\frac{dx}{dt} = m\,\frac{d^2x}{dt^2}$$

Using the value $m = 1,000$, we rewrote this equation as a second-order linear homogeneous constant-coefficient differential equation:

$$\frac{d^2x}{dt^2} + \frac{1}{100}\,\frac{dx}{dt} + \frac{1}{2}x = 0$$

We set up the characteristic equation:

$$r^2 + \frac{1}{100}r + \frac{1}{2} = 0$$

$$r = \frac{-0.01 \pm i1.414}{2}$$

We set up the solution:

$$x = e^{-0.005t}(A \sin 0.707t + B \cos 0.707t)$$

Using the initial conditions ($x = 20$ and $dx/dt = 0$ when $t = 0$), we could solve for A and B:

$$20 = B, \qquad 0.14 = A$$

we obtained the final solution:

$$x = e^{-0.005t}(0.14 \sin 0.707t + 20 \cos 0.707t)$$

30. The professor was bothered by what a unit of work was, but we realized we didn't have time to worry about that now. The integral was straightforward, with the result (work) = $9k/100$.

31. We set up the partial fractions:

$$\frac{x-1}{(x-2)(x-3)} = \frac{A}{x-2} + \frac{B}{x-3}$$

$$x - 1 = A(x-3) + B(x-2)$$

coefficients of x:

$$1 = A + B$$

constant terms:

$$-1 = -3A - 2B$$

The result was that $A = -1, B = 2$.

$$\frac{x-1}{(x-2)(x-3)} = \frac{2}{x-3} - \frac{1}{x-2}$$

$$\int_4^5 \frac{x-1}{(x-2)(x-3)} dx = 2 \ln (x-3) \Big|_4^5 - \ln (x-2) \Big|_4^5$$

$$= 2 \ln 2 - 2 \ln 1 - \ln 3 + \ln 2$$

$$= 3 \ln 2 - \ln 3$$

$$\int_4^5 \frac{x-1}{(x-2)(x-3)} dx = 0.98$$

32. This was easy:

$$(4)(\tfrac{1}{3})\, x^3 \,\Big|_0^{10} = 1{,}333$$

33. The king suggested the substitution $u = 2x$.

$$\int_0^{\pi/6} \tan 2x \; dx = \frac{1}{2} \int_0^{\pi/3} \tan u \; du$$

$$= -\frac{1}{2} \ln \cos u \;\Big|_0^{\pi/3}$$

$$= -\frac{1}{2} \ln \cos \left(\frac{\pi}{3} \right) + \frac{1}{2} \ln \cos 0$$

$$= -\frac{1}{2} \ln \frac{1}{2}$$

$$\int_0^{\pi/6} \tan 2x \; dx = 0.35$$

34. $e^x \Big|_0^{10} - \dfrac{1}{4} x^4 \Big|_0^{10} = 22{,}026 - 1 - 2{,}500 = 19{,}525$

35. This looked hard until the king remembered a perfect substitution:

Let $u = x^5$, $\qquad du = 5x^4 \; dx$.

$$\int_1^2 x^4 \sin (x^5) \; dx = \frac{1}{5} \int_1^{32} \sin u \; du = \frac{-1}{5}(\cos 32 - \cos 1)$$

$$= -0.059$$

36. Just as we were ready to try integration by parts, Trigonometeris remembered a formula to make matters much simpler:

$$\sin x \cos x = \frac{1}{2}\sin 2x$$

$$\int_0^{\pi/2} \sin x \cos x \; dx = \frac{1}{2} \int_0^{\pi/2} \sin 2x \; dx = -\frac{1}{4} \cos 2x \;\Big|_0^{\pi/2}$$

$$= -\frac{1}{4}(\cos \pi - \cos 0) = \frac{1}{2}$$

37. We realized that we had no choice but integration by parts. We decided to treat the integral as if it were an indefinite integral first, and then use the limits of integration after we had found the antiderivative function:

Let $u = x^2$. Also,

$$dv = \sin x \, dx$$

$$du = 2x \, dx$$

$$v = -\cos x$$

$$\int x^2 \sin x \, dx = -x^2 \cos x + \int \cos x \, 2x \, dx$$

We used integration by parts again:
Let $u = x$. Also,

$$du = dx$$

$$dv = \cos x \, dx$$

$$v = \sin x$$

$$\int x^2 \sin x \, dx = -x^2 \cos x + 2x \sin x - \int 2\sin x \, dx$$

$$\int_2^3 x^2 \sin x \, dx = (-x^2 \cos x + 2x \sin x + 2\cos x) \Big|_2^3$$

$$= -7 \cos 3 + 6 \sin 3 + 2 \cos 2 - 4 \sin 2$$

$$= 3.3$$

38. We defined a new function equal to the distance between the two curves:
$y = y_1 - y_2 = \sin x - x^2 + \pi x$.

"Now we need to know what limits of integration to use," Recordis said.

After making a quick sketch of both curves, it was clear that we wanted to integrate between the two points where the curves intersected. One intersection point came at $x = 0$, because then both y_1 and y_2 were equal to zero. We found that the other intersection point occurred where $x = \pi$, so we integrated from $x = 0$ to $x = \pi$ to find the total area between the curves:

$$A = \int_0^\pi (\sin x - x^2 + \pi x) \, dx = (-\cos x - \frac{1}{3}x^3 + \frac{1}{2}\pi x^2) \Big|_0^\pi$$

$$= 1 + 1 - \frac{1}{3}\pi^3 + \frac{1}{2}\pi^3 = 7.17$$

39. Realizing that the balancing point must lie along the central line, we used Builder's formula for the center of mass:

$$x_{com} = \frac{\int x \, dm}{\rho h a^2}$$

Then we calculated what the differential mass element dm would be (Figure 18–6):

$$dm = \rho \, dV$$

$$= 2\rho h y \, dx$$

From a similar triangle relation:

$$\frac{y}{a} = \frac{a - x}{a}$$

$$y = a - x$$

FIGURE 18–6.

We put these into the formula and did the integration:

$$x_{com} = \int_0^a \frac{2\rho h(a - x)x \, dx}{\rho h a^2}$$

$$= \frac{2}{a^2} \int_0^a (ax - x^2) \, dx$$

$$= \frac{2}{a^2} \left(\frac{ax^2}{2} - \frac{x^3}{3} \right) \Big|_0^a$$

$$x_{com} = \frac{2}{a^2} \left(\frac{a^3}{6} \right) = \frac{a}{3}$$

40. We divided the football into a collection of little cylinders:

$$dV = \pi y^2 \, dx$$

$$y = \sin x$$

Now all we had to do was integrate from $x = 0$ to $x = \pi$:

$$V = \pi \int_0^\pi \sin^2 x \, dx$$

$$= \pi \int_0^\pi \frac{1}{2}(1 - \cos 2x) \, dx$$

$$= \frac{1}{2}\pi^2$$

41. We found the first derivative to look for horizontal tangents:

$$y = e^{-x^2}$$

$$\frac{dy}{dx} = -2xe^{-x^2}$$

"Where does that equal zero?" Recordis asked.

"The only point is where $x = 0$," the professor said. "That means that this curve has only one horizontal tangent."

We calculated the second derivative:

$$\frac{d^2y}{dx^2} = (4x^2 - 2)(e^{-x^2})$$

The second derivative will be zero if

$$4x^2 - 2 = 0$$

$$x^2 = \frac{1}{2}$$

$$x = \pm \frac{1}{\sqrt{2}}$$

"That means that there are two points of inflection," the professor noted.

"One more thing," the king said. "We need to check whether the second derivative is positive or negative at the point where the horizontal tangent occurs."

$$x = 0, \qquad \frac{d^2y}{dx^2} = -2$$

"That means the point is a maximum," the king said.

"I remember," Trigonometeris stated. "If the second derivative is negative, the curve spills water."

We could easily see that $y = 1$ when $x = 0$. We could also see that the function was always greater than 0 but less than 1, so Igor was able to sketch the graph (Figure 18–7).

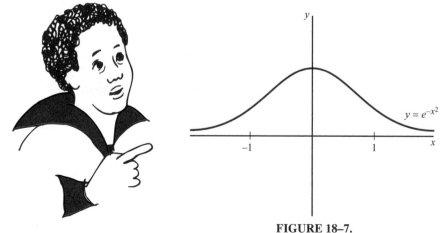

FIGURE 18–7.

42. First we found the homogeneous solution:

$$\frac{d^2y}{dx^2} + 2\frac{dy}{dx} + 2y = 0$$

We set up the characteristic equation:

$$r^2 + 2r + 2 = 0$$

$$r = \frac{-2 \pm \sqrt{4 - 8}}{2}$$

$$= -1 + i, \qquad = -1 - i$$

Since we had a complex answer, we set up an exponential and a trigonometric solution:

$$y = e^{-x}(A \cos x + B \sin x)$$

Now we needed to guess a particular solution to the nonhomogeneous equation. We spent hours trying to find a solution that would work. After each guess that failed we came up with a new guess that was even more complicated than the preceding one. No guess, no matter how complicated, worked.

"If only we just had $2y = 10$," Recordis moaned. "Then we could say $y = 5$. If only we knew a function that would allow us to ignore those derivatives."

"That wouldn't help," the professor said. "We know that the only kind of function that has a derivative of zero is a constant function."

"That's it!" the king exclaimed. "We'll use a constant function! Let's guess $y = 5$."

"We can't use a constant function!" Recordis said. "This is a differential equation. Differential equations are hard, and they always have complicated solutions."

We tried the king's guess anyway, and found that it did work:

$$\frac{d^2}{dx^2}5 + \frac{d}{dx}5 + 2 \times 5 = 10$$

"That is the particular solution we need," the professor said. We found the complete solution by adding the particular solution to the homogeneous solution:

$$y = 5 + e^{-x}(A \cos x + B \sin x)$$

43. We realized that we would have to use the curve length integral. (See Figure 18–8.) First we found dy/dx:

$$\frac{dy}{dx} = \tfrac{1}{2}(e^{x/36} - e^{-x/36})$$

$$\left(\frac{dy}{dx}\right)^2 = \tfrac{1}{4}(e^{x/18} - 2 + e^{-x/18})$$

$$1 + \left(\frac{dy}{dx}\right)^2 = \tfrac{1}{4}(e^{x/18} + 2 + e^{-x/18})$$

$$\sqrt{1 + (dy/dx)^2} = \tfrac{1}{2}(e^{x/36} - e^{-x/36})$$

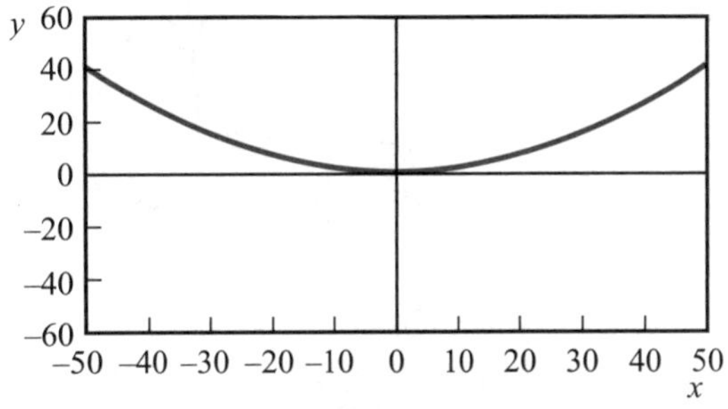

FIGURE 18–8.

Now we set up the length integral:

$$L = \frac{1}{2} \int_{-50}^{50} (e^{x/36} + e^{-x/36})\, dx$$

$$= 18(e^{x/36} - e^{-x/36}) \Big|_{-50}^{50}$$

$$= 135.4$$

44. We set up the function $y = y_2 - y_1 = x - x^2$, and integrated it from $x = 0$ to $x = 1$ in order to find the total area between the two curves:

$$A = \int_0^1 (x - x^2)\, dx = \frac{1}{2}x^2 - \frac{1}{3}x^3 \Big|_0^1 = \frac{1}{2} - \frac{1}{3} = \frac{1}{6}$$

45. Our sand was just about ready to run out. We had finished all the problems except this one.

"We've got to hurry!" Recordis said. "We have only that much sand left!" He held his fingers a tiny distance apart.

The professor desperately guided Igor through the required calculations.

"It can't be done," Recordis moaned. "There is no answer!"

"That's the answer!" the professor cried.

"What's the answer?"

"There is no answer! Look at the graph (Figure 18–9). There is no way to find a derivative for the absolute-value function at that point, because there is no real tangent line. In fact, I bet that's true for any function with a cusp in it—you would have to say that the function is nondifferentiable at the point where the cusp is. So the answer must be that there is no answer! The gremlin threw an impossible problem into the test to confuse us."

"That gremlin!" Recordis cried. "I hate him!" He whipped out his pen and wrote "no answer" on the large scroll where he had been keeping track of the answers to the test.

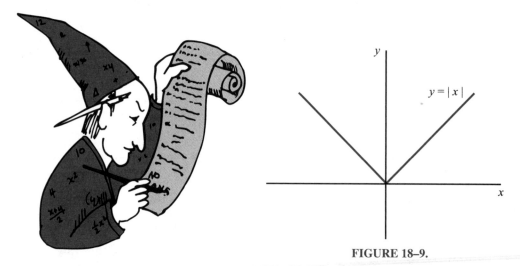

FIGURE 18–9.

At the instant he wrote the last letter, the very last grain of sand hesitated for a moment at the vertex of the hourglass cone, and then dropped to the pile of sand below. There was a trumpet call from the hallway and in marched the gremlin, wearing an impressive royal robe and followed by two short gremlin attendants, each carrying a trumpet.

"I want you to know that I inspected the palace and I like it very much, although there are a few parts that will have to be changed to suit my taste. It's been nice knowing you. I hope you are all packed." We all looked at him in terror. "Well, which one did you in? If you really want to give me the satisfaction, I will look over your answers and see where you went wrong."

"I think we got them all," the professor said timidly.

"Are you sure?" Recordis asked. "We were in such a hurry that we may have misplaced one or two problems."

"We did get them all, you vile creature," the king said, standing straight up and staring the gremlin in the eye. He took the answer scroll from Recordis and handed it to the gremlin. He remained standing there while the gremlin slowly read through the entire scroll. The next few minutes seemed to take forever, but the king did not budge. A look of consternation grew on the gremlin's face as he read the answer to each problem that he was sure would have ruined us.

"You can't have solved the last one!" he said desperately, but when he saw Recordis' writing he fell to the floor sobbing. "Curses! I'm ruined!" he moaned. "All that effort wasted! All is lost! You know everything worth knowing about single-variable calculus!"

A large puddle of tears developed around the prostrate gremlin. Recordis felt so sorry for him that he patted him on the back. "It can't be that bad," he said consolingly. "I'm sure you'll find other complications. You'll think of something like multi-variable calculus or scientifically applied problems or . . ."

"Shut up, Recordis!" the professor cried. "What are you saying?"

"Enough of that!" the gremlin said, suddenly standing up. "I shouldn't have lost control of myself like that. I can make things more complicated, and I will. It is just a matter of time before you discover another subject, and then I shall win and rule Carmorra!" He grabbed his two attendants and whooshed out the window.

"We surely got rid of him," Recordis said, and we all laughed, for the first time in a long while.

The party the next day was very festive and bright. The shining ribbons decorated the grounds. The roses in the garden were all in bloom, and Spike Rock was safely covered. The children loved the doughnuts, the ice cream cones, and all the rides.

That night the king asked me how long I could stay. "I will have to return home when I regain my memory," I said, "but I will be glad to stay in Carmorra for the present."

"We deeply appreciate your services," the king said. "If you ever want to return home, we'll do our best to help you, even if we need to invent a new subject."

As we looked out at the moon, Farmer Floran said, "I wonder whether you could figure out how the moon moves."

"We wouldn't have had a chance before," the king said. "Maybe we could now."

This brings to an end my part of the story. There were indeed more adventures, and we were constantly amazed at the applications of the subject of calculus. I'll let the professor summarize what we did in her book, which she graciously granted me permission to reprint here as Chapter 19. "It's been great having another woman in the Main Conference Room," she said wistfully.

I hope that this entire account has been as much fun for the reader as it was for me, and that it will be beneficial to anyone who is interested in discovering the mysteries of calculus.

The Professor's Guide to Calculus

The Complete, Authoritative, Summary Guide to the Mysterious Subject of Calculus

by Professor A. A. A. Stanislavsky, Ph.D., etc., etc.
Royal Institute of Carmorra

Calculus begins by trying to solve the following problem: What is the slope of the tangent line for a particular curve at a given point? It turns out that this problem is exactly the same problem as finding the speed of an object if we are given its position function.

The slope of the tangent line to the curve $y = f(x)$ is given by the following expression, known as the derivative:

$$(\text{slope of tangent line}) = (\text{derivative}) = f'(x) = \frac{dy}{dx}$$

$$= \lim_{\Delta x \to 0} \frac{f(x + \Delta x) - f(x)}{\Delta x}$$

The operation "lim" means to take the limit of the expression as Δx moves very close to zero, but we don't ever let Δx actually equal zero.

We can derive a set of rules that make it possible to find the derivatives of different functions. (We call this process differentiating the function.) Let c represent any constant number.

$$y = c \qquad\qquad y' = 0$$
$$y = cx \qquad\qquad y' = c$$

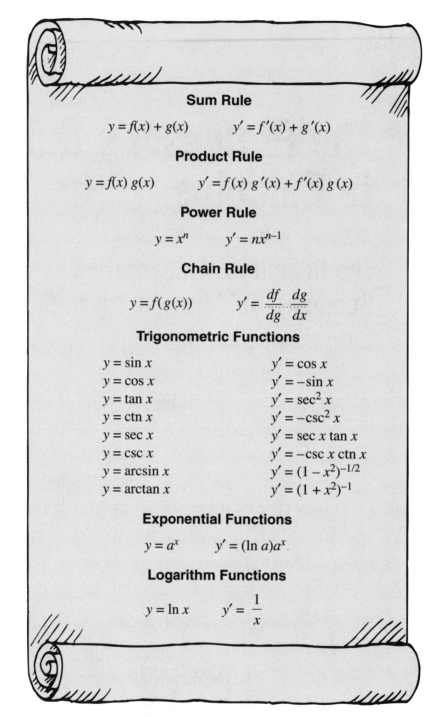

Sum Rule

$$y = f(x) + g(x) \qquad y' = f'(x) + g'(x)$$

Product Rule

$$y = f(x)\, g(x) \qquad y' = f(x)\, g'(x) + f'(x)\, g(x)$$

Power Rule

$$y = x^n \qquad y' = nx^{n-1}$$

Chain Rule

$$y = f(g(x)) \qquad y' = \frac{df}{dg}\,\frac{dg}{dx}$$

Trigonometric Functions

$y = \sin x$	$y' = \cos x$
$y = \cos x$	$y' = -\sin x$
$y = \tan x$	$y' = \sec^2 x$
$y = \operatorname{ctn} x$	$y' = -\csc^2 x$
$y = \sec x$	$y' = \sec x \tan x$
$y = \csc x$	$y' = -\csc x \operatorname{ctn} x$
$y = \arcsin x$	$y' = (1 - x^2)^{-1/2}$
$y = \arctan x$	$y' = (1 + x^2)^{-1}$

Exponential Functions

$$y = a^x \qquad y' = (\ln a)a^x.$$

Logarithm Functions

$$y = \ln x \qquad y' = \frac{1}{x}$$

It is often useful to reverse the process of differentiation. We call this process *integration.* We make this definition for indefinite integral:

$$\int f(x)\, dx = F(x) + C \quad \text{if and only if } \frac{dF}{dx} = f(x)$$

We can then make a set of rules to calculate integrals:

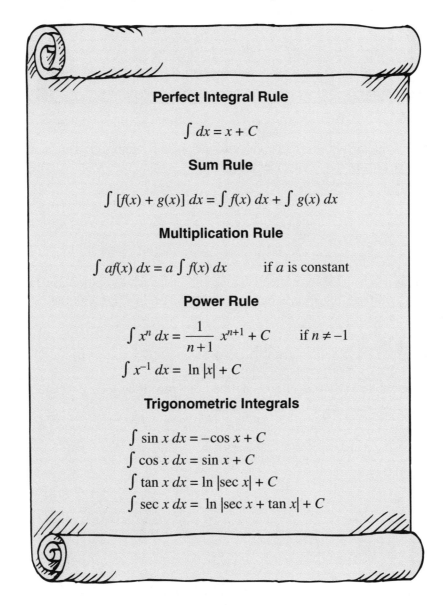

Perfect Integral Rule

$$\int dx = x + C$$

Sum Rule

$$\int [f(x) + g(x)]\, dx = \int f(x)\, dx + \int g(x)\, dx$$

Multiplication Rule

$$\int af(x)\, dx = a \int f(x)\, dx \qquad \text{if } a \text{ is constant}$$

Power Rule

$$\int x^n\, dx = \frac{1}{n+1} x^{n+1} + C \qquad \text{if } n \neq -1$$

$$\int x^{-1}\, dx = \ln |x| + C$$

Trigonometric Integrals

$$\int \sin x\, dx = -\cos x + C$$
$$\int \cos x\, dx = \sin x + C$$
$$\int \tan x\, dx = \ln |\sec x| + C$$
$$\int \sec x\, dx = \ln |\sec x + \tan x| + C$$

An integral that contains the quotient of two polynomials can be solved by the method of partial fractions. An integral that contains an expression such as $\sqrt{1 + x^2}$ or $(1 - x^2)^{-1}$ can be solved by using a trigonometric substitution, based on the identity

$$\sin^2 \theta + \cos^2 \theta = 1 \qquad \text{or} \qquad 1 + \tan^2 \theta = \sec^2 \theta$$

An integral containing the product of two different types of functions, or some other hard integrals, can be solved by using the method of integration by parts.

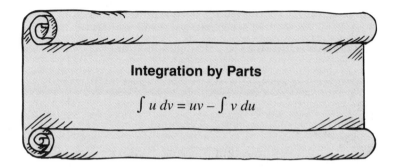

Integration by Parts

$$\int u \, dv = uv - \int v \, du$$

The amazing feature of integral calculus is that an integral represents not only the antiderivative but also the area under a curve. This is often written as a definite integral.

$$(\text{area under } f(x), \text{ from } x = a \text{ to } x = b) = \int_a^b f(x) \, dx$$

$$= F(b) - F(a)$$

where $F(x)$ is a function such that $dF/dx = f(x)$.

Integrals can be used for more general problems, such as volume, surface areas, or the center of mass, by comparing the formula for an integral with the formula for a continuous sum:

$$\int_a^b f(x) \, dx = \lim_{\Delta x \to 0} \sum_{i=1}^n f(x_i) \, \Delta x \qquad (x_1 = a, x_n = b)$$

The natural logarithm function is defined in terms of definite integrals:

$$\ln x = \int_1^x \frac{1}{t} \, dt$$

The base of the natural logarithm function is the mysterious number e, approximately equal to 2.71:

$$e^{\ln x} = x$$

Appendix 1: Answers to Worksheets and Exercises

Chapter 1

WORKSHEET

	x_1	y_1	x_2	y_2	$\Delta x = x_2 - x_1$	$\Delta y = y_2 - y_1$	slope $= \Delta y/\Delta x$
1.	10	100	15.0000	225.000000	5.0000	125.000000	25.0000
2.	10	100	11.0000	121.000000	1.0000	21.000000	21.0000
3.	10	100	10.1000	102.010000	0.1000	2.010000	20.1000
4.	10	100	10.0100	100.200100	0.0100	0.200100	20.0100
5.	10	100	10.0010	100.020001	0.0010	0.020001	20.0010
6.	10	100	10.0001	100.00200001	0.0001	0.00200001	20.0001

EXERCISES

1. and **2.** See Table A–1 and Figure A–1.

Table A–1

x	$y = x^2$	Slope of Secant Line
3	9	5
2.5	6.25	4.5
2.3	5.29	4.3
2.1	4.41	4.1
2.05	4.2025	4.05
1.95	3.8025	3.95
1.9	3.61	3.9
1.8	3.24	3.8
1.7	2.89	3.7
1.5	2.25	3.5
1	1	3

Note: The slope column gives the slope of the secant line between the point (2,4) and the point (x,y).

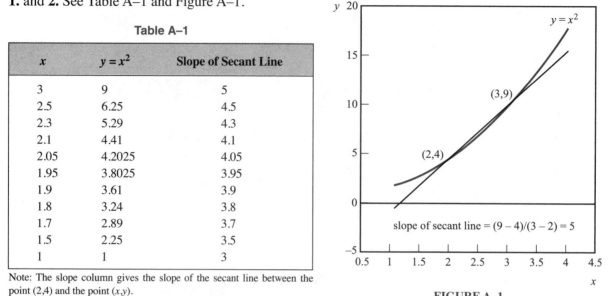

slope of secant line = $(9 - 4)/(3 - 2) = 5$

FIGURE A–1.

3. (Slope) = $2 \cdot 2 = 4$. Point-slope formula: $(y - 4)/(x - 2) = 4$. $y = 4x - 4$.

4. (Slope) = $2 \cdot 7 = 14$. $y = 14x - 49$. When $y = 50$, the x coordinate of the tangent line is $x = 99/14$ = 7.0714. (This is reasonably close to the best decimal approximation, which is $\sqrt{50} \approx 7.0711$.)

5. $f(0)$ is undefined, since at $x = 0$ the expression for $f(x)$ is $0/0$. On the graph this is signified by drawing an open circle at the point $x = 0$. It is clear from the graph that $\lim_{x \to 0} f(x) = 8$. (See Figure A–2.)

6. If $f(t)$ represents the position of an object as a function of time, then the speed of the object between time t_1 and time t_2 is $[f(t_2) - f(t_1)]/(t_2 - t_1)$. Note that this is the same formula that is used to find the slope of a curve.

8.

x	y	slope of tangent
−6	36	−12
−4	16	−8
−2	4	−4
0	0	0
2	4	4
4	16	8
6	36	12

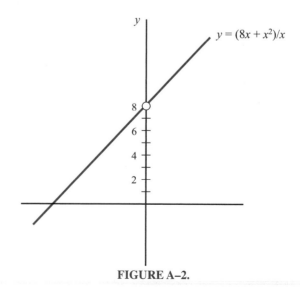

FIGURE A–2.

Chapter 2

WORKSHEET

		$f(x)$	$f'(x)$
1.		6	0
2.		32	0
3.		68	0
4.		$3x + 125$	3
5.		$-8x - 643$	-8
6.		$-15x - 549$	-15
7.		$8x^2 + 4x + 419$	$16x + 4$
8.		$7x^2 - 7x - 7{,}652$	$14x - 7$
9.		$3x^2 + 6x + 16$	$6x + 6$
10.		$10x^3 + 2x^2 - 18x - 86$	$30x^2 + 4x - 18$
11.		$6x^3 - 6x^2 + 12x + 788$	$18x^2 - 12x + 12$
12.		$4x^3 + 9x^2 + 20x + 604$	$12x^2 + 18x + 20$
13.		$2x^4 + 3x^3 + 12x^2 + 16x - 13$	$8x^3 + 9x^2 + 24x + 16$
14.		$4x^4 + 5x^3 - 8x^2 - 44x + 42$	$16x^3 + 15x^2 - 16x - 44$
15.		$8x^4 + 7x^3 + 5x^2 + 25x + 121$	$32x^3 + 21x^2 + 10x + 25$
16.		$h - \dfrac{1}{2}gx^2$	$-gx$

Note: Question 16 describes the motion of a falling object, as mentioned in Chapter 1. In this case x represents time, h represents the height from which the object falls, and g represents the acceleration of gravity. The velocity of the object is $-gx$, which is negative because the object is going down. Note how the object moves faster and faster as time (x) increases. See also Exercise 16 at the end of this Chapter; Chapter 3, Exercises 6 and 7; and Chapter 7, Exercise 48.

EXERCISES

1. $y' = 9x^2 + 4x + 1.$ $x = 3, y' = 94$

2. $y' = 20x^4 + 2x.$ $x = 10, y' = 200{,}020$

3. $y' = 35x^{34}.$ $x = 1, y' = 35$

4. $y' = x^2 + x + 1.$ $x = 6.42, y' = 48.636$

5. $y' = 8.68x + 0.98.$ $x = -4, y' = -33.74$

6. $f(t) = 6t^2 - 15t + 8t - 20.$ $f'(t) = 12t - 7.$ $t = \frac{1}{2} f'(t) = -1$

7. $y' = a$

8. $y' = 2ax - 1$

9. $y' = 3ax^2 + 2bx + c$

10. $y' = bax^{b-1}$

11. $y = ab + (a + b)x + x^2; y' = 2x + a + b$

12. $\dfrac{dy}{dx} = \lim\limits_{\Delta x \to 0} \dfrac{cu(x + \Delta x) - cu(x)}{\Delta x} = c \lim\limits_{\Delta x \to 0} \dfrac{u(x + \Delta x) - u(x)}{\Delta x} = c\dfrac{du}{dx}$

13. $f(x + \Delta x) = ax^2 + 2ax\,\Delta x + a\,\Delta x^2 + bx + b\,\Delta x + c.$

    $f(x + \Delta x) - f(x) = 2ax\,\Delta x + a\,\Delta x^2 + b\,\Delta x.$

    $f'(x) = \lim\limits_{\Delta x \to 0} \dfrac{2ax\,\Delta x + a\,\Delta x^2 + b\,\Delta x}{\Delta x} = \lim\limits_{\Delta x \to 0} (2ax + a\,\Delta x + b) = 2ax + b.$

14. $y + \Delta y = f(x + \Delta x) + g(x + \Delta x) + h(x + \Delta x).$

    $\dfrac{\Delta y}{\Delta x} = \dfrac{f(x + \Delta x) - f(x) + g(x + \Delta x) - g(x) + h(x + \Delta x) - h(x)}{\Delta x}$

    $\dfrac{dy}{dx} = \lim\limits_{\Delta x \to 0} \left[\dfrac{f(x + \Delta x) - f(x)}{\Delta x} + \dfrac{g(x + \Delta x) - g(x)}{\Delta x} + \dfrac{h(x + \Delta x) - h(x)}{\Delta x} \right]$

    $= f'(x) + g'(x) + h'(x).$

15. (A) $v(t) = dh/dt = -gt + v_0.$ (b) $v(0) = v_0.$ (c) $dh/dt = 0$ when $0 = -gt + v_0. t = v_0/g.$

16. (a) $v(t) = -gt.$ The velocity is negative, which means that h is becoming smaller (the ball is going down). (b) $h = 0$ when $0 = 64 - \frac{1}{2}gt^2. t = 8\sqrt{2}/\sqrt{g}.$ (c) When $t = 8\sqrt{2}/\sqrt{g}, v(t) = -8\sqrt{2g}.$ (d) $3.61 = 8\sqrt{2}/\sqrt{g}. g = 9.8.$

17. $y' = x^2 - 2x + 3.$ Now set $y' = 3$ and solve for x: $3 = x^2 - 2x + 3. x^2 - 2x = 0. x = 0$ or $x = 2.$

18. The slope of the curve is $y' = 3x^2.$ The slope of the line is $y' = 8.$ Set those two slopes equal, and solve for $x. x = \pm\sqrt{8/3}.$ The line $y = 8x + b$ will be tangent to the curve if it passes through the point $((8/3)^{1/2}, (8/3)^{3/2})$ or the point $(-(8/3)^{1/2}, -(8/3)^{3/2}).$ This will happen if $b = 8.7$ or $b = -8.7.$

19. First, we need the slope of the line between (2, 1) and (6, 9) (call the slope m). $m = (9 - 1)/(6 - 2)$ = 2. Now we need to find the point where the slope of the curve $f(x) = -x^2 + 10x - 15$ equals 2. $f'(x) = -2x + 10$. $f'(x)$ equals 2 when $2 = -2x + 10$. $x = 4$.

20. Calculate the derivative: $f'(x) = 3x^2$. The formula now becomes:

$$x_{i+1} = x_i = \frac{x_i^3 - 7}{3x_i^2}.$$

For $x_1 = 2$, $x_2 = 2 - 1/12 = 1.917$. $x_3 = 1.917 - 0.041/11.02$. $x_3 = 1.913$. $x_4 = 1.913 - 0.00076/10.978$ = 1.913. This result is close to the best decimal approximation for $\sqrt[3]{7}$, which is 1.9129.

Chapter 3

WORKSHEET

1. distance $= D = \sqrt{(x - 82)^2 + (60 - 12)^2}$

$$z = D^2 = x^2 - 2 \times 82x + 82^2 + 48^2$$

$$\frac{dz}{dx} = 2x - 164$$

Set the derivative to zero:

$$0 = 2x - 164$$

$$164 = 2x$$

$$x = \frac{164}{2} = 82, y = 60$$

2. distance $= D = \sqrt{(x - 15)^2 + [(100 - x) - 45]^2}$

$$z = D^2 = x^2 - 2 \times 15x + 15^2 + (55 - x)^2$$

$$z = x^3 - 30x + 15^2 + 55^2 - 2 \times 55x + x^2$$

$$\frac{dz}{dx} = 2x - 30 - 110 + 2x$$

Set the derivative to zero:

$$0 = 4x - 140$$

$$140 = 4x$$

$$x = \frac{140}{4} = 35, y = 100 - 1 \times 35 = 65$$

3. distance $= D = \sqrt{(x - 18)^2 + [(30 + 2x) - 86]^2}$

$$z = D^2 = x^2 - 2 \times 18x + 18^2 + (-56 + 2x)^2$$

$$z = x^2 - 36x + 18^2 + (-56)^2 + 2 \times (-56) \times 2x + 2^2 x^2$$

$$\frac{dz}{dx} = 2x - 36 - 224 + 2 \times 4x$$

Set the derivative to zero:

$$0 = 10x - 260$$

$$260 = 10x$$

$$x = \frac{260}{10} = 26, y = 30 + 2 \times 26 = 82$$

4. distance $= D = \sqrt{(x - 402)^2 + [(2 + 100x) - 198]^2}$

$$z = D^2 = x^2 - 2 \times 402x + 402^2 + (-196 + 100x)^2$$

$$z = x^2 - 804x + 402^2 + (-196)^2 + 2 \times (-196) \times 100x + 100^2 x^2$$

$$\frac{dz}{dx} = 2x - 804 - 39{,}200 + 2 \times 10{,}000x$$

Set the derivative to zero:

$$0 = 20{,}002x - 40{,}004$$

$$40{,}004 = 20{,}002x$$

$$x = \frac{40{,}004}{20{,}002} = 2, y = 2 + 100 \times 2 = 202$$

5. distance $= D = \sqrt{(x - 27)^2 + [(72 - 12x) - 38]^2}$

$$z = D^2 = x^2 - 2 \times 27x + 27^2 + (34 - 12x)^2$$

$$z = x^2 - 54x + 27^2 + 34^2 + 2 \times 34 \times (-12)x + 12^2 x^2$$

$$\frac{dz}{dx} = 2x - 54 - 816 + 2 \times 144x$$

Set the derivative to zero:

$$0 = 290x - 870$$

$$870 = 290x$$

$$x = \frac{870}{290} = 3, y = 72 - 12 \times 3 = 36$$

6. distance $= D = \sqrt{(x - 21)^2 + [(32 + 0.1250x) - 384]^2}$

$$z = D^2 = x^2 - 2 \times 21x + 21^2 + (-352 + 0.1250x)^2$$

$$z = x^2 - 42x + 21^2 + (-352)^2 + 2 \times (-352) \times 0.1250x + 0.1250^2 x^2$$

$$\frac{dz}{dx} = 2x - 42 - 88 + 2 \times 0.0156x$$

Set the derivative to zero:

$$0 = 2.0313x - 130$$

$$130 = 2.0313x$$

$$x = \frac{130}{2.0313} = 64, y = 32 - 0.1250 \times 64 = 40$$

7. distance $= D = \sqrt{(x-h)^2 + [(mx+b)-k]^2}$

$z = x^2 - 2xh + h^2 + [mx + (b-k)]^2$

$z = x^2 - 2xh + h^2 + (mx)^2 + 2mx(b-k) + (b-k)^2$

$\dfrac{dz}{dx} = 2x - 2h + 2m^2x + 2m(b-k)$

$0 = (2 + 2m^2)x - 2h + 2m(b-k)$

$0 = (1 + m^2)x - h + m(b-k)$

$h - m(b-k) = (1 + m^2)x$

$x = \dfrac{h - m(b-k)}{1 + m^2}$

WORKSHEET

1. $dy/dx = 4x + 44, x = -44/(4) = -11.000, y = 2(-11.000)^2 + 44 \times (-11.000) + 6 = -236.000,$ $d^2y/dx^2 = 4$, min

2. $dy/dx = 12x - 84, x = 84/(12) = 7.000, y = 6(7.000)^2 - 84 \times 7.000 + 3 = -291.000, d^2y/dx^2 = 12$, min

3. $dy/dx = 16x - 80, x = 80/(16) = 5.000, y = 8(5.000)^2 - 80 \times 5.000 + 4 = -196.000, d^2y/dx^2 = 16$, min

4. $dy/dx = -24x + 184, x = -184/(-24) = 7.667, y = -12(7.667)^2 + 184 \times 7.667 + 9 = 714.333,$ $d^2y/dx^2 = -24$, max

5. $dy/dx = -20x + 36, x = -36/(-20) = 1.800, y = -10(1.800)^2 + 36 \times 1.800 + 18 = 50.400, d^2y/dx^2 = -20$, max

6. $dy/dx = 40x + 23, x = -23/(40) = -0.575, y = 20(-0.575)^2 + 23 \times (-0.575) + 34 = 27.388, d^2y/dx^2 = 40$, min

7. $dy/dx = -30x - 18, x = 18/(-30) = -0.600, y = -15(0.600)^2 - 18 \times (-0.600) + 72 = 77.400,$ $d^2y/dx^2 = -30$, max

8. $dy/dx = 60x - 9, x = 9/(60) = 0.150, y = 30(0.150)^2 - 9 \times 0.150 + 65 = 64.325, d^2y/dx^2 = 60$, min

9. $dy/dx = -84x - 96, x = 96/(-84) = -1.143, y = -42(-1.143)^2 - 96 \times 1.143) + 112 = 166.857,$ $d^2y/dx^2 = -84$, max

10. $dy/dx = -44x - 82, x = 82/(-44) = -1.864, y = -22(-1.864)^2 - 82 \times (-1.864) + 360 = 436.409,$ $d^2y/dx^2 = -44$, max

11. $dy/dx = 32x + 42, x = -42/(32) = -1.313, y = 16(-1.313)^2 + 42 \times (-1.313) + 94 = 66.438, d^2y/dx^2 = 32$, min

12. $dy/dx = -22x + 18, x = -18/(-22) = 0.818, y = -11(0.818)^2 + 18 \times 0.818 + 178 = 185.364, d^2y/dx^2 = -22$, max

13. $dy/dx = -8x - 16, x = 16/(-8) = -2.000, y = -4(-2.000)^2 - 16 \times (-2.000) + 3 = 19.000, d^2y/dx^2 = -8, \text{max}$

14. $dy/dx = 18x + 20, x = -20/(18) = -1.111, y = 9(-1.111)^2 + 20 \times (-1.111) + 5 = -6.111, d^2y/dx^2 = 18, \text{min}$

WORKSHEET

1. $y = 2x^3 - 33x^2 + 168x + 12$

$y' = 6x^2 - 66x + 168$

Factor this expression:

$y' = 6(x - 4)(x - 7)$

List the two values of x that make the first derivative zero: $x = 4$ and $x = 7$.

Find the second derivative:

$y'' = 12x - 66$

Value of the second derivative at $x = 4$:

$y'' = (12 \times 4) - 66 = -18$

Value of the second derivative at $x = 7$:

$y'' = (12 \times 7) - 66 = 18$

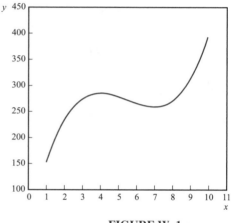

FIGURE W–1.

2. $y = 2x^3 - 51x^2 + 360x - 164$

$y' = 6x^2 - 102x + 360$

Factor this expression:

$y' = 6(x - 5)(x - 12)$

List the two values of x that make the first derivative zero: $x = 5$ and $x = 12$.

Find the second derivative:

$y'' = 12x - 102$

Value of the second derivative at $x = 5$:

$y'' = (12 \times 5) - 102 = -42$

Value of the second derivative at $x = 12$:

$y'' = (12 \times 12) - 102 = 42$

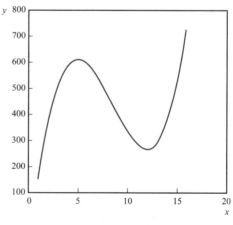

FIGURE W–2.

3. $y = 2x^3 - 72x^2 + 810x + 17$

$y' = 6x^2 - 144x + 810$

Factor this expression:

$y' = 6(x - 9)(x - 15)$

List the two values of x that make the first derivative zero: $x = 9$ and $x = 15$.

Find the second derivative:

$y'' = 12x - 144$

Value of the second derivative at $x = 9$:

$y'' = (12 \times 9) - 144 = -36$

Value of the second derivative at $x = 15$:

$y'' = (12 \times 15) - 144 = 36$

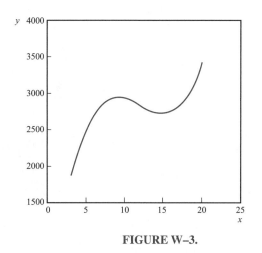

FIGURE W–3.

4. $y = 2x^3 + 48x^2 + 378x - 124$

$y' = 6x^2 + 96x + 378$

Factor this expression:

$y' = 6(x + 7)(x + 9)$

List the two values of x that make the first derivative zero: $x = -7$ and $x = -9$. Find the second derivative:

$y'' = 12x + 96$

Value of the second derivative at $x = -7$:

$y'' = (12 \times (-7)) + 96 = 12$

Value of the second derivative at $x = -9$:

$y'' = (12 \times (-9)) + 96 = -12$

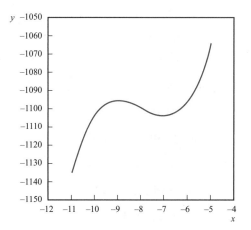

FIGURE W–4.

5. $y = 2x^3 + 6x^2 - 480x - 14$

$y' = 6x^2 + 12x - 480$

Factor this expression:

$y' = 6(x - 8)(x + 10)$

List the two values of x that make the first derivative zero: $x = 8$ and $x = -10$.

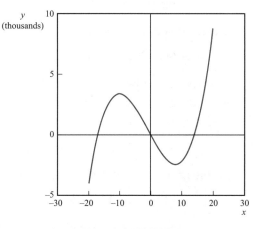

FIGURE W–5.

Find the second derivative:

$$y'' = 12x + 12$$

Value of the second derivative at $x = 8$:

$$y'' = (12 \times 8) + 12 = 108$$

Value of the second derivative at $x = -10$:

$$y'' = (12 \times (-10)) + 12 = -108$$

6.　　$y = 2x^3 - 51x^2 - 360x + 96$

$$y' = 6x^2 - 102x - 360$$

Factor this expression:

$$y' = 6(x + 3)(x - 20)$$

List the two values of x that make the first derivative zero: $x = -3$ and $x = 20$.

Find the second derivative:

$$y'' = 12x - 102$$

Value of the second derivative at $x = -3$:

$$y'' = (12 \times (-3)) - 102 = -138$$

Value of the second derivative at $x = 20$:

$$y'' = (12 \times 20) - 102 = 138$$

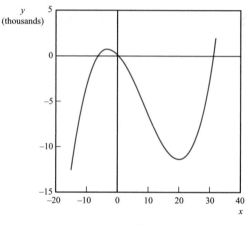

FIGURE W–6.

7.　　$y = 2x^3 + 9x^2 - 60x + 84$

$$y' = 6x^2 + 18x - 60$$

Factor this expression:

$$y' = 6(x - 2)(x + 5)$$

List the two values of x that make the first derivative zero: $x = 2$ and $x = -5$. Find the second derivative:

$$y'' = 12x + 18$$

Value of the second derivative at $x = 2$:

$$y'' = (12 \times 2) + 18 = 42$$

Value of the second derivative at $x = -5$:

$$y'' = (12 \times (-5)) + 18 = -42$$

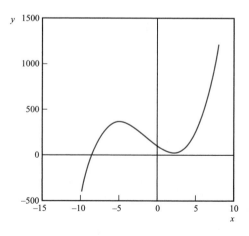

FIGURE W–7.

8. $y = 2x^3 + 60x^2 + 384x - 43$

$y' = 6x^2 + 120x + 384$

Factor this expression:

$y' = 6(x + 16)(x + 4)$

List the two values of x that make the first derivative zero: $x = -16$ and $x = -4$.

Find the second derivative:

$y'' = 12x + 120$

Value of the second derivative at $x = -16$:

$y'' = (12 \times (-16)) + 120 = -72$

Value of the second derivative at $x = -4$:

$y'' = (12 \times (-4)) + 120 = 72$

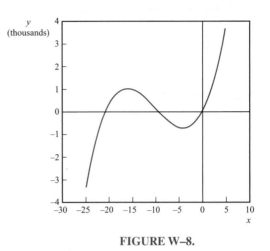

FIGURE W–8.

9. $y = 2x^3 - 24x^2 + 96x + 44$

$y' = 6x^2 - 48x + 96$

Factor this expression:

$y' = 6(x - 4)(x - 4)$

There is only one value of x that make the first derivative zero: $x = 4$. Find the second derivative:

$y'' = 12x - 48$

Value of the second derivative at $x = 4$:

$y'' = 12 \times 4 - 48 = 0$

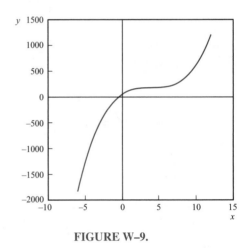

FIGURE W–9.

Since the second derivative is zero, you can't tell from this test whether this point is a local maximum, a local minimum, or neither. Here is the graph:

10. $y = 2x^3 - 9x^2 + 204x + 16$

$y' = 6x^2 - 18x + 204$

Try to solve for $y' = 0$ using the quadratic formula:

$$x = \frac{-(-18) \pm \sqrt{(-18)^2 - 4 \times 6 \times 204}}{2 \times 6} = \frac{18 \pm \sqrt{324 - 4{,}896}}{12} = \frac{18 \pm \sqrt{-4{,}572}}{12}$$

Since the quadratic formula requires you to find the square root of a negative number, there are no real number values of x that make the first derivative zero. This means this curve has no horizontal tangents. Here is the graph:

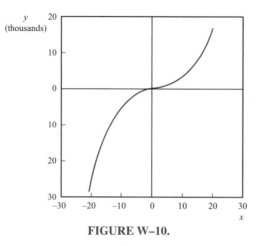

FIGURE W–10.

EXERCISES

1. $y' = -6x + 5$; $y'' = -6$; $x = 5/6$ is a maximum

2. $y' = 26x + 4$; $y'' = 26$; $x = -2/13$ is a minimum

3. $y' = 34x + 11$; $y'' = 34$; $x = -11/34$ is a minimum

4. $y' = 38x - 10$; $y'' = 38$; $x = 5/19$ is a minimum

5. $y' = -18x + 16$; $y'' = -18$; $x = 8/9$ is a maximum

6. (Acceleration) $= h'' = -g$. The acceleration is negative.

7. $h'' = -g$. Note that the acceleration due to gravity is the same whether Pal drops the ball or throws it into the air.

8. $x'' = 0$. Pal is traveling with a constant velocity. Any object traveling with a constant velocity has zero acceleration.

9. $y' = 2x - 3$. $y' = 0$ when $x = 3/2$. $y'' = 2$. Since y'' is positive, the curve has a local minimum at $x = 3/2$. The curve never turns down, so the absolute minimum occurs at $x = 3/2$, $y = -9/4$. Since there are no other points where the curve has horizontal tangents, the maximum value of the function must occur at one of the end points of the interval ($x = 0$ or $x = 5$). Such a case is known as a corner solution. In this case the maximum value occurs at $x = 5$, $y = 10$.

10. $y' = -4x^3 + 16x$. $y'' = -12x^2 + 16$. $y' = 0$ when $x = 0$, $x = 2$, or $x = -2$. At $x = 0$ the curve has a local minimum. At $x = 2$ and $x = -2$ it has local maxima. It is concave upward when $16 - 12x^2$ is positive, which means that $-2/\sqrt{3} < x < 2/\sqrt{3}$.

11. $y' = 2x - x^2$. The curve is rising when $0 < x < 2$. $y'' = -2x + 2$. The curve is concave downward when $x > 1$.

12. $d^4y/dx^4 = 0$.

13. $d^3y/dx^3 = 6a$.

14. $d^ny/dx^n = n!$ (*n* factorial)

15. See the book's Web page: *http://myhome.spu.edu/ddowning/easycalc.html*

16. $y = x^3 - 15x^2 + 48x + 12$; $y' = 3x^2 - 30x + 48$; $y'' = 6x - 30$; $y = 0$ when $x = -0.2328$, $x = 5.0741$, or $x = 10.1587$; $y' = 0$ when $x = 2$ (local maximum) and $x = 8$ (local minimum). $y'' = 0$ when $x = 5$. See Figure A–3.

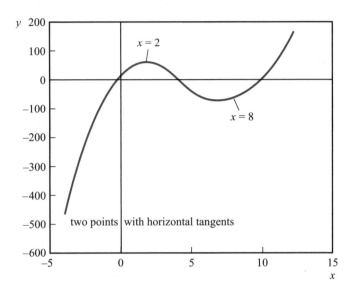

two points with horizontal tangents

FIGURE A–3.

17. $y = x^3 - 36x^2 + 432x - 37$; $y' = 3x^2 - 72x + 432$; $y'' = 6x - 72$. $y = 0$ when $x = 0.0863$; $y' = 0$ when $x = 12$; $y'' = 0$ when $x = 12$. See Figure A–4.

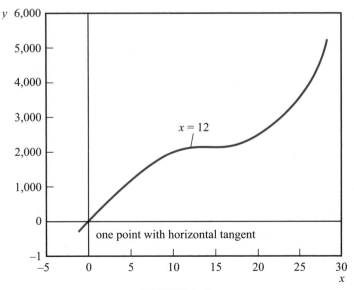

one point with horizontal tangent

FIGURE A–4.

18. $y = x^3 + 3x^2 + 243x + 600$; $y' = 3x^2 + 6x + 243$; $y'' = 6x + 6$; $y = 0$ when $x = -2.4823$; y' is never zero; $y'' = 0$ when $x = -1$. See Figure A–5.

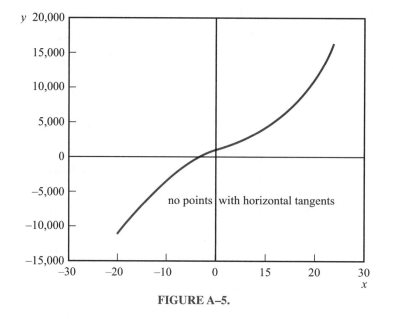

FIGURE A–5.

19. $y = x^4 - 11x^3 - 7x^2 + 155x + 150$; $y' = 4x^3 - 33x^2 - 14x + 155$; $y'' = 12x^2 - 66x - 14$; $y = 0$ when $x = -3$, $x = 1$, $x = 5$, or $x = 10$; $y' = 0$ when $x = -2.1103$ (local minimum), $x = 2.2696$ (local maximum), or $x = 8.0906$ (absolute minimum); $y'' = 0$ when $x = -0.2045$ or $x = 5.7045$. See Figure A–6.

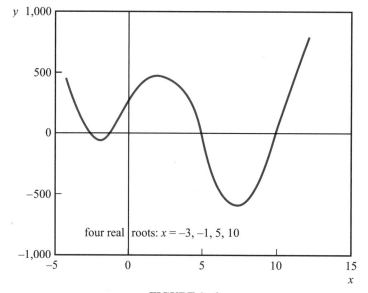

FIGURE A–6.

20. $y = x^4 - 19x^3 + 114x^2 - 256x + 160$; $y' = 4x^3 - 57x^2 + 228x - 256$; $y'' = 12x^2 - 114x + 228$; $y = 0$ when $x = 1$, $x = 4$, or $x = 10$; $y' = 0$ when $x = 1.9210$ (local minimum), $x = 4$ (local maximum), $x = 8.3290$ (absolute minimum); $y'' = 0$ when $x = 2.8625$ or $x = 6.6375$. See Figure A–7.

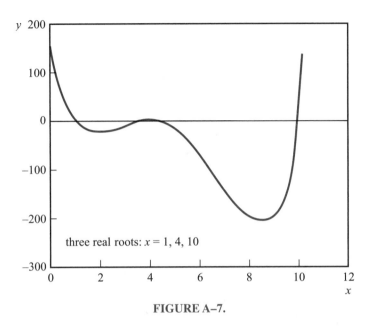

three real roots: $x = 1, 4, 10$

FIGURE A–7.

21. $y = x^4 - 4x^3 - 5x^2 - 2x + 10$; $y' = 4x^3 - 12x^2 - 10x - 2$; $y'' = 12x^2 - 24x - 10$; $y = 0$ when $x = 1$ or $x = 5$; $y' = 0$ when $x = 3.7102$ (absolute minimum); $y'' = 0$ when $x = -0.3540$ or $x = 2.3540$. See Figure A–8.

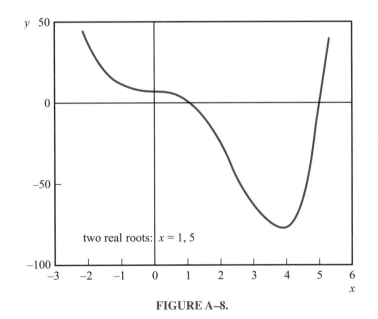

two real roots: $x = 1, 5$

FIGURE A–8.

22. $y = x^4 - 6x^3 + 2x^2 + 16x + 32$; $y = 4x^3 - 18x^2 + 4x + 16$; $y'' = 12x^2 - 36x + 4$; $y = 0$ when $x = 4$; $y' = 0$ when $x = -0.7808$ (local minimum), $x = 1.2808$ (local maximum), $x = 4$ (absolute minimum); $y'' = 0$ when $x = 0.1156$ or $x = 2.8844$. See Figure A–9.

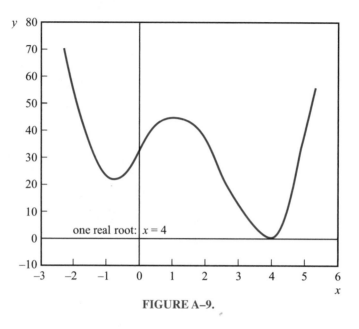

FIGURE A–9.

23. $y = x^4 - 11x^3 - 7x^2 + 155x + 800$; $y' = 4x^3 - 33x^2 - 14x + 155$; $y'' = 12x^2 - 66x - 14$; y does not ever equal zero; $y' = 0$ when $x = -2.1103$ (local minimum), $x = 2.2696$ (local maximum), or $x = 8.0906$ (absolute minimum); $y'' = 0$ when $x = -0.2045$ or $x = 5.7045$. See Figure A–10. (Compare this answer with Exercise 18.)

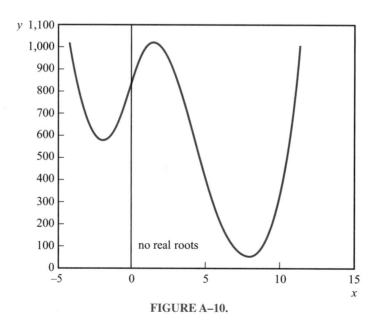

FIGURE A–10.

Chapter 4

WORKSHEET

	u	dv/dx	v	du/dx	$f'(x)$
1.	ax	b	bx	a	$axb + bxa = 2abx$
2.	$a + x$	1	$b + x$	1	$a + x + b + x = a + b + 2x$
3.	$a + bx$	k	$h + kx$	b	$(a + bx)k + (h + kx)b = 2bkx + ak + bh$
4.	ax^m	nbx^{n-1}	bx^n	max^{m-1}	$ab(m + n)x^{m+n-1}$
5.	$(a + x)^m$	$n(b + x)^{n-1}$	$(b + x)^n$	$m(a + x)^{m-1}$	$n(a + x)^m(b + x)^{n-1} + m(b + x)^n(a + x)^{m-1}$
6.	x	$(0.5)(1 - x^2)^{-.5}(-2x)$	$\sqrt{1 - x^2}$	1	$-x^2/\sqrt{1 - x^2} + \sqrt{1 - x^2}$

WORKSHEET

1. $y = \dfrac{a + x}{b + x} = (a + x)(b + x)^{-1}$

Let $u = (a + x)$; let $v = (b + x)^{-1}$; $du/dx = 1$; $dv/dx = (-1)(b + x)^{-2}$.

$\dfrac{dy}{dx} = (a + x)(-1)(b + x)^{-2} + (b + x)^{-1} \times 1 = \dfrac{b + x - (a + x)}{(b + x)^2} = \dfrac{b - a}{(b + x)^2}$

2. $y = \dfrac{a + bx}{h + kx} = (a + bx)(h + kx)^{-1}$

Let $u = (a + bx)$; let $v = (h + kx)^{-1}$; $du/dx = b$; $dv/dx = (-1)(h + kx)^{-2}k$.

$\dfrac{dy}{dx} = (a + bx)(-1)(h + kx)^{-2}k + (h + kx)^{-1}b = \dfrac{b(h + kx) - k(a + bx)}{(h + kx)^2}$

3. $y = \dfrac{(a + bx)^m}{(h + kx)^n} = (a + bx)^m(h + kx)^{-n}$

Let $u = (a + bx)^m$; let $v = (h + kx)^{-n}$; $du/dx = m(a + bx)^{m-1}b$; $dv/dx = (-n)(h + kx)^{-n-1}k$.

$\dfrac{dy}{dx} = (a + bx)^m(-n)(h + kx)^{-n-1}k + (h + kx)^{-n}m(a + bx)^{m-1}b$

4. $y = \dfrac{x}{\sqrt{1 - x^2}} = x(1 - x^2)^{-1/2}$

Let $u = x$; $v = (1 - x^2)^{-1/2}$; $du/dx = 1$; $dv/dx = (-1/2)(1 - x^2)^{-3/2}(2x) = -x(1 - x^2)^{-3/2}$.

$\dfrac{dy}{dx} = x(-x)(1 - x^2)^{-3/2} + (1 - x^2)^{-1/2} \times 1 = \dfrac{-x^2 + 1 - x^2}{(1 - x^2)^{3/2}} = \dfrac{1 - 2x^2}{(1 - x^2)^{3/2}}$

EXERCISES

1. $f'(t) = 2(3t + 4) + 3(2t - 5)$.

2. $y' = \dfrac{2}{3x - 1} - \dfrac{3(2x + 5)}{(3x - 1)^2}$.

3. $y' = \dfrac{10}{4x - 3} - \dfrac{40x}{(4x - 3)^2}$.

4. $y' = \dfrac{a}{cx + d} - \dfrac{c(ax + b)}{(cx + d)^2}$.

5. $y' = (x^2)(3x^2) + (2x)(x^3) = 3x^4 + 2x^4 = 5x^4$.

6. $y' = uv\dfrac{dw}{dx} + w\dfrac{d}{dx}(uv) = uv\dfrac{dw}{dx} + w\left(u\dfrac{dv}{dx} + v\dfrac{du}{dx}\right) = uv\dfrac{dw}{dx} + uw\dfrac{dv}{dx} + wv\dfrac{du}{dx}$.

   $y = x^3, y' = x \times x \times 1 + x \times x \times 1 + x \times x \times 1 = 3x^2$.

7. $y' = u(-1)v^{-2}\dfrac{dv}{dx} + \dfrac{du}{dx}\dfrac{1}{v} = \dfrac{v\, du/dx}{v^2} - \dfrac{u\, dv/dx}{v^2}$.

8. $dy/du = (3/2)u^{1/2}$. $du/dx = 2x$. $dy/dx = (3/2)(x^2 + 3)^{1/2}(2x)$. $dy/dx = 3x\sqrt{x^2 + 3}$.

9. $dy/du = \frac{1}{2}u^{-1/2}$. $du/dx = 2x$. $dy/dx = x(1 + x^2)^{-1/2}$.

10. $dy/du = (3/2)u^{1/2}$. $du/dx = 4$. $dy/dx = 6\sqrt{3 + 4x}$.

11. $dy/du = \frac{1}{2}u^{-1/2}$. $du/dx = 2ax + b$. $dy/dx = \frac{1}{2}(ax^2 + bx + c)^{-1/2}(2ax + b)$.

12. $dy/dv = \frac{1}{2}v^{-1/2}$. $dv/du = -(1/u^2)$. $du/dx = 2x$.

$$\frac{dy}{dx} = \frac{-x}{(x^2 + 4)^2\sqrt{1 + 1/(x^2 + 4)}}.$$

13. $y = \Delta y = \sqrt{x + \Delta x}$.

$$\frac{\Delta y}{\Delta x} = \frac{\sqrt{x + \Delta x} - \sqrt{x}}{\Delta x}.$$

This expression can be simplified by multiplying the top and bottom by $(\sqrt{x + \Delta x} + \sqrt{x})$.

$$\frac{\Delta y}{\Delta x} = \frac{x + \Delta x - x}{\Delta x(\sqrt{x + \Delta x} + \sqrt{x})} = \frac{1}{\sqrt{x + \Delta x} + \sqrt{x}}.$$

$$\frac{dy}{dx} = \lim_{\Delta x \to 0} \frac{1}{\sqrt{x + \Delta x} + \sqrt{x}} = \frac{1}{2\sqrt{x}}.$$

14. $y + \Delta y = 1/(x + \Delta x)$.

$$\frac{\Delta y}{\Delta x} = \frac{\dfrac{1}{x + \Delta x} - \dfrac{1}{x}}{\Delta x} = \frac{\dfrac{x}{(x + \Delta x)x} - \dfrac{x + \Delta x}{(x + \Delta x)x}}{\Delta x}.$$

$$= -\frac{1}{(x + \Delta x)x}.$$

$$\frac{dy}{dx} = \lim_{\Delta x \to 0} -\frac{1}{(x + \Delta x)x} = -\frac{1}{x^2}.$$

15. $dy/dx = nu^{n-1}(du/dx)$.

16. $y' = (4 + 2x)^{-1/2}$.

17. $y' = 9.9(4 + 3x)^{2.3}$.

18. $y' = \frac{1}{2}(3x^3 + 2x^2 + x)^{-1/2} (9x^2 + 4x + 1)$.

19. $y' = -3x(x^2 - 1)^{-3/2}$.

20. $y' = \frac{1}{3}x^{-2/3}$.

21. $(2x/3)(x^2 + 4)^{-2/3}$.

22. $y' = (-1/2x^2)(1 + 1/x)^{-1/2}$.

23. $y' = \frac{1}{2}(x + 1/x)^{-1/2}(1 - 1/x^2)$.

24. $y' = \dfrac{-2x}{(x^2 + 4)^2}$.

25. $\dfrac{d^2 A}{dw^2} = -\dfrac{w^3}{(r^2 - w^2)^{1.5}} - \dfrac{3w}{(r^2 - w^2)^{0.5}}$.

When $w = r/\sqrt{2}$, $d^2A/dw^2 = -4$. Since the second derivative is negative, the curve has a local maximum at this point. Recordis is building the largest possible house.

26. (a) $2x + 2y(dy/dx) = 0$. $dy/dx = -x/y$.

(b) $y = (r^2 - x^2)^{1/2}$. $dy/dx = -\frac{1}{2}(r^2 - x^2)^{-1/2}(2x) = -x/y$.

27. Using the implicit method:

$$\frac{2}{9}(y - 1)\frac{dy}{dx} - \frac{1}{2}(x - 3) = 0.$$

$$\frac{dy}{dx} = \frac{\frac{1}{2}(x - 3)}{\frac{2}{9}(y - 1)}.$$

Chapter 5

WORKSHEET

		$f(x)$	u	dv/dx	v	du/dx	$f'(x)$
1.		$\sin x \cos x$	$\sin x$	$-\sin x$	$\cos x$	$\cos x$	$\cos^2 x - \sin^2 x = \cos (2x)$
2.		$\sin (mx) \cos (nx)$	$\sin (mx)$	$-n \sin (nx)$	$\cos (nx)$	$m \cos (mx)$	$-n \sin (mx) \sin (nx) + m \cos (nx) \cos (mx)$

		$f(x)$	$f'(x)$
3.		$\sin^2 x$	$2 \sin x \dfrac{d \sin x}{dx} = 2 \sin x \cos x$
4.		$\cos^2 x$	$2 \cos x \dfrac{d \sin x}{dx} = -2 \cos x \sin x$
5.		$\sin^2 x + \cos^2 x$	0 (combine the previous two)
6.		$\sin (x^n)$	$\cos (x^n) \dfrac{dx^n}{dx} = nx^{n-1} \cos (x^n)$
7.		$(\sin x)^n$	$n(\sin x)^{n-1} \dfrac{d \sin x}{dx} = n (\sin x)^{n-1} \cos x$
8.		$\sec^2 x$	$2 \sec x(\tan x \sec x) = 2 \sec^2 x \tan x$
9.		$\tan^2 x$	$2 \tan x \sec^2 x$
10.		$\sec^2 x - \tan^2 x$	0 (combine the previous two)

EXERCISES

1. $y' = \cos (x^2)2x$.

2. $y' = 2 \sin x \cos x$.

3. $y' = (-1) \sin^{-2} x \cos x = -\text{ctn } x \csc x$.

4. $y' = x \cos x + \sin x$.

5. $y' = (\sin x)(-\sin x) + (\cos x)(\cos x) = \cos^2 x - \sin^2 x = \cos 2x$.

6. (a) $y = \sin (x^2) \cos x + \sin x \cos (x^2)$.

   $y' = \sin (x^2)(-\sin x) + (\cos x)(2x) \cos (x^2) + \sin x(2x)(-\sin x^2) + \cos (x^2) \cos x$

   $= (2x + 1)(\cos x \cos x^2 - \sin x \sin x^2)$

   $= (2x + 1) \cos (x^2 + x)$.

   (b) When the chain rule is used, the answer comes directly: $y' = \cos (x^2 + x)(2x + 1)$.

7. If A measured in degrees and θ in radians, then $A = 180\theta/\pi$. $y = \sin(180\theta/\pi)$. $y'(180/\pi)\cos(180\theta/\pi) = (180/\pi)\cos A$.

8. $y' = \cos x$. The curve will have horizontal tangents when $\cos x = 0$. This happens when $x = n\pi + \pi/2$, where n is any integer. When n is even, there is a maximum (as at $x = \pi/2$). When n is odd, there is a minimum (as at $x = 3\pi/2$). $y'' = -\sin x$. The curve is concave downward from $x = 0$ to $x = \pi$; it is concave upward from $x = \pi$ to $x = 2\pi$; etc.

9. $y' = \pi\cos(2\pi t/10)$. $y'' = -(\pi^2/5)\sin(2\pi t/10)$.

10. $32° = 0.559$. $x = \pi/6$, $y' = \cos(\pi/6) = 0.866$. Using the point-slope formula for the tangent line gives: $(y - \frac{1}{2})/(x - \pi/6) = 0.866$. $y = 0.866x + 0.047$. When $x = 0.559$, $y = 0.531$. The best decimal approximation for $\sin 32° = 0.5299$.

11. $y' = \frac{1}{2}(1 + \tan^2\theta)^{-1/2}(2\tan\theta\sec^2\theta) = \dfrac{\tan\theta\,\sec^2\theta}{\sec\theta} = \tan\theta\sec\theta$. Remember that $\sec\theta = (1 + \tan^2)^{1/2}$.

12. $y' = \sec^2\theta = 1/\cos^2\theta$; y' will never be zero, so this curve has no horizontal tangents. $y'' = 2\sec\theta\sec\theta\tan\theta$; $y'' = 2\sec^2\theta\tan\theta$. y'' will be positive when tan is positive, so the curve will be concave upward in the interval $\theta = 0$ to $\theta = \pi/2$. It will be concave downward in the interval $\theta = -\pi/2$ to $\theta = 0$.

13. See Table A–2.

Table A–2

x	$(\sin x)/x$
0.785	0.9004
0.5	0.9589
0.3	0.9851
0.1	0.9983
0.05	0.9996

14. $dx/dt = (0.8)(3)\cos(3t)$. $d^2x/dt^2 = -(0.8)(9)\sin(3t) = -9x$. From the equation of motion: $-kx = m\,d^2x/dt^2$. $-kx = m(-9x)$. $k = 9m$. $k = 18$.

15. midway between low tide and high tide.

16. (a) $dV/dt = \omega A\cos(\omega t)$.

    (b) $\omega = 2\pi \times 60$.

17. $dA/dt = wA_0\cos(wt + k)$.
    dA/dt will be the greatest when $(wt + k) = 0$ or 2π or other values coterminal with these. These are the values where $\sin(wt + k)$ is 0—in other words, when the pendulum is hanging straight down.

18. (a) $v = \omega A\cos(\omega t)$; kinetic energy $= KE = \omega^2A^2\cos^2(\omega t)/2$.

    (b) $dKE/dt = -m\omega^3A^2\cos(\omega t)\sin(\omega t)$.

    (c) potential energy $= PE = m\omega^2A^2\sin^2(\omega t)/2$

(d) $dPE/dt = m\omega^3 A^2 \sin(\omega t) \cos(\omega t)$.

(e) total energy $= TE = KE + PE = \omega^2 A^2/2$.

(f) $dTE/dt = 0$; that is, total energy is constant.

Chapter 6

EXERCISES

1. Let r equal the radius of the cone formed by the water. $r = ha/2b$. $V = \frac{1}{3}\pi r^2 h = \pi a^2 h^3/12b^2$.
    $dV/dh = \pi a^2 h^2/4b^2$. $dV/dt = u$. $dh/dt = 4b^2 u/\pi a^2 h^2 = 0.11$ meter/minute.

2. Let x be the distance from the wall to the bottom of the ladder, and let y be the distance up
    the wall to the top of the ladder. $x^2 + y^2 = L^2$. $dx/dt = u$. $dy/dx = -x/\sqrt{L^2 - x^2}$. $dy/dt = -ux/\sqrt{L^2 - x^2}$. (The result is negative because y is becoming smaller as the ladder slides down.)

3. Let D be the distance from the point to the curve:

    $$D = \sqrt{(x - 4)^2 + (y - 2)}.$$

    It will be easier to minimize D^2 than to attempt to minimize D:

    $$S = D^2 = (x^2 - 8x + 16) + (y^2 - 4y + 4)$$
    $$= x^2 - 8x + y^2 - 4y + 20.$$

    Solving for y, using the equation of the curve, gives:

    $$S = x^2 - 8 \cdot 2^{1/2} x^{1/2} + 20.$$
    $$\frac{dS}{dx} = 2x - 4\sqrt{2}x^{-1/2} = 0.$$
    $$x = 2; y = 4.$$

4. Maximize $A = xy$ subject to $2x + 2y = k$: $A = \frac{1}{2}kx - x^2$. $dA/dx = \frac{1}{2}k - 2x = 0$. $x = \frac{1}{4}k$; $y = \frac{1}{4}k$. The shape with maximum area will be a square.

5. Minimize $z = 2x + 2y$ subject to $xy = A$: $z = 2x + 2A/x$. $dz/dx = 2 - 2A/x^2 = 0$. $x = \sqrt{A}$. The shape with minimum perimeter will be a square.

6. $dY/dQ = P - (dC/dQ) = 0$. $P = dC/dQ = MC$. To maximize profits the firm must choose Q so that price is equal to marginal cost.

7. Set marginal cost equal to price:

    $$\frac{dTC}{dQ} = MC = 4.5Q^2 - 16Q + 32 = 1{,}858$$

    Rewrite the equation:

    $$4.5Q^2 - 16Q = 1{,}826 = 0$$

Use the quadratic formula:

$$Q = \frac{16 \pm \sqrt{256 - (-32{,}868)}}{9}$$

First, use the plus sign:

$$Q = \frac{16 + \sqrt{33{,}124}}{9} = \frac{16 + 182}{9} = \frac{198}{9}$$

$$Q = 22$$

If you check the second derivative for $Q = 22$ and $Q = (16 - \sqrt{33{,}124})/9$, you can verify that $Q = 22$ is the profit maximum.

8. $24Q^2 - 240Q + 780 = 2{,}124$; $Q = 14$

9. $6Q^2 - 60Q + 210 = 2{,}004$; $Q = 23$

10. $6Q^2 - 90Q + 365 = 3{,}929$; $Q = 33$

11. $9Q^2 - 120Q + 800 = 4{,}496$; $Q = 28$

12. $AC = TC/Q = TCQ^{-1}$

Use the product rule for derivatives to find the derivative of the average cost with respect to quantity:

$$\frac{dAC}{dQ} = TC\frac{d(Q^{-1})}{dQ} + \frac{1}{Q}\frac{dTC}{dQ}$$

$$= TC(-Q^{-2}) + Q^{-1}MC$$

$$= \left(\frac{TC}{Q}\right)\left(\frac{-1}{Q}\right) + \frac{MC}{Q}$$

$$= \frac{MC - AC}{Q}$$

If $MC > AC$, then the derivative of the average cost with respect to quantity is positive, so average cost will increase if the quantity increases.

13.
$$P = a/b - Q/b$$
$$TR = a/bQ - Q^2/b$$
$$MR = a/b - 2Q/b$$

Set $MR = 0$:
$$a/b = 2Q/b$$
$$Q = a/2$$

14. $dQ/dP = -b$

elasticity $|-bp/Q| = |-b(a/b - Q/b)/Q| = |-a/Q + 1|$

when $Q = a/2$, then the elasticity $= 1$

15. $TR = PQ$

To find the marginal revenue, use the product rule for derivatives:

$$\frac{dTR}{dQ} = P\,\frac{dQ}{dQ} + Q\,\frac{dP}{dQ}$$

$$= P + Q\,\frac{dp}{dQ}$$

$$= P\left(1 + \frac{Q}{P}\,\frac{dP}{dQ}\right)$$

$\dfrac{Q}{P}\,\dfrac{dP}{dQ}$ is equal to $-1/e$, where e is the elasticity.

$MR = P(1 - 1/e)$

The revenue is maximized where $MR = 0$, which means $P(1 - 1/e) = 0$, or the elasticity is equal to one. If the elasticity is less than one, then the marginal revenue is negative; if the elasticity is greater than one, then the marginal revenue is positive. This means that if the elasticity of demand is less than one it is possible to increase your revenue by raising your price and reducing quantity. However, if the elasticity is greater than one, then raising the price will cause the revenue to decline.

16.   Use the formula from the previous exercise for marginal revenue, and set that equal to marginal cost:

$MC = MR$
$MC = P(1 - 1/e)$
$P = MC/(P(1 - 1/e) = eMC/(e - 1)$

Solve for the markup:

$$\frac{P - MC}{MC} = \frac{eMC/(e - 1) - MC}{MC} = \frac{e}{e - 1} - 1 = \frac{e - (e - 1)}{e - 1}$$

$$\frac{P - MC}{MC} = \frac{1}{e - 1}$$

If the elasticity is larger, then the markup is smaller. This means that the buyers are more flexible, so the seller has less ability to charge higher prices.

17. $\dfrac{dQ}{dP} = -eaP^{-e-1}$

$$\text{elasticity} = \left|\frac{dQ}{dP}\,\frac{P}{Q}\right| = \left|\frac{-eaP^{-e-1}P}{Q}\right|$$

$$= \left|\frac{-eaP^{-e}}{aP^{-e}}\right| = e$$

This demand curve has the special property that the elasticity is the same at all points on the curve.

18. $V = hxy.\ y = V/xh.\ (\text{cost}) = C = 3rxy + 2rxh + 2ryh.\ dC/dx = 2rh - 2rV/x^2 = 0.\ x^2 = V/h.\ x = y.$

19. $C = 2rxy + 2ryh + rxh + 2rxh$

$= 2rV/h + 2rV/x + 3rxh.$

$dC/dx = -2rV/x^2 + 3rh.\ x = (2V/3h)^{1/2};\ y = (3V/2h)^{1/2}.$

20. $V = \text{volume} = \pi r^2 h.$

$S = (\text{surface area}) = 2\pi rh + 2\pi r^2$

$= 2V/r + 2\pi r^2.$

$dS/dr = -2V/r^2 + 4\pi r = 0.\ r^3 = V/2\pi;\ r = (V/2\pi)^{1/3}.$

21. $y' = 4x^3 - 9x^2 + 5.\ y'' = 12x^2 - 18x.\ x = 0, x = 3/2.\ x = 0$ is a local maximum. $x = 3/2$ is a local minimum. The absolute maximum will occur at plus infinity.

22.   Let the origin be at the point 20 meters due north of South Beach. (Notice that this point is also 20 meters due west of East Beach.) Let x equal the distance from Pal's boat to the origin, and y be the distance from Recordis' boat to the origin. Then $y = 2,000 - 5t$, and $x = 2,000 - 7t$. Let D be the distance between the two boats. $D = \sqrt{x^2 + y^2}.\ D^2 = 2,000^2 \times 2 + 74t^2 - 48,000t.$ To minimize $D^2, t = 48,000/148 = 324.32.$ When $t = 324.32, D = 465$ cm.

23.   Let y equal the distance that the weight has moved off the ground, and x be the distance from Pal to the point on the ground directly under the pulley. $dx/dt = u.\ h^2 + x^2 = (L - h + y)^2.\ dy/dx = x/(L - h + y).\ dy/dt = ux/(L - h + y).$

24.   Let f be the amount of fuel that the car uses per hour. Then the total fuel used on the trip will be $F = Tf$, where T is the total time that the trip takes. $T = D/v.$

$F = (D/v)(10v^2 - 100v + 290)$

$= D(10v - 100 + 290/v).$

$dF/dv = D(10 - 290/v^2) = 0.\ 10 = 290/v^2.\ v = \sqrt{29} = 5.39.$

25.   Distance from a point on the line $y = mx + b$ to the point (h, k):

$$z = \sqrt{(x-h)^2 + [(mx+b)-k]^2}.$$

Let $u = z^2.$

$$u = (x - h)^2 + [(mx + b) - k]^2$$

$$u = x^2 - 2xh + h^2 + (mx + b)^2 - 2(mx + b)k + k^2$$

$$u = x^2 - 2xh + h^2 + (mx)^2 + 2mxb + b^2 - 2(mx + b)k + k^2$$

$$u = x^2 - 2xh + h^2 + m^2x^2 + 2mxb + b^2 - 2mxk - 2bk + k^2$$

$$\frac{du}{dx} = 2x - 2h + 2m^2x + 2mb - 2mk$$

$$0 = (2 + 2m^2)x - 2h + 2mb - 2mk$$

$$0 = (1 + m^2)x - h + mb - mk$$

$$h - mb + mk = (1 + m^2)x$$

$$x = \frac{h - mb + mk}{1 + m^2}$$

26. If the ball is thrown at a steep angle, it goes very high, but it does not travel far. If it is thrown at too shallow of an angle, then it doesn't stay in the air too long so it doesn't go very far.

Solve this equation to find the time when the height will be zero:

$$0 = -0.5gt^2 + tv \sin \theta$$

One solution is $t = 0$, but that just means the height is zero at the moment the ball is thrown. (We're measuring the height above the level of your arm, not the height above the ground.) To find the other solution (for the time when the ball's height is again at arm level), factor out t:

$$0 = t(-0.5gt + v \sin \theta)$$

Set the factor in parentheses equal to zero:

$$0 = -0.5gt_{air} + v \sin \theta$$

Solve for t_{air}:

$$0.5gt_{air} = v \sin \theta$$

$$t_{air} = \frac{2v \sin \theta}{g}$$

Put this expression into the formula for distance:

$$z = v_h t_{air} = (v \cos \theta)t = (v \cos \theta)\left(\frac{2v \sin \theta}{g}\right)$$

$$z = \frac{v^2}{g}(2 \sin \theta \cos \theta)$$

Remember the formula for sin 2θ:

$$z = \frac{v^2}{g} \sin (2\theta)$$

$$\frac{dz}{d\theta} = \frac{v^2}{g}[2 \cos (2\theta)]$$

Set the derivative equal to zero:

$$0 = \frac{v^2}{g}[2 \cos (2\theta)]$$

$$0 = \cos (2\theta)$$

$$2\theta = \arccos (0) = \pi/2$$

$$\theta = \pi/4 \text{ radians} = 45 \text{ degrees}$$

Chapter 7

WORKSHEET

	$f(x)$	$\int f(x)\,dx$
1.	6	$6x + C$
2.	32	$32x + C$
3.	68	$68x + C$
4.	$4x + 12$	$2x^2 + 12x + C$
5.	$-8x - 64$	$-4x^2 + 64x + C$
6.	$-16x - 54$	$-8x^2 - 54x + C$
7.	$6x^2 + 4x + 49$	$2x^3 + 2x^2 + 49x + C$
8.	$9x^2 - 8x - 762$	$3x^3 - 4x^2 - 762x + C$
9.	$12x^2 + 6x + 16$	$4x^3 + 3x^2 + 16x + C$
10.	$12x^3 - 21x^2 - 18x - 86$	$3x^4 - 7x^3 - 9x^2 - 86x + C$
11.	$8x^3 - 30x^2 + 12x + 78$	$2x^4 - 10x^3 + 6x^2 + 78x + C$
12.	$32x^3 + 18x^2 + 20x + 604$	$8x^4 + 6x^3 + 10x^2 + 604x + C$
13.	$10x^4 + 16x^3 + 12x^2 + 8x - 13$	$2x^5 + 4x^4 + 4x^3 + 4x^2 - 13x + C$
14.	$20x^4 + 24x^3 + 24x^2 + 44x + 42$	$4x^5 + 6x^4 + 8x^3 + 22x^2 + 42x + C$
15.	$75x^4 + 20x^3 + 15x^2 + 24x + 121$	$15x^5 + 5x^4 + 5x^3 + 12x^2 + 121x + C$
16.	$y = x^4 - 3x^3 + 5x^2 - 7x + 6$	$\dfrac{x^5}{5} - \dfrac{3x^4}{4} + \dfrac{5x^3}{3} - \dfrac{7x^2}{2} + 6x + C$
17.	$y = x^3 - 7x^2 + 9x + 4$	$\dfrac{x^4}{4} - \dfrac{7x^3}{3} + \dfrac{9x^2}{2} + 4x + C$
18.	$y = 5x^3 + 4x^2 - 11x + 17$	$\dfrac{5x^4}{4} + \dfrac{4x^3}{3} - \dfrac{11x^2}{2} + 17x + C$

WORKSHEET

1. $y = \int \sin(mx)\,dx$. Let $u = mx$, $du/dx = m$, $du = m\,dx$, $dx = du/m$

   $y = \int \sin u\,du/m = (1/m) \int \sin u\,du = -(1/m)\cos u + C = -(1/m)\cos(mx) + C$

2. $y = \int \sqrt{a + x}\,dx$. Let $u = a + x$, $du/dx = 1$, $du = dx$.

   $y = \int \sqrt{u}\,du = \dfrac{u^{1.5}}{1.5} = \dfrac{(a + x)^{1.5}}{1.5}$

3. $y = \int \sqrt{ax + b}\,dx$. Let $u = ax + b$, $du/dx = a$, $du = du/a$.

   $y = \int \sqrt{u}\,du/a = \dfrac{u^{1.5}}{1.5a} = \dfrac{(ax + b)^{1.5}}{1.5a}$

4. $y = \int (ax + b)^n\,dx$. Let $u = ax + b$, $du/dx = a$, $dx = du/a$.

   $y = \int u^n\,du/a = \dfrac{u^{n+1}}{(n + 1)a} = \dfrac{(ax + b)^{n+1}}{(n + 1)a}$

5. $y = \int \cos x(\sin x)^3\,dx$. Let $u = \sin x$.

   $du = \cos x\,dx$; $y = \int u^3\,du = \dfrac{u^4}{4} = \dfrac{(\sin x)^4}{4} + C$

6. $y = \int \sin x(\cos x)^4\,dx$. Let $u = \cos x$.

   $du = -\sin x\,dx$; $y = -\int u^4\,du = -\dfrac{u^5}{5} = -\dfrac{(\cos x)^5}{5} + C$

7. $y = \int \cos x(\sin x)^n\,dx$. Let $u = \sin x$.

   $du = \cos x\,dx$; $y = \int u^n\,du = \dfrac{u^{n+1}}{n+1} = \dfrac{(\sin x)^{n+1}}{n+1} + C$

8. $y = \int x^{n-1} \sin(x^n)\,dx$. Let $u = x^n$.

   $du = nx^{n-1}\,dx$; $y = \dfrac{1}{n} \int \sin u\,du = -\dfrac{1}{n}\cos u = -\dfrac{1}{n}\cos(x^n) + C$

9. $y = \int \dfrac{\sin(\sqrt{x})}{\sqrt{x}}\,dx$. Let $u = \sqrt{x}$.

   $du = \dfrac{1}{2}x^{-1/2}\,dx$; $y = 2 \int \sin u\,du = -2\cos u + C = -2\cos\sqrt{x} + C$

10. $y = \int \cos x(a + \sin x)^n\,dx$. Let $u = a + \sin x$.

   $du = \cos x\,dx$; $y = \int u_n\,du = \dfrac{u^{n+1}}{n+1} = \dfrac{(a + \sin x)^{n+1}}{n+1} + C$

11. $y = \int \cos x \sqrt{a + \sin x}\ dx.$ Let $u = a + \sin x.$

$$du = \cos x\ dx;\ y = \int \sqrt{u}\ du = \frac{2}{3}u^{3/2} = \frac{2}{3}(a + \sin x)^{3/2} + C$$

12. $y = \int (ax^2 + bx)^n(2ax + b)\ dx.$ Let $u = ax^2 + bx.$

$$du = (2ax + b)\ dx;\ y = \int u^n\ du = \frac{u^{n+1}}{n+1} = \frac{(ax^2 + bx)^{n+1}}{n+1} + C$$

13. $y = \int 2ax(ax^2 + b)^n\ dx.$ Let $u = ax^2 + b.$

$$du = 2ax\ dx;\ y = \int u^n\ du = \frac{u^{n+1}}{n+1} = \frac{(ax^2 + b)^{n+1}}{n+1} + C$$

14. $y = \int x(ax^2 + b)^2\ dx.$ Let $u = ax^2 + b.$
$du = 2ax\ dx$

$$\frac{1}{2a} \int u^2\ du = \frac{1}{2a}\frac{1}{3}u^3 = \frac{1}{6a}(ax^2 + b)^3$$

$$= \frac{1}{6a}(a^3x^6 + 3a^2x^4b + 3ax^2\,b^2 + b^3) = \frac{1}{6}(a^2x^6 + 3ax^4b + 3x^2b^2 + \frac{b^3}{a})$$

$$= \frac{1}{6}a^2x^6 + \frac{1}{2}ax^4b + \frac{1}{2}x^2b^2 + C\ (\text{since } \frac{b^3}{a} \text{ is a constant})$$

15. $y = \int (a^2x^5 + 2ax^3b + b^2x)\ dx$

$$y = \frac{1}{6}a^2x^6 + \frac{2}{4}ax^4b + \frac{1}{2}b^2x^2$$

The answers for Questions 14 and 15 are the same except for the constants.

EXERCISES

1. $(1/3)x^3 + (3/2)x^2 + 5x + C.$

2. $(a/3)x^3 + (b/2)x^2 + cx + C.$

3. $(9/2)x^2 + 10x + C.$

4. $14x + C.$

5. $\frac{1}{4}x^4 + x + C.$

6. $x^6 + (5/2)x^4 + (3/2)x^2 + C.$

7. $x^4 + x^3 + x^2 + x + C.$

8. $\frac{1}{12}x^4 + \frac{1}{6}x^3 + \frac{1}{2}x^2 + C.$

9. $x^{m+1}/(m+1) + x^{n+1}/(n+1) + C$.

10. $x^{m+1} + x^{n+1} + x^{p+1} + C$.

11. $(1/101)x^{101} + C$.

12. $-\cos\theta + C$.

13. $\sin\theta + C$.

14. $\tan\theta + C$ (remember what the derivative of the tangent function is).

15. $(-\frac{1}{3})x^{-3} - \frac{1}{2}x^{-2} + C$.

16. $\frac{1}{3}x^{-3} - x^{-1} + C$.

17. $\operatorname{ctn}\theta + C$.

18. $\sec\theta + C$.

19. $\csc\theta + C$.

20. $\frac{1}{2}\theta - \frac{1}{4}\sin 2\theta + C$.

21. Use this identity: $\sin^2\theta = \frac{1}{2}(1 - \cos 2\theta)$. The answer is the same as for Exercise 20: $\frac{1}{2}\theta - \frac{1}{4}\sin 2\theta + C$.

22. $x\int x^2\,dx = \frac{1}{3}x^4 + Cx$. $\int x^3\,dx = \frac{1}{4}x^4 + C$. The two results are different, showing that a variable such as x cannot be moved across an integral sign.

23. $x = 5$. This is an object that is not moving at all, such as Hasselbluff Mountain.

24. $x = 4t + 2$. This object moves with a constant velocity, such as an ice skater on a frictionless lake.

25. $x = 55t - 265$. If t is measured in hours and x in miles, this object is a driver who leaves home at 8 o'clock in the morning, enters the freeway at mile 175, and drives at a constant speed of 55 miles per hour.

26. $x = \frac{1}{2}at^2$. This object moves with a constant acceleration, as the train did when Pal pushed it.

27. $x = \frac{1}{2}at^2 + x_0$. This equation represents the time Pal started pushing the train at the point where $x = x_0$.

28. $x = (1/a)\sin(at) + \frac{1}{2} - (1/a)$. This object is a weight attached to a spring. It keeps moving back and forth.

29. $x = \frac{1}{5}t^5 + \frac{4}{5}$. I don't know any objects that move like this.

30. Integrate once to get dx/dt: $dx/dt = \int a\,dt = at + C$. Now use the initial condition $dx/dt = v_0$ when $t = 0$ to show that $dx/dt = at + v_0$. Now integrate again to find x: $x = \int(at + v_0)\,dt$. $x = \frac{1}{2}at^2 + v_0t + C$.

Use the initial condition $x = x_0$ when $t = 0$ to show that $x = \frac{1}{2}at^2 + v_0t + x_0$. This represents an object that starts at x_0 with velocity v_0 and then moves with constant acceleration, as Pal's ball does when he throws it into the air with initial velocity v_0 from height x_0.

31. Integrating once gives $dx/dt = C$. Using the initial condition gives $dx/dt = v_0$. Integrating again gives $x = v_0t + x_0$. This object is another ice skater.

32. $dy = 2x\,dx.$

33. $dy = \cos x\,dx.$

34. $dy = c\,dx.$

35. $dy = \frac{1}{2}x^{-1/2}\,dx.$

36. $dy = x(1 + x^2)^{-1/2}\,dx.$

37. $du = -2x\,dx.\ y = -\frac{1}{2}\int u^{1/2}\,du = -\frac{1}{3}u^{3/2} + C = -\frac{1}{3}(1 - x^2)^{3/2} + C.$

38. $du = 3bx^2\,dx.\ y = (1/3b)\int u^{1/2}\,du = (2/9b)\,(a + bx^3)^{3/2} + C.$

39. $du = 2x\,dx.\ y = \int \frac{1}{2}\sin u\,du = -\frac{1}{2}\cos(x^2) + C.$

40. $du = \cos x\,dx.\ y = \int u^3\,du = \frac{1}{4}\sin^4 x + C.$

41. Let $u = 5 + 6x^2.\ du = 12x\,dx.\ y = \frac{1}{12}\int u^{-2}\,du.\ y = -\frac{1}{12}(5 + 6x^2)^{-1}.$

42. Let $u = a + x^n.\ du = nx^{n-1}\,dx.\ y = (1/n)\int u^{1/2}\,du.\ y = (2/3n)\,(a + x^n)^{3/2} + C.$

43. Let $u = \tan x.\ du = \sec^2 x\,dx.\ y = \int u^3\,du = \frac{1}{4}\tan^4 x + C.$

44. Let $u = x^{10}.\ du = 10x^9\,dx.\ y = \int \sin u\,du\ (\frac{1}{10}).\ y = -\frac{1}{10}\cos x^{10} + C.$

45. $y = [u^{n+1}/(n + 1)] + C.$

46. $y = -\cos \theta.$

47. (a) Integrate once: $v = \int -g\,dt = -gt + C.$ Since $v = v_0$ when $t = 0,\ v = -gt + v_0.$

   (b) Integrating again gives $h = \int(-gt + v_0)\,dt.\ h = -\frac{1}{2}gt^2 + v_0 t + C.\ h = 0$ when $t = 0,$ so $C = 0.$
   $h = -\frac{1}{2}gt^2 + v_0 t.$

48. $v = -gt.\ h = -\frac{1}{2}gt^2 + 64.$

49. $\int \frac{dy}{dx}\,dx = y + C;\ \frac{d}{dx}\int y\,dx = y.$

Chapter 8

WORKSHEET

1. $\int_{10}^{20} 17\,dx = 17x\Big|_{10}^{20} = (17 \times 20) - (17 \times 10) = 340 - 170 = 170$

2. $\int_{3}^{8}(12x + 20)\,dx = (6x^2 + 20x)\Big|_{3}^{8} = (6 \times 8^2) + (20 \times 8) - [(6 \times 3^2) + (20 \times 3)] = 544 - 114 = 430$

3. $\int_{2}^{10}(9x^2 + 8x + 60)\,dx = (3x^3 + 4x^2 + 60x)\Big|_{2}^{10}$

   $= (3 \times 10^3) + 4 \times (10^2) + (60 \times 10) - [(3 \times 2^3) + (4 \times 2^2) + (60 \times 2)] = 4{,}000 - 160 = 3{,}840$

4. $\int_1^5 (24x^3 + 21x^2 + 6x + 30)\, dx = (6x^4 + 7x^3 + 3x^2 + 30x)\Big|_1^5$

    $= (6 \times 5^4) + 7 \times 5^3 + 3 \times 5^2 + 30 \times 5) - [(6 \times 1) + (7 \times 1) + (3 \times 1) + (30 \times 1)]$

    $= (6 \times 625) + (7 \times 125) + (3 \times 25) + 150 - (6 + 7 + 3 + 30) = 4{,}850 - 46 = 4{,}804$

5. $\int_0^3 (25x^4 + 16x^3 + 6x^2 + 14x + 10)\, dx = (5x^5 + 4x^4 + 2x^3 + 7x^2 + 10x)\Big|_0^3$

    $= (5 \times 3^5) + (4 \times 3^4) + (2 \times 3^3) + (7 \times 3^2) + (10 \times 3) - [(5 \times 0^5) + (4 \times 0^4) + (2 \times 0^3) + (7 \times 0^2)$
    $+ (10 \times 0)]$

    $= (5 \times 243) + (4 \times 81) + (2 \times 27) + (7 \times 9) + 30 - 0 = 1{,}686$

EXERCISES

1. Area $= \int_{-\pi/2}^{\pi/2} \cos x\, dx = \sin x\, \Big|_{-\pi/2}^{\pi/2} = 1 - (-1) = 2.$

2. Area $= \int_0^\pi \frac{1}{2}(1 - \cos 2x)\, dx = \frac{1}{2}x\Big|_0^\pi - \frac{1}{4}\sin 2x\Big|_0^\pi = \pi/2.$

3. Area $= \int_0^{\pi/5} 3 \sin 5x\, dx = (-3/5) \cos 5x\, \Big|_0^{\pi/5} = (-3/5)(\cos \pi - \cos 0) = 6/5.$

4. $\int_{-2}^2 (2x^2 - 8)\, dx = (2/3)x^3\Big|_{-2}^2 - 8x\Big|_{-2}^2 = 16/3 + 16/3 - 16 - 16 = -21.3.$

    The area is 21.3. (Note that the value of the definite integral is negative, because the function is negative everywhere in the interval where the integration is carried out.)

5. $(1/3) + (3/2) + 5.$

6. $(9/2)x^2\Big|_{-3}^5 + 10x\Big|_{-3}^5 = 152.$

7. $(a/3)(d_2^3 - d_1^3) + (b/2)(d_2^2 - d_1^2) + c(d_2 - d_1).$

8. $(a/3)(d_2^3 - d_1^3) + (b/2)(d_2^2 - d_1^2) + c(d_2 - d_1).$

9. $14a.$

10. $8/3.$

11. $(5/6) + (5/2) + (3/2) = 4.83.$

12. $4.$

13. $\frac{1}{12}a^4 + \frac{1}{6}a^3 + \frac{1}{2}a^2.$

14. $\dfrac{d_2^{m+1} - d_1^{m+1}}{m + 1} + \dfrac{d_2^{n+1} - d_1^{n+1}}{n + 1}$

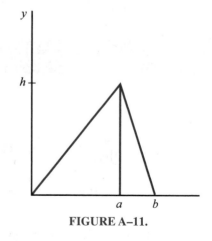

FIGURE A–11.

15. 3.

16. 2/101.

17. $\int_a^b \sin\theta\, d\theta = -\cos\theta\Big|_a^b = -(\cos b - \cos a) = -\cos b + \cos a.$

    (a) $\int_0^{\pi/4} \sin\theta\, d\theta = -\cos(\pi/4) + \cos(0) = 2^{-1/2} + 1 = 0.293.$

    (b) $\int_0^{\pi/2} \sin\theta\, d\theta = -\cos(\pi/2) + \cos(0) = 0 + 1 = 1.$

    (c) $\int_0^{\pi} \sin\theta\, d\theta = -\cos(\pi) + \cos(0) = -(-1) + 1 = 2.$

    (d) $\int_0^{2\pi} \sin\theta\, d\theta = -\cos(2\pi) + \cos(0) = -1 + 1 = 0.$

    (In this case the positive arch of the curve from $\theta = 0$ to $\theta = \pi$ cancels out the negative arch of the curve from $\theta = \pi$ to $\theta = 2\pi$, so the final result for the definite integral is zero.)

    (e) $\int_{-\pi}^{\pi} \sin\theta\, d\theta = \cos(\pi) + \cos(-\pi) = -(-1) - 1 = 0.$

18. $\sin(\pi/4) - \sin(\pi/4) = 2/\sqrt{2} = \sqrt{2} = 1.414.$

19. $(-1/3)x^{-3}\Big|_1^2 - \frac{1}{2}x^{-2}\Big|_1^2 = \frac{2}{3}.$

20. $\frac{1}{3}x^3\Big|_1^2 - x^{-1}\Big|_1^2 = 17/6.$

21. The area of the triangle will be the same wherever the triangle is located, so we may as well choose the most convenient location. We can put one vertex of the triangle at the origin, and we can let one side coincide with the x axis. The base of the triangle we can call b, and the height we can call h (Figure A–2). We need to solve two integrals to find the total area of the triangle. The first integral represents the area under the line $y = hx/a$, from $x = 0$ to $x = a$. The second integral represents the area under the line $y = hx/(a - b) + bh/(b - a)$, from $x = a$ to $x = b$.

$$A = \int_0^a \frac{hx}{a}\, dx + \int_a^b \left(\frac{hx}{a - b} + \frac{bh}{b - a} \right) dx$$

$$= \frac{h}{a} \frac{1}{2} x^2 \Big|_0^a + \frac{h}{a-b} \frac{1}{2} x^2 \Big|_a^b + \frac{bh}{b - a} x \Big|_a^b$$

$$= \frac{1}{2} ha + \frac{h}{a-b} \frac{1}{2} (b^2 - a^2) + bh$$

$$= \frac{1}{2} ha + \frac{h}{a-b} \frac{1}{2} (b - a)(b + a) + bh$$

$$= \frac{1}{2}ha - \frac{1}{2}h(b + a) + bh$$

$$= \frac{1}{2}ha - \frac{1}{2}hb - \frac{1}{2}ha + bh$$

$$A = \frac{1}{2}bh.$$

22. $\int_a^b f(x)\, dx = F(b) - F(a) = -[F(a) - F(b)] = -\int_b^a f(x)\, dx.$

23. $y = -\cos \theta \Big|_0^{2\pi} = -\cos (2\pi) + \cos 0 = -1 + 1 = 0.$

24. (a) $\Delta x = \pi/6.\, A \simeq (\pi/6)(0.5 + 0.8660) = 0.7152.$

   (b) $\Delta x = \pi/18.\, A \simeq (\pi/18)\, S = 0.9102.$

   (Here $S = \sum_{i=1}^{9} \sin (\pi i/18) = 5.21,$ which is the sum of the first eight elements in Table 8–1.)

   Evaluating the definite integral shows that the true area is 1.00.

25. $y_{av} = [1/(3 - 1)] \int_1^3 x^2\, dx = 13/3.$

26. $y_{av} = \frac{1}{4}.$

27. $V_{av}^2 = \frac{1}{2}A^2.\, V_{rms} = A2^{-1/2} = 0.707A.$ Since $V_{pp} = 2A,$ it follows that $V_{pp} = 2.8V_{rms}.$ When $V_{rms} = 1,$
   $V_{pp} = 2.8.$

28. Since the space from $x = 0$ to $x = a$ is divided into n rectangles of uniform width, we can see that $\Delta x = a/n.$ We can also tell that $x_i = ia/n.$ That makes the area equal to:

$$A = \lim_{n \to \infty} \sum_{i=1}^{n} \frac{i^2 a^2}{n^2} \frac{a}{n}.$$

Since a^3/n^3 is constant, we can pull it across the summation sign:

$$A = \lim_{n \to \infty} \frac{a^3}{n^3} \sum_{i=1}^{n} i^2.$$

Now we use the formula:

$$A = a^3 \lim_{n \to \infty} \frac{1}{n^3} \frac{n}{6}(2n^2 + 3n + 1)$$

$$= a^3 \lim_{n \to \infty} \left(\frac{1}{3} + \frac{1}{2n} + \frac{1}{6n^2} \right)$$

$$= \frac{1}{3}a^3$$

(This result is the same as the result of the definite integral.)

29. The area between the curves is the difference of the areas under the curves:

$$A = \int_a^b f(x)\, dx - \int_a^b g(x)\, dx$$

$$= \int_a^b f(x)\, dx + \int_a^b (-1)g(x)\, dx$$

$$= \int_a^b [f(x) - g(x)]\, dx$$

Chapter 9

WORKSHEET

1. $y = \ln x$, $dy/dx = 1/x$

2. $y = \ln (a + x)$, $dy/dx = 1/(a + x)$

3. $y = \ln (bx)$, $dy/dx = \dfrac{1}{bx} \times b = \dfrac{1}{x}$

4. $y = \ln (a + bx)$, $dy/dx = b/(a + bx)$

5. $y = \ln (x^n)$, $y = n \ln x$, $dy/dx = n/x$

6. $\int \left(\dfrac{1}{x}\right) dx = \ln x + C$

7. $\int \left(\dfrac{1}{a + x}\right) dx = \ln (a + x) + C$

8. $\int \left(\dfrac{1}{bx}\right) dx = (\ln x)/b$

9. $I = \int \left(\dfrac{2ax}{ax^2 + b}\right) dx$, let $u = ax^2 + b$, $du = 2ax\,dx$, $I = \int \frac{1}{u}\,du = \ln u + C = \ln (ax^2 + b) + C$

10. $I = \int \left(\dfrac{x}{ax^2 + b}\right) dx$

    Let $u = ax^2 + b$, $du = 2ax\,dx$.

    $I = \int \left(\dfrac{\frac{1}{2a} \times 2ax}{ax^2 + b}\right) dx = \left(\dfrac{1}{2a}\right) \int \left(\dfrac{1}{u}\right) du = \dfrac{\ln u}{2a} + C = \dfrac{\ln (ax^2 + b)}{2a} + C$

11. $\int_1^2 x^{-1}\,dx = \ln 2 - \ln 1 = \ln 2 - 0 = 0.6931$

12. $\int_1^2 x^{-.9999}\,dx = \dfrac{x^{0.0001}}{0.0001} \Big|_1^2 = \dfrac{2^{0.0001} - 1^{0.0001}}{0.0001} = \dfrac{1.00006932 - 1}{0.0001} = 0.6932$

    Compare this result with Question 11.

EXERCISES

1. $dn/n = 3\,dt$. $\ln n = 3t + C$. $n = 10$ when $t = 0$, so $C = \ln 10$. $t = \frac{1}{3} \ln (n/10)$.

    $n = 1{,}000$ when

    $t = \frac{1}{3} \ln 100 = 1.5$ hours.

2. $y = 2 \ln x$. $y' = 2/x$.

3. $y' = 1/x$. (Notice that this function is defined only when $x < 0$!)

4. $y' = (\sin x)^{-1} \cos x = \text{ctn } x$.

5. $y' = (\tan x)^{-1} \sec^2 x$.

6. $y = \frac{1}{2} \ln (x^2 + 4)$. $y' = \frac{1}{2}(x^2 + 4)^{-1}(2x) = x/(x^2 + 4)$.

7. $y' = (ax^2 + bx + c)^{-1}(2ax + b)$.

8. $y = \ln x + \frac{1}{2} \ln (x + 1)$. $y' = x^{-1} + \frac{1}{2}(x + 1)^{-1}$.

9. $y' = \lim_{\Delta x \to 0} \dfrac{\log (x + \Delta x) - \log x}{\Delta x} = \lim_{\Delta x \to 0} \dfrac{\log [(x + \Delta x)/x]}{\Delta x}$

$\qquad = \lim_{\Delta x \to 0} (x^{-1})(x/\Delta x) \log (1 + \Delta x/x)$

$\qquad = x^{-1} \log \lim_{\Delta x \to 0} (1 + \Delta x/x)^{x/\Delta x}$

$\qquad = x^{-1} \log e$

$y' = 0.434x^{-1}$. (Notice that in this problem $\log x$ has been taken to mean $\log_{10} x$.)

10. $y' = x^{-2}(1 - \ln x)$. This curve will have a horizontal tangent where $x = e$. $y'' = x^{-3}(3 + 2 \ln x)$. This curve will have a point of inflection where $x = e^{3/2} = 4.48$.

11. The logarithm of 2 is equal to the area under the curve $y = x^{-1}$ from $x = 1$ to $x = 2$. The width of each of the 10 rectangles is 0.1, and the height is given by the second column of Table 9–3. Adding together the areas of all the rectangles gives the approximate value $\ln 2 = 0.6688$. The best decimal approximation is $\ln 2 = 0.6931$.

12. $\ln (a/b) = \int_1^{a/b} t^{-1}\, dt = \int_1^a t^{-1}\, dt + \int_a^{a/b} t^{-1}\, dt$

$\qquad = \ln a + I_2$.

$\qquad I_2 = \int_a^{a/b} t^{-1}\, dt$. Let $u = bt/a$. $du = (b/a)\, dt$.

$\qquad I_2 = \int_b^1 \dfrac{ab\, du}{ab\, u} = -\int_1^b u^{-1}\, du = -\ln b$.

$\ln \left(\dfrac{a}{b}\right) = \ln a - \ln b$.

13. $\ln (x^n) = \int_1^{x^n} t^{-1}\, dt$. Let $u = t^{1/n}$. $du = (1/n)t^{(1-n)/n}\, dt$.

$\ln (x^n) = \int_1^x u^{-n} n u^{n-1}\, du = n \int_1^x u^{-1}\, du = n \ln x$.

14. $y = \ln |x + 4| + C.$

15. $y = \frac{1}{3} \ln |3x + 4| + C.$

16. $y = \int (x + 1)^{-2} \, dx = -(x + 1)^{-1} + C.$

17. $y = (1/a) \ln |ax + b| + C.$

18. $y = -(ax + b)^{-1}(1/a).$

19. Let $u = 3x^2 + 6$. $du = 6x \, dx$. $y = \frac{1}{6} \int u^{-1} \, du$. $y = \frac{1}{6} \ln |3x^2 + 6| + C.$

20. Let $u = x^2 + 4x + 6$. $du = (2x + 4) \, dx$. $y = \int u^{-1} \, du$. $y = \ln |x^2 + 4x + 6| + C.$

21. Let $u = \cos \theta$. $du = -\sin \theta \, d\theta$. $y = -\int u^{-1} \, du = -\ln |\cos \theta| + C.$

22. Let $u = \sin \theta + 4$. $du = \cos \theta \, d\theta$. $y = \int u^{-1} \, du = \ln |\sin \theta + 4| + C.$

23. See Table A–3. Since $f(0)$ is undefined, it should be marked on the graph with an open circle (Figure A–12). It is clear that $\lim_{x \to 0} f(x)$ is somewhere close to 2.8, and of course the actual value of the limit is the mysterious number e. As x goes to infinity, the value of the function goes to 1. Note that the curve is not defined for $x < -1$.

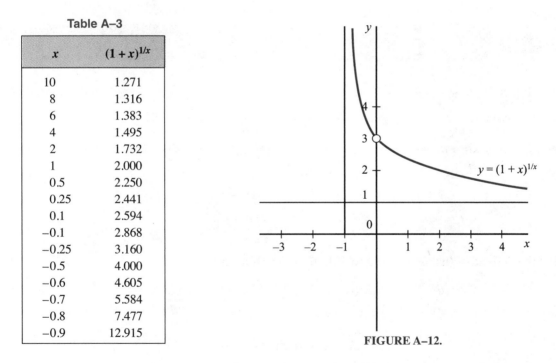

Table A–3

x	$(1 + x)^{1/x}$
10	1.271
8	1.316
6	1.383
4	1.495
2	1.732
1	2.000
0.5	2.250
0.25	2.441
0.1	2.594
−0.1	2.868
−0.25	3.160
−0.5	4.000
−0.6	4.605
−0.7	5.584
−0.8	7.477
−0.9	12.915

FIGURE A–12.

24. Let $K = PV$. Then the work is given by $W = -\int_{V_1}^{V_2} KV^{-1} \, dV$. $W = K \ln (V_1/V_2)$. If $V_1 > V_2$, the gas is being compressed and positive work by some outside force is required to accomplish this. If $V_1 = 2V_2$, then $W = K \ln 2$.

25. Let $L = \int_1^e t^n \, dt = \dfrac{t^{n+1}}{n+1} \Big|_1^e$

$$= \dfrac{e^{n+1}}{n+1} - \dfrac{1^{n+1}}{n+1}$$

$$= \dfrac{e^{n+1} - 1}{n+1}$$

$$(n+1)L = e^{n+1} - 1$$

$$e^{n+1} = (n+1)L + 1$$

$$e = [(n+1)L + 1]^{1/(n+1)}$$

As n approaches -1, L will approximately be equal to the natural logarithm of e (which is 1), so:

$$e = \lim_{n \to -1} [(n+1) + 1]^{1/(n+1)}$$

Let $w = n + 1$:

$$e = \lim_{w \to 0} (w+1)^{1/w}$$

which is the same formula that was found in the chapter.

Chapter 10

WORKSHEET

1.
$$y = \int x \sin x \, dx$$

Let $u = x$, $dv = \sin x \, dx$, $du = dx$

$v = \int \sin x \, dx = -\cos x$

$$y = \int u \, dv = uv - \int v \, du = -x \cos x - \int (-\cos x) \, dx$$

$$y = -x \cos x + \int \cos x \, dx$$

$$y = -x \cos x + \sin x + C$$

Find the derivative (using the product rule for the first term):

$$\frac{dy}{dx} = (-x)\left(\frac{d \cos x}{dx}\right) + (\cos x)\left(\frac{d(-x)}{dx}\right) + \frac{d \sin x}{dx}$$

$$\frac{dy}{dx} = (-x)(-\sin x) + (\cos x)(-1) + \cos x = x \sin x$$

2.
$$y = \int x^2 \sin x \, dx$$

Let $u = x^2$, $dv = \sin x \, dx$, $du = 2x \, dx$

$v = \int \sin x \, dx = -\cos x$

$$y = \int u \, dv = uv - \int v \, du = -x^2 \cos x - \int (-\cos x)\, 2x \, dx$$

$$y = -x^2 \cos x + 2 \int x \cos x \, dx$$

Let $y_2 = \int x \cos x \, dx$

Let $u = x$, $dv = \cos x \, dx$, $du = dx$.

$$v = \int \cos x \, dx = \sin x$$
$$y_2 = \int u \, dv = uv - \int v \, du = x \sin x - \int (\sin x) \, dx$$
$$y_2 = x \sin x + \cos x$$
$$y = -x^2 \cos x + 2y_2$$
$$y_2 = x \sin x + \cos x$$
$$y = \int x^2 \sin x \, dx = -x^2 \cos x + 2x \sin x + 2 \cos x$$

Find the derivative:

$$\frac{dy}{dx} = [(-x^2)(-\sin x) + \cos x(-2x)] + [2x \cos x + \sin x(2)] - 2 \sin x$$

The last four terms cancel out:

$$\frac{dy}{dx} = x^2 \sin x$$

3. $$y = \int x^3 \sin x \, dx$$

Let $u = x^3$, $dv = \sin x \, dx$, $du = 3x^2 \, dx$

$$v = \int \sin x \, dx = -\cos x$$
$$y = \int u \, dv = uv - \int v \, du = -x^3 \cos x - \int (-\cos x)3x^2 \, dx$$
$$y = -x^3 \cos x + 3 \int x^2 \cos x \, dx$$

Let $y_3 = \int x^2 \cos x \, dx$. Evaluate this integral with integration by parts. Let $u = x^2$, $dv = \cos x \, dx$.

$$du = 2x \, dx$$
$$v = \int \cos x \, dx = \sin x$$
$$y_3 = \int u \, dv = uv - \int v \, du = x^2 \sin x - \int (\sin x)2x \, dx$$
$$y_3 = x^2 \sin x - 2 \int x \sin x \, dx$$

The answer to the last integral comes from Question 1:

$$\int x \sin x \, dx = -x \cos x + \sin x$$

Therefore:

$$y_3 = \int x^2 \cos x \, dx = x^2 \sin x - 2(-x \cos x + \sin x)$$
$$= x^2 \sin x + 2x \cos x - 2 \sin x$$

Putting this result back in our original integral:

$$y = \int x^3 \sin x \, dx = -x^3 \cos x + 3y_3$$
$$y = \int x^3 \sin x \, dx = -x^3 \cos x + 3(x^2 \sin x + 2x \cos x - 2 \sin x)$$
$$y = \int x^3 \sin x \, dx = -x^3 \cos x + 3x^2 \sin x + 6x \cos x - 6 \sin x)$$

Find the derivative:

$$\frac{dy}{dx} = [(-x^3)(-\sin x) + \cos x(-3x^2)] + [(3x^2)(\cos x) + \sin x\,(6x)] + [(6x)(-\sin x) + \cos x(6)] - 6 \cos x$$

The last six terms cancel out, leaving

$$\frac{dy}{dx} = (-x^3)(-\sin x) = x^3 \sin x$$

EXERCISES

1. $y' = ae^{ax}$.

2. $y' = -2xe^{-x^2}$.

3. $y' = 2axe^{ax^2+b}$.

4. $y' = mae^{\max}$.

5. $y' = (\ln 10)10^x$.

6. $y' = (e^{-ax}) \cos x - ae^{-ax} \sin x$.

7. $\ln y = x^x \ln a$. $(1/y)(dy/dx) = \ln a(x^x + x^x \ln x)$. $dy/dx = ax^{x^x} \ln a (x^x + x^x \ln x)$.

8. $\ln y = \frac{1}{2} \ln(x - a) + \frac{1}{2} \ln (x - b)$.

    $dy/dx = (x - a)^{1/2}(x - b)^{1/2}[\frac{1}{2}(x - a)^{-1} + (x - b)^{-1}]$.

9. $\dfrac{dy}{dx} = y\left[\dfrac{1}{x + 1} - \dfrac{3}{2}\dfrac{2ax + b}{ax^2 + bx + c}\right]$

10. $\ln y = \ln (x - 5) - \ln (x + 2) - \ln (x + 3)$.

    $y' = y[(x - 5)^{-1} - (x + 2)^{-1} - (x + 3)^{-1}]$.

11. $y' = (2\pi)^{-1/2} e^{-[(x - \mu)2/\sigma 2]} [-2(x - \mu)/\sigma^2]$.

12. $y = x^n$. $\ln y = n \ln x$. $(1/y)(dy/dx) = nx^{-1}$. $dy/dx = nx^n x^{-1}$. $dy/dx = nx^{n-1}$.

13. $y = uv$. $\ln y = \ln u + \ln v$.

    $$\frac{1}{y}\frac{dy}{dx} = \frac{1}{u}\frac{du}{dx} + \frac{1}{v}\frac{dv}{dx}$$

    $$\frac{dy}{dx} = uv\left(\frac{1}{u}\frac{du}{dx} + \frac{1}{v}\frac{dv}{dx}\right) = v\frac{du}{dx} + u\frac{dv}{dx}.$$

14. $$\frac{f(x + \Delta x) - f(x)}{\Delta x} = \frac{e^{x+\Delta x} - e^x}{\Delta x} = \frac{e^x e^{\Delta x} - e^x}{\Delta x} = \frac{e^x(e^{\Delta x} - 1)}{\Delta x}$$

    $$= \frac{e^x(1 + \Delta x + \Delta x^2 / 2! + \Delta x^3 / 3! + \cdots - 1)}{\Delta x}.$$

    $$\frac{dy}{dx} = \lim_{\Delta x \to 0} e^x(1 + \frac{\Delta x}{2!} + \frac{\Delta x^2}{3!} + \cdots)$$

    $$= e^x.$$

15. $(1/a)e^{ax} + C$.

16. $(1/a)e^{ax+b} + C$.

17. Let $u = x^2$. $du = 2x\, dx$. $y = \int \frac{1}{2}e^u\, du = \frac{1}{2}e^{x^2} + C$.

18. Let $u = e^x + 5$. $du = e^x\, dx$. $y = \int u^{-1}\, du = \ln |e^x + 5| + C$.

19.   Let $u = \ln x. \, du = x^{-1} \, dx. \, y = \int u \, du = \frac{1}{2}(\ln|x|)^2 + C.$

20. $y' = \cos x - (\cos x - x \sin x) = x \sin x.$

21.   19.1.

22. $y' = (x^2)(-e^{-x}) + (e^{-x})(2x) = xe^{-x}(2 - x).$ This curve has horizontal tangents where $x = 0$ or $x = 2.$

23.   (a) Let $u = \sin x. \, dv = \cos x \, dx. \, du = \cos x \, dx. \, v = \sin x. \, y = \sin^2 x - \int \sin x \cos x \, dx = \sin^2 x - y + C.$
   $2y = \sin^2 x + C. \, y = \frac{1}{2} \sin^2 x + C. \, y = \frac{1}{4}(1 - \cos 2x) + C = -\frac{1}{4} \cos 2x + C.$ (Note that we can
   consider the $\frac{1}{4}$ term to be part of the constant.)

   (b) $u = \sin x. \, du = \cos x \, dx. \, y = \int u \, du = \frac{1}{2}u^2 + C. \, y = \frac{1}{2} \sin^2 x + C = -\frac{1}{4} \cos 2x + C.$

   (c) $y = \frac{1}{2} \int \sin 2x \, dx = -\frac{1}{4} \cos 2x + C.$

   (Remember that $\sin 2x = 2 \sin x \cos x.$) The trigonometric identity method seems to be easiest in
   this case.

24. $3t = \ln (n/10). \, e^{3t} = n/10. \, n = 10e^{3t}.$ When $t = 5,$ the number of bacteria will be $n = 10e^{15} = 3.3 \times 10^7.$

25.   A problem of this type, where two different types of functions are multiplied together, calls for
   integration by parts. Let $u = x \ln x. \, dv = dx. \, du = (1 + \ln x) \, dx. \, v = x. \, y = x^2 \ln x - \int x \, dx$
   $- \int x \ln x \, dx.$ Our first reaction on seeing that last integral was to become discouraged, since
   it is no simpler than the original integral. In fact, it is exactly the same as the original integral.
   But that means we can substitute: $y = \int x \ln x \, dx.$ Then $y = x^2 \ln x - \frac{1}{2}x^2 - y. \, 2y = x^2 \ln x - \frac{1}{2}x^2.$
   $y = \frac{1}{2}x^2 \ln x - \frac{1}{4}x^2 + C.$ (Remember this trick, because it can be used often.)

26. $y' = \frac{1}{2}x^2(1/x) + x \ln x - \frac{1}{2}x = x \ln x.$

27.   Using integration by parts: Let $u = x. \, dv = e^x \, dx. \, du = dx. \, v = e^x. \, y = xe^x - \int e^x \, dx = xe^x - e^x + C.$

28. $y' = xe^x + e^x - e^x = xe^x.$

29.   Let $u = x^2. \, dv = e^x \, dx. \, du = 2x \, dx. \, v = e^x. \, y = x^2e^x - 2 \int xe^x \, dx.$ Now we have to apply integration
   by parts again, or we can use the results from Exercise 27. $y = x^2e^x - 2xe^x + 2e^x + C.$

30. $y' = x^2e^x + 2xe^x - 2xe^x - 2e^x + 2e^x = x^2e^x.$

31.   Let $u = x^2 \ln x. \, dv = dx. \, du = x^2x^{-1} + 2x \ln x. \, v = x. \, y = x^3 \ln x - \int (x^2 + 2x^2 \ln x) \, dx.$
   $y = x^3 \ln x - \frac{1}{3}x^3 - 2y.$ (Here we used the same trick as in Exercise 25.) $y = \frac{1}{3}x^3 \ln x - \frac{1}{9}x^3 + C.$

32.   Let $u = x. \, dv = (1 + x)^{1/2} \, dx. \, du = dx. \, v = \frac{2}{3}(1 + x)^{3/2}. \, y = (2x/3)(1 + x)^{3/2} - \frac{2}{3} \int (1 + x)^{3/2} \, dx.$
   $y = (2x/3)(1 + x)^{3/2} - (\frac{2}{3})(\frac{2}{5})(1 + x)^{5/2} + C.$

33.   Let $u = x^2. \, dv = \sin x \, dx. \, du = 2x \, dx. \, v = -\cos x. \, y = -x^2 \cos x + 2 \int x \cos x \, dx.$ Use integration by
   parts again: $u = x. \, dv = \cos x \, dx. \, du = dx. \, v = \sin x. \, y = -x^2 \cos x + 2(x \sin x - \int \sin x \, dx).$
   $y = -x^2 \cos x + 2x \sin x + 2 \cos x + C.$

34.   Let $u = e^x. \, dv = \cos x \, dx. \, du = e^x \, dx. \, v = \sin x. \, y = e^x \sin x - \int e^x \sin x \, dx.$ Use integration by parts
   again: $u = e^x. \, dv = \sin x \, dx. \, du = e^x \, dx. \, v = \cos x. \, y = e^x \sin x + e^x \cos x - \int \cos xe^x \, dx.$ Now use
   the Exercise 25 trick again: $y = e^x \sin x + e^x \cos x - y. \, y = \frac{1}{2}e^x (\sin x + \cos x).$

35. Let $u = \sin^{m-1} x$. $dv = \sin x\, dx$. $du = (m-1)\sin^{m-2} x \cos x\, dx$. $v = -\cos x$.

$$y = -\sin^{m-1} x \cos x + (m-1)\int \cos^2 x \sin^{m-2} x\, dx$$
$$= -\sin^{m-1} x \cos x + (m-1)\int (1 - \sin^2 x)\sin^{m-2} x\, dx$$
$$= -\sin^{m-1} x \cos x + (m-1)\int \sin^{m-2} x\, dx - (m-1)\int \sin^m x\, dx$$
$$y = -\sin^{m-1} x \cos x + (m-1)\int \sin^{m-2} x\, dx - (m-1)y$$
$$my = -\sin^{m-1} x \cos x + (m-1)\int \sin^{m-2} x\, dx$$
$$y = \int \sin^m x\, dx = -\frac{\sin^{m-1} x \cos x}{m} + \frac{m-1}{m}\int \sin^{m-2} x\, dx$$

Chapter 11

WORKSHEET

$$y = \int \frac{1}{\sqrt{m^2 - n^2 x^2}}\, dx$$

Factor m^2 out of the expression in the square root sign:

$$y = \int \frac{1}{\sqrt{m^2\left(1 - \left(\frac{nx}{m}\right)^2\right)}}\, dx$$

Move the m^2 outside the square root sign, then outside the integral:

$$y = \left(\frac{1}{m}\right)\int \frac{1}{\sqrt{1 - \left(\frac{nx}{m}\right)^2}}\, dx$$

Let $u = nx/m$.

$$\frac{du}{dx} = \frac{n}{m}$$

$du = (n/m)\, dx$; $dx = (m/n)\, du$

Substitute the expressions for u and du into the integral:

$$y = \left(\frac{1}{m}\right)\int \frac{1}{\sqrt{1 - u^2}}\left(\frac{m}{n}\right) du$$

Move the constants in front of du outside the integral:

$$y = \left(\frac{1}{m}\right)\left(\frac{m}{n}\right)\int \frac{1}{\sqrt{1 - u^2}}\, du$$

$$y = \left(\frac{1}{n}\right)\int \frac{1}{\sqrt{1 - u^2}}\, du$$

Let $u = \sin \theta$.

$$\frac{du}{d\theta} = \cos \theta$$

$$du = \cos \theta \, d\theta$$

Substitute the expressions for u and du into the integral:

$$y = \left(\frac{1}{n}\right) \int \frac{1}{\sqrt{1 - \sin^2 \theta}} \cos \theta \, d\theta$$

Simplify the expression under the square root sign:

$$y = \left(\frac{1}{n}\right) \int \frac{1}{\sqrt{\cos^2 \theta}} \cos \theta \, d\theta$$

$$y = \left(\frac{1}{n}\right) \int \frac{1}{\cos \theta} \cos \theta \, d\theta$$

Cancel $\cos \theta / \cos \theta$:

$$y = \left(\frac{1}{n}\right) \int d\theta$$

Evaluate the perfect integral:

$$y = \left(\frac{1}{n}\right) \theta$$

From the definitions of θ and u:

$$\theta = \arcsin u = \arcsin \left(\frac{nx}{m}\right)$$

$$y = \left(\frac{1}{n}\right) \arcsin \left(\frac{nx}{m}\right)$$

Verify this result by finding the derivative:

$$\frac{dy}{dx} = \left(\frac{1}{n}\right) \frac{1}{\sqrt{1 - n^2 x^2 / m^2}} \left(\frac{n}{m}\right)$$

$$\frac{dy}{dx} = \left(\frac{1}{m}\right) \frac{1}{\sqrt{(m^2 - n^2 x^2)/m^2}}$$

$$\frac{dy}{dx} = \frac{1}{m\sqrt{(m^2 - n^2 x^2)/m^2}}$$

$$\frac{dy}{dx} = \frac{1}{\sqrt{m^2(m^2 - n^2 x^2)/m^2}}$$

$$\frac{dy}{dx} = \frac{1}{\sqrt{m^2 - n^2 x^2}}$$

EXERCISES

1. Let $x = \tan \theta$. $dx = \sec^2 \theta \, d\theta$. $\theta = \arctan x$.

$$y = \int \frac{\sec^2 \theta \, d\theta}{\tan^2 \theta \sec \theta} = \int \frac{\cos \theta}{\sin^2 \theta} \, d\theta = \int \operatorname{ctn} \theta \csc \theta \, d\theta$$

$$= -(\sin \theta)^{-1} + C. \; \sin \theta = x / \sqrt{1 + x^2}.$$

$$= -\frac{\sqrt{1 + x^2}}{x} + C.$$

2. Let $x = \sec \theta$. $dx = \sec \theta \tan \theta \, d\theta$. $\theta = \operatorname{arcsec} x$.

$$y = \int \frac{\sec \theta \, \tan \theta \, d\theta}{\sec \theta \, \tan \theta} = \int d\theta = \theta + C$$

$$= \operatorname{arcsec} x + C.$$

3. Let $u = e^x$. $du = e^x \, dx$.

$$y = \int \frac{1}{1 + u^2} \, du = \arctan u + C$$

$$= \arctan e^x + C.$$

4. Let $x = (a/b) \sin \theta$. $dx = (a/b) \cos \theta \, d\theta$. $\theta = \arcsin (bx/a)$.

$$y = \int_0^{\pi/2} \sqrt{a^2 - a^2 \sin^2 \theta} \, (a/b) \cos \theta \, d\theta = (a^2/b) \int_0^{\pi/2} \cos^2 \theta \, d\theta$$

$$= \frac{\pi a^2}{4b}.$$

5. $y' = (-\frac{1}{3})(1 - x^2)^{3/2} + C$. To evaluate y_2, use the substitution $x = \sin \theta$. $dx = \cos \theta \, d\theta$. $\theta = \arcsin x$.

   $y_2/x = \int \cos^2 \theta \, d\theta = \frac{1}{2}\theta + \frac{1}{2} \sin 2\theta + C = \frac{1}{2} \arcsin x + \frac{1}{2} \sin \theta \cos \theta + C$. We know that $\sin \theta = x$,

   and $\cos \theta = \sqrt{1 - \sin^2 \theta} = \sqrt{1 - x^2}$. Therefore $y_2 = \frac{1}{2}x \arcsin x + \frac{1}{2}x^2 \sqrt{1 - x^2} + C$.

6. To find the area of one quarter of the circle, set up the integral: $A = \int_0^r \sqrt{r^2 - x^2} \, dx$. Let

   $x = r \sin \theta$. $dx = r \cos \theta \, d\theta$. $\theta = \arcsin (x/r)$. $A = \int_0^{\pi/2} r^2 \cos^2 \theta \, d\theta = \pi r^2/4$. (You can use the

   result of Exercise 4, with $a = r$ and $b = 1$.) The area of the entire circle will therefore be πr^2.

7. The area will be given by the definite integral $A = 2b \int_{a/2}^a \sqrt{1 - x^2/a^2} \, dx$. Make the substitution

   $x = a \sin \theta$ and the integral becomes:

$$A = 2ab \int_{\pi/6}^{\pi/2} \cos^2 \theta \, d\theta = 2ab \left(\frac{\pi}{6} - \frac{\sqrt{3}}{8} \right) = 0.61ab.$$

8.   Make this substitution: $x = a \sin \theta$. $z = 4ab \int \cos^2 \theta \, d\theta$.

$$z = 2ab \int (1 + \cos 2\theta) \, d\theta$$
$$= 2ab(\theta + \tfrac{1}{2} \sin 2\theta)$$
$$= 2ab\left[\arcsin\left(\frac{x}{a}\right) + \left(\frac{x}{a}\right)\sqrt{1 - x^2 a^2} \right].$$

$$\frac{dz}{dx} = 2ab\left[\frac{1}{a}\left(1 - \frac{x^2}{a^2}\right)^{-1/2} - \frac{x^2}{a^3}\left(1 - \frac{x^2}{a^2}\right)^{-1/2} + \frac{1}{a} \times \sqrt{1 - x^2 a^2} \right]$$

$$= 2b\left(1 - \frac{x^2}{a^2}\right)^{-1/2} - 2bx^2 a^{-2}\left(1 - \frac{x^2}{a^2}\right)^{-1/2} + 2b\left(1 - \frac{x^2}{a^2}\right)^{1/2}$$

$$= 2b\left(1 - \frac{x^2}{a^2}\right)^{-1/2}\left(1 - \frac{x^2}{a^2}\right) + 2b\left(1 - \frac{x^2}{a^2}\right)^{1/2}$$

$$\frac{dz}{dx} = 2b\left(1 - \frac{x^2}{a^2}\right)^{1/2} + 2b\left(1 - \frac{x^2}{a^2}\right)^{1/2} = 4b\sqrt{1 - x^2 a^2} \; .$$

9.   Let $u = x/a$. $dx = a \, du$. $y = a \int \frac{1}{u^2 + 1} \, du$. $y = a \arctan (x/a) + C$.

10. $y = \frac{1}{a^2} \int \frac{1}{(x/a)^2 + 1} \, dx = (1/a) \arctan (x/a) + C$. (Use the result from Exercise 9.)

11. $y = \frac{1}{5} \arctan (x/5) + C$. (Use the result from Exercise 10.)

12. $y = \int \frac{1}{u^2 + 9} \, du = \frac{1}{3} \arctan [(x + 2)/3] + C$.

13. $y = \int \frac{1}{u^2 + c^2 - b^2} \, du$. Define d so that $d^2 = c^2 - b^2$.

$y = (1/d) \arctan [(x + b)/d] + C$.

14. Let $u = x + \frac{1}{2}$. $y = \int \frac{1}{u^2 + 19/4} \, du$. Let $d = \sqrt{19/4}$. $y = (1/d) \arctan [(x + \frac{1}{2})/d] + C$.

15. (a) $x = \sin \theta$. $dx = \cos \theta \, d\theta$. $\theta = \arcsin x$. $y = \int \cos^2 \theta \, d\theta$. $y = \frac{1}{2}\theta + \frac{1}{4} \sin 2\theta + C = \frac{1}{2} \arcsin x + \frac{1}{2}x\sqrt{1 - x^2} + C$.

(b) $x = \cos \theta$. $dx = -\sin \theta \, d\theta$. $\theta = \arccos x$, $y = -\int \sin^2 \theta \, d\theta = -\frac{1}{2}\theta + \frac{1}{4} \sin 2\theta = -\frac{1}{2} \arccos x + \frac{1}{2}x(1 - x^2)^{1/2} + C$.

You can decide for yourself which method is easier.

16. There is one last hope for an integral that seems to defy any other means of solution: integration by parts. Let $u = \arcsin x$. $dv = dx$. $du = (1 - x^2)^{-1/2} \, dx$. $v = x$. $y = x \arcsin x - \int \frac{x}{\sqrt{1 - x^2}} \, dx$. This integral can be solved with a regular substitution: Let $u = 1 - x^2$. $y = x \arcsin x + \sqrt{1 - x^2} + C$. Differentiating gives:

$$y' = x(1 - x^2)^{-1/2} + \arcsin x + \frac{1}{2}(1 - x^2)^{-1/2}(-2x) = \arcsin x.$$

17. Using integration by parts: Let $u = \arctan x$. $dv = dx$. $du = [1/(1 + x^2)]\, dx$. $v = x$.

$$y = x \arctan x - \int \frac{x}{\sqrt{1 + x^2}}\, dx.$$

Let $u = 1 + x^2$. $y = x \arctan x - \frac{1}{2} \ln (1 + x^2) + C$.

$$\frac{dy}{dx} = \arctan x + \frac{x}{1 + x^2} - \frac{1}{2}(1 + x^2)^{-1}(2x) = \arctan x.$$

Chapter 12

WORKSHEET

1.
$$\frac{102x + 126}{(3x + 9)(21x + 18)} = \frac{H}{3x + 9} + \frac{K}{21x + 18}$$

$$= \frac{H(21x + 18) + K(3x + 9)}{(3x + 9)(21x + 18)}$$

$$= \frac{21xH + 18H + 3xK + 9K}{(3x + 9)(21x + 18)}$$

$$= \frac{x(21H + 3K) + (18H + 9K)}{(3x + 9)(21x + 18)}$$

Looking only at the numerators:

$$102x + 126 = x(21H + 3K) + (18H + 9K)$$

This equation comes from the coefficients of x:

$$102 = 21H + 3K$$

This equation comes from the constant terms:

$$126 = 18H + 9K$$

The solution is:

$$H = \frac{\begin{vmatrix} 102 & 3 \\ 126 & 9 \end{vmatrix}}{\begin{vmatrix} 21 & 3 \\ 18 & 9 \end{vmatrix}} = \frac{(102) \times (9) - (126) \times (3)}{(21) \times (9) - (18) \times (3)} = \frac{540}{135} = 4$$

$$K = \frac{\begin{vmatrix} 21 & 102 \\ 18 & 126 \end{vmatrix}}{\begin{vmatrix} 21 & 3 \\ 18 & 9 \end{vmatrix}} = \frac{(21) \times (126) - (18) \times (102)}{(21) \times (9) - (18) \times (3)} = \frac{810}{135} = 6$$

The final result is:

$$\frac{102x + 126}{(3x + 9)(21x + 18)} = \frac{4}{3x + 9} + \frac{6}{21x + 18}$$

2.
$$\frac{36x + 57}{(2x + 4)(6x + 9)} = \frac{H}{2x + 4} + \frac{K}{6x + 9}$$

$$= \frac{H(6x + 9) + K(2x + 4)}{(2x + 4)(6x + 9)}$$

$$= \frac{6xH + 9H + 2xK + 4K}{(2x + 4)(6x + 9)}$$

$$= \frac{x(6H + 2K) + (9H + 4K)}{(2x + 4)(6x + 9)}$$

Looking only at the numerators:

$$36x + 57 = x(6H + 2K) + (9H + 4K)$$

This equation comes from the coefficients of x:

$$36 = 6H + 2K$$

This equation comes from the constant terms:

$$57 = 9H + 4K$$

The solution is:

$$H = \frac{\begin{vmatrix} 36 & 2 \\ 57 & 4 \end{vmatrix}}{\begin{vmatrix} 6 & 2 \\ 9 & 4 \end{vmatrix}} = \frac{(36) \times (4) - (57) \times (2)}{(6) \times (4) - (9) \times (2)} = \frac{30}{6} = 5$$

$$K = \frac{\begin{vmatrix} 6 & 36 \\ 9 & 57 \end{vmatrix}}{\begin{vmatrix} 6 & 2 \\ 9 & 4 \end{vmatrix}} = \frac{(6) \times (57) - (9) \times (36)}{(6) \times (4) - (9) \times (2)} = \frac{18}{6} = 3$$

The final result is:

$$\frac{36x + 57}{(2x + 4)(6x + 9)} = \frac{5}{2x + 4} + \frac{3}{6x + 9}$$

3. $$\frac{107x + 106}{(10x + 11)(3x + 2)} = \frac{H}{10x + 11} + \frac{K}{3x + 2}$$

$$= \frac{H(3x + 2) + K(10x + 11)}{(10x + 11)(3x + 2)}$$

$$= \frac{3xH + 2H + 10xK + 11K}{(10x + 11)(3x + 2)}$$

$$= \frac{x(3H + 10K) + (2H + 11K)}{(10x + 11)(3x + 2)}$$

Looking only at the numerators:

$$107x + 106 = x(3H + 10K) + (2H + 11K)$$

This equation comes from the coefficients of x:

$$107 = 3H + 10K$$

This equation comes from the constant terms:

$$106 = 2H + 11K$$

The solution is:

$$H = \frac{\begin{vmatrix} 107 & 10 \\ 106 & 11 \end{vmatrix}}{\begin{vmatrix} 3 & 10 \\ 2 & 11 \end{vmatrix}} = \frac{(107) \times (11) - (106) \times (10)}{(3) \times (11) - (2) \times (10)} = \frac{117}{13} = 9$$

$$K = \frac{\begin{vmatrix} 3 & 107 \\ 2 & 106 \end{vmatrix}}{\begin{vmatrix} 3 & 10 \\ 2 & 11 \end{vmatrix}} = \frac{(3) \times (106) - (2) \times (107)}{(3) \times (11) - (2) \times (10)} = \frac{104}{13} = 8$$

The final result is:

$$\frac{107x + 106}{(10x + 11)(3x + 2)} = \frac{9}{10x + 11} + \frac{8}{3x + 2}$$

4.
$$\frac{408x + 208}{(20x + 15)(24x + 4)} = \frac{H}{20x + 15} + \frac{K}{24x + 4}$$

$$= \frac{H(24x + 4) + K(20x + 15)}{(20x + 15)(24x + 4)}$$

$$= \frac{24xH + 4H + 20xK + 15K}{(20x + 15)(24x + 4)}$$

$$= \frac{x(24H + 20K) + (4H + 15K)}{(20x + 15)(24x + 4)}$$

Looking only at the numerators:

$$408x + 208 = x(24H + 20K) + (4H + 15K)$$

This equation comes from the coefficients of x:

$$408 = 24H + 20K$$

This equation comes from the constant terms:

$$208 = 4H + 15K$$

The solution is:

$$H = \frac{\begin{vmatrix} 408 & 20 \\ 208 & 15 \end{vmatrix}}{\begin{vmatrix} 24 & 20 \\ 4 & 15 \end{vmatrix}} = \frac{(408) \times (15) - (208) \times (20)}{(24) \times (15) - (4) \times (20)} = \frac{1,960}{280} = 7$$

$$K = \frac{\begin{vmatrix} 24 & 408 \\ 4 & 208 \end{vmatrix}}{\begin{vmatrix} 24 & 20 \\ 4 & 15 \end{vmatrix}} = \frac{(24) \times (208) - (4) \times (408)}{(24) \times (15) - (4) \times (20)} = \frac{3,360}{280} = 12$$

The final result is:

$$\frac{408x + 208}{(20x + 15)(24x + 4)} = \frac{7}{20x + 15} + \frac{12}{24x + 4}$$

EXERCISES

1.
$$\frac{x + 4}{(x - 3)(x - 2)} = \frac{A}{x - 3} + \frac{B}{x - 2} = \frac{A(x - 2) + B(x - 3)}{(x - 3)(x - 2)}.$$

$x + 4 = x(A + B) - 2A - 3B.\ 1 = A + B.\ 4 = -2A - 3B.\ A = 7, B = -6.$

$$\frac{x + 4}{(x - 3)(x - 2)} = \frac{7}{x - 3} - \frac{6}{x - 2}.$$

2. $3x = Ax - \frac{1}{2}A + Bx + \frac{1}{2}B.$

$$\frac{3x}{(x + \frac{1}{2})(x - \frac{1}{2})} = \frac{3}{2x + 1} + \frac{3}{2x - 1}.$$

3. $2x - 4 = A(4x + 3) + B(2x - 2).$

$$\frac{2x - 4}{(2x - 2)(4x + 3)} = -\frac{2}{7(2x - 2)} + \frac{11}{7(4x + 3)}.$$

4. $1 = A(x + 3) + Bx.$

$$\frac{1}{x(x + 3)} = \frac{1}{3x} - \frac{1}{3(x + 3)}.$$

5. $\dfrac{x^2 + x + 1}{(x + 1)^3} = \dfrac{A}{x + 1} + \dfrac{B}{(x + 1)^2} + \dfrac{C}{(x + 1)^3}.$

$$x^2 + x + 1 = A(x + 1)^2 + B(x + 1) + C.$$

The three-equation system becomes: $A = 1. \, 2A + B = 1. \, A + B + C = 1.$ The result is:

$$\frac{x^2 + x + 1}{x^3 + 3x^2 + 3x + 1} = \frac{1}{x + 1} + \frac{1}{(x + 1)^2} + \frac{1}{(x + 1)^3}.$$

6. $\dfrac{(7.1)(9 - 0.4z) - 6.4(10 - 0.6z)}{(10 - 0.6z)(0 - 0.4z)} = \dfrac{63.9 - 2.84z - 64 + 3.84z}{(10 - 0.6z)(0 - 0.4z)}$

$$\simeq \frac{z - 0.1}{(10 - 0.6z)(0 - 0.4z}.$$

7. $\dfrac{2 + 2x + 2 - 2x}{2(1 - x)2(1 + x)} = \dfrac{4}{4(1 - x)(1 + x)} = \dfrac{1}{(1 - x)(1 + x)}.$

8. $\dfrac{2(x^2 + 2x + 2) + (x - 1)(x - 1)}{(x - 1)(x^2 + 2x + 2)} = \dfrac{3x^2 + 2x + 5}{(x - 1)(x^2 + 2x + 2)}.$

9. $y = \int \dfrac{2x - 7}{(x - 4)(x - 3)} \, dx.$ Partial fractions are called for in a case like this.

$$\frac{2x - 7}{(x - 4)(x - 3)} = \frac{A}{(x - 4)} + \frac{B}{(x - 3)}.$$

$2x - 7 = Ax - 3A + Bx - 4B. \, A = 1. \, B = 1.$
$y = \int [(x - 4)^{-1} + (x - 3)^{-1}] \, dx = \ln|x - 4| + \ln|x - 3| + C.$

10. $\dfrac{2x + 9}{(\frac{1}{2}x + 5)(x - 1)} = \dfrac{A}{\frac{1}{2}x + 5} + \dfrac{B}{x - 1}.$

$B = 2. \, A = 1. \, y = 2\ln|\frac{1}{2}x + 5| + 2\ln|x - 1| + C.$

11. $y = (-2 - 3\sqrt{3}) \ln |x - \sqrt{3}| + (3 + 2\sqrt{3}) \ln |x - 1|$.

12. $y = \frac{1}{2} \ln |x - 1| + \frac{1}{2} \ln |x + 1| + C$.

13. $y = \frac{1}{2} \ln (1 + x) - \frac{1}{2} \ln (1 - x)$.

$$y' = \frac{1}{2(1 + x)} + \frac{1}{2(1 - x)} = \frac{1}{(x - 1)(x + 1)}.$$

14. $y' = (\sec x + \tan x)^{-1} (\sec x \tan x + \sec^2 x)$
$= (\sec x + \tan x)^{-1} (\sec x + \tan x) \sec x = \sec x$.

15. $y' = -(\cos x)^{-1} (-\sin x) = \tan x$.

16. $y' = \frac{1}{2} (x^2 + 2x + 2)^{-1} (2x + 2) - \frac{2}{1 + (x + 1)^2} + \frac{2}{x - 1}$

$$= \frac{x + 1}{x^2 + 2x + 2} - \frac{2}{x^2 + 2x + 2} + \frac{2}{x - 1}$$

$$= \frac{2}{x - 1} + \frac{x - 1}{x^2 + 2x +}.$$

17. This integral requires integration by parts. It is hard to tell at first glance what the best parts are, but after some trial and error it turns out that the best strategy is to set $u = \sec x$. $dv = \sec^2 x \, dx$. $du = \sec x \tan x \, dx$. $v = \tan x$. $y = \sec x \tan x - \int \tan^2 x \sec x \, dx$.

Using the secant-tangent identity gives:

$$y = \sec x \tan x - \int \sec^3 x \, dx + \int \sec x \, dx$$
$$= \sec x \tan x - y + \ln |\sec x + \tan x| + C$$
$$= \frac{1}{2} \sec x \tan x + \frac{1}{2} \ln |\sec x + \tan x| + C.$$

Dffferentiating, we obtain:

$$y' = \frac{1}{2} (\sec x \sec^2 x + \sec x \tan^2 x) + \frac{1}{2} \sec x$$
$$= \frac{1}{2} (\sec^3 x + \sec^3 x - \sec x + \sec x)$$
$$= \sec^3 x.$$

18. The area is given by $A = 2 \int_a^{2a} b\sqrt{x^2 / a^2 - 1} \, dx$. Make the substitution $x = a \sec \theta$. The final result is $2.14ab$.

19. Let $x = \tan \theta$. $y = \int \sec^3 \theta \, d\theta$. Use the result of Exercise 17:

$$y = \frac{1}{2} \sec \theta \tan \theta + \frac{1}{2} \ln |\sec \theta + \tan \theta| + C$$
$$= \frac{1}{2} x \sqrt{1 + x^2} + \frac{1}{2} \ln |x + \sqrt{1 + x^2}| + C.$$

20. Let $x = \tan \theta$. $y = \int \sec \theta \, d\theta$. $y = \ln |\sec \theta + \tan \theta| + C$. $y = \ln |x + \sqrt{1 + x^2}| + C$.

21. Let $x = \sin \theta$.

$$y = \int \cos^2 \theta \, (\sin \theta)^{-1} \, d\theta = \int (\sin \theta)^{-1} \, d\theta - \int \sin \theta \, d\theta$$

$$= \cos \theta - \ln |\csc \theta + \text{ctn } \theta| + C$$

$$= \sqrt{1 - x^2} - \ln |x^{-1} + (1 - x^2)^{1/2} \, x^{-1}| + C.$$

22. Let $x = \tan \theta$.

$$y = \int (\sin \theta)^{-1} \, d\theta = -\ln |\csc \theta + \text{ctn } \theta| + C$$

$$= \ln |x^{-1} + (1 - x^2)^{1/2} \, x^{-1}| + C.$$

23. Let $x = \tan \theta$.

$$y = \int \tan^2 \theta \sec \theta \, d\theta = \int \sec^3 \theta \, d\theta - \int \sec \theta \, d\theta$$

$$= \frac{1}{2} \sec \theta \tan \theta - \frac{1}{2} \ln |\sec \theta + \tan \theta| + C$$

$$= \frac{1}{2} x \sqrt{x^2 + 1} + \frac{1}{2} \ln |x + \sqrt{x^2 + 1}| + C.$$

24. These problems can be done by using algebraic division.

$$
\begin{array}{r}
x^2 + x \\
2x^2 + 3x + 4 \overline{) 2x^4 + 5x^3 + 7x^2 + 5x} \\
\underline{-2x^4 + 3x^3 + 4x^2} \\
2x^3 + 3x^2 + 5x \\
\underline{-2x^3 + 3x^2 + 4x} \\
x
\end{array}
$$

The remainder is x, so the final result is

$$x^2 + x + \frac{x}{2x^2 + 3x + 4}.$$

25. $5x + [1/(3x^3 - 3)]$.

26. $5x^2 + 4x + [1/(x^2 - 1)]$.

27. $3x^2 + 2x + 5 + [(4x - 1)/(x^2 - x - 12)]$.

28. (a) When $b^2 - 4ac$ is positive, the integral is of the same general family as $\int [1/(x^2 - 1)] \, dx$. The integral can be simplified with the substitution $u = x + \frac{1}{2} b/a$. $(dx = du.)$ The result:

$$y = \int \frac{1}{au^2 - \frac{1}{4} D^2 / a}.$$

$$du = \frac{4a}{D^2} \int \frac{1}{[(2au)^2 / D^2] - 1} \, du.$$

Now let $v = 2au/D$. $y = -\dfrac{2}{D} \int \dfrac{1}{1 - v^2}\, dv$.

This integral can be solved with partial fractions, as was done in the chapter.

$$y = -\dfrac{2}{D} \int \left[\dfrac{1}{2(1 - v)} + \dfrac{1}{2(1 + v)} \right] dv$$

$$= \dfrac{1}{D} \ln\left| \dfrac{v - 1}{v + 1} \right|.$$

Solving for v in terms of x yields $v = (2ax + b)/D$. So the final result is

$$y = \dfrac{1}{D} \ln\left| \dfrac{2ax + b - D}{2ax + b + D} \right| + C.$$

(b) When $b^2 - 4ac$ is negative, the integral is of the same general family as $\int [1/(1 + x^2)]\, dx$. Start by making the same substitution: $v = (2ax + b)/D$. After some algebra the integral turns out to be equal to $(2/D) \int [1/(1 + v^2)]\, dv$. That integral is instantly recognizable, so the final answer is

$$y = \dfrac{2}{D} \arctan\left(\dfrac{2ax + b}{D} \right) + C.$$

You can see how a table of integrals would spare you from repeatedly performing some of these calculations.

29. $b^2 - 4ac = -4$. Use Exercise 28, case (b). $y = \arctan(x + 1) + C$.

30. $b^2 - 4ac = 13$. Case (a): $y = 13^{-1/2} \ln\left| \dfrac{2x - 3 - \sqrt{13}}{2x - 3 + \sqrt{13}} \right| + C$.

31. $b^2 - 4ac = -3$. Case (b): $y = (2/\sqrt{3}) \arctan[(2x + 1)\sqrt{3}] + C$.

32. $b^2 - 4ac = -3$. Case (b): $y = (2/\sqrt{3}) \arctan[(2x - 1)\sqrt{3}] + C$.

33. Factor the denominator: $y = \int \dfrac{1}{(x - 1)(x^2 + x + 1)}\, dx$.

Set up the partial fractions:

$$\dfrac{1}{(x - 1)(x^2 + x + 1)} = \dfrac{A}{x - 1} + \dfrac{Bx + C}{x^2 + x + 1}.$$

$1 = Ax^2 + Ax + A + Bx^2 - Bx + Cx - C$. $A + B = 0$. $A - B + C = 0$.

$A - C = 1$. $A = \frac{1}{3}$. $B = -\frac{1}{3}$. $C = -\frac{2}{3}$.

$$y = \frac{1}{3} \ln|x - 1| - \int \dfrac{x/3 + \frac{2}{3}}{x^2 + x + 1}\, dx$$

$$= \frac{1}{3} \ln|x - 1| - \frac{1}{6} \ln|x^2 + x + 1| - \left(\frac{2}{3} - \frac{1}{6}\right)\left(\dfrac{2}{\sqrt{3}}\right) \times \arctan\left(\dfrac{2x + 1}{\sqrt{3}}\right) + C.$$

(Use the result from Exercise 31.)

34. Factor the denominator: $y = \int \dfrac{1}{(x+1)(x^2-x+1)}\,dx.$

After solving the partial fractions:

$$y = \int \frac{1}{3(x+1)}\,dx + \int \frac{(-\frac{1}{3})x + \frac{2}{3}\,dx}{x^2-x+1}$$

$$= \tfrac{1}{3}\ln|x+1| - \tfrac{1}{6}\ln|x^2-x+1| + (\tfrac{2}{3}+\tfrac{1}{6})\left(\frac{2}{\sqrt{3}}\right) \times \arctan\left(\frac{2x-1}{\sqrt{3}}\right) + C.$$

35. $b^2 - 4ac = 4.$ Case (a): $y = \tfrac{1}{2}\ln\left|\dfrac{4x}{4x+4}\right| + C.$

36. $b^2 - 4ac = 13.$ Case (a): $y = 13^{-1/2}\ln\left|\dfrac{6x+5-\sqrt{13}}{6x+5+\sqrt{13}}\right| + C.$

37. Factor the denominator: $y = \int \dfrac{x}{(x^2+1)(x-1)(x+1)}\,dx.$

After solving the partial fractions the result is:

$$y = \int\left[\frac{-x}{2(x^2+1)} + \frac{1}{4(x-1)} + \frac{1}{4(x+1)}\right]dx$$

$$= \tfrac{1}{4}\ln|x^2+1| - \tfrac{1}{4}\ln|x-1| + \tfrac{1}{4}\ln|x+1| + C.$$

38. After solving for partial fractions:

$$y = \int\left[\frac{1}{2(x-1)} + \frac{x+1}{2(x^2+1)}\right]dx$$

$$= \tfrac{1}{2}\ln|x-1| - \tfrac{1}{4}\ln|x^2+1| - \tfrac{1}{2}\arctan x + C.$$

39. $y' = \left(\dfrac{2}{D}\right)\dfrac{D^2}{1 + \left(\dfrac{2ax+b}{D}\right)^2}\left(\dfrac{2a}{D}\right)$

$$= \frac{4a}{D^2}\cdot\frac{D^2}{D^2 + 4a^2x^2 + 4axb + b^2}$$

$$= \frac{4a}{4ac - b^2 + 4a^2x^2 + 4axb + b^2}$$

$$= \frac{1}{ax^2 + bx + c}$$

40. Let $x_1 = x + b/2a$. Then:

$$Z = \int \frac{1}{\sqrt{a(x_1^2 - x_1 b/a + b^2/4a^2) + bx_1 - b^2/2a + c}}\, dx_1$$

$$= \int \frac{1}{\sqrt{ax_1^2 + (4ac - b^2)/4a}}\, dx_1$$

Let $K = (4ac - b^2)/4a$. Then:

$$Z = \int \frac{1}{\sqrt{K(ax_1^2/K + 1)}}\, dx_1$$

Let $\cos \theta = \sqrt{\dfrac{-a}{K}}\, x_1$. (Note that because of the assumptions in the problem, $a < 0$, $K > 0$, and $\sqrt{a^2} = -a$.)

$$dx_1 = \sqrt{\frac{K}{-a}}\, \sin \theta\, d\theta$$

$$Z = \frac{1}{\sqrt{K}} \int \left(\frac{1}{\sqrt{1 - \cos^2 \theta}} \right) \left(-\sqrt{\frac{K}{-a}}\, \sin \theta \right) d\theta$$

$$Z = -\frac{1}{\sqrt{-a}} \int d\theta = -\frac{1}{\sqrt{-a}}\, \theta = -\frac{1}{\sqrt{-a}}\, \arccos \left(\sqrt{\frac{-a}{K}}\, x_1 \right)$$

$$Z = -\frac{1}{\sqrt{-a}}\, \arccos \left[\left(\frac{-2a}{\sqrt{b^2 - 4ac}} \right) \left(x + \frac{b}{2a} \right) \right]$$

$$Z = -\frac{1}{\sqrt{-a}}\, \arccos \left(\frac{-(2ax + b)}{\sqrt{b^2 - 4ac}} \right)$$

Since $\arccos(-x) = \pi - \arccos x$, this can be rewritten:

$$Z = \frac{1}{\sqrt{-a}}\, \arccos \left(\frac{2ax + b}{\sqrt{b^2 - 4ac}} \right) + C$$

where the π term is absorbed into the constant of integration. Also, since $\arcsin x = \pi/2 - \arccos x$, the result can also be written:

$$Z = \frac{-1}{\sqrt{-a}}\, \arcsin \left(\frac{2ax + b}{\sqrt{b^2 - 4ac}} \right) + C$$

(then $\pi/2$ disappears into the constant of integration).

41. $$Z = \frac{a(1 - \cos \theta)a \sin \theta\, d\theta}{\sqrt{2a^2(1 - \cos \theta) - a^2(1 - \cos \theta)^2}}$$

$$= \frac{a(1 - \cos \theta)a \sin \theta\, d\theta}{\sqrt{2a^2 - 2a^2 \cos \theta - a^2 + 2a^2 \cos \theta - a^2 \cos^2 \theta}}$$

$$= \frac{a(1 - \cos \theta)a \sin \theta\, d\theta}{\sqrt{a^2(1 - \cos^2 \theta)}}$$

$Z = a\theta - a \sin \theta$

$\theta = \arccos (1 - x/a); \sin \theta = \sqrt{1 - (1 - x/a)^2}$

$Z = a \arccos (1 - x/a) - \sqrt{2ax - x^2}$

Chapter 13

WORKSHEET

1. $\int_0^h s^2 \, dy = s^2 y \Big|_0^h = s^2(h - 0) = s^2 h$

2. $\int_0^h \pi r^2 \, dy = \pi r^2 y \Big|_0^h = \pi r^2(h - 0) = \pi r^2 h$

3. $\int_0^h \pi x^2 \, dy = \int_0^h \pi (Ry/h)^2 \, dy = \dfrac{\pi R^2}{h^2} \int_0^h y^2 \, dy = \dfrac{\pi R^2}{h^2} \times \dfrac{y^3}{3}\Big|_0^h = \dfrac{\pi R^2}{h^2} \times \dfrac{h^3 - 0^3}{3}\Big|_0^h = \dfrac{\pi R^2 h^3}{3h^2} = \dfrac{\pi R^2 h}{3}$

4. $\int_0^h (sy/h)^2 \, dy = \dfrac{s^2}{h^2} \int_0^h y^2 \, dy = \dfrac{s^2}{h^2} \times \dfrac{y^3}{3}\Big|_0^h = \dfrac{s^2}{h^2} \times \dfrac{h^3 - 0^3}{3}\Big|_0^h = \dfrac{s^2 h^3}{3h^2} = \dfrac{s^2 h}{3}$

5. $I = \int_0^R 2\pi x \sqrt{R^2 - x^2} \, dx$

Let $u = R^2 - x^2$, $du = -2x \, dx$, $dx = -du/2x$

$\qquad I = 2\pi \int_{u=R^2}^{u=0} x \sqrt{u} \, (-du/2x) \, du$

Cancel the 2 and x:

$\qquad I = -\pi \int_{u=R^2}^{u=0} \sqrt{u} \, du$

$\qquad I = -\pi \left(\dfrac{2}{3}\right) u^{1.5} \Big|_{u=R^2}^{u=0}$

$\qquad I = -\pi \left(\dfrac{2}{3}\right) [0^{1.5} - (R^2)^{1.5}]$

$\qquad I = -\pi \left(\dfrac{2}{3}\right) (0 - R^3)$

$\qquad I = \dfrac{2\pi R^3}{3}$

This integral shows how to use the method of cylindrical shells to find the volume of the top half of a sphere, using cylinders of radius x and height $\sqrt{R^2 - x^2}$, where R is the radius of the sphere.

EXERCISES

1. $dV = \pi r^2\, dx.\; r = y.\; y^2 = b^2(1 - x^2 a^2).\; V = (4/3)\pi ab^2.$

2. $dV = A\, dy.\; A = \frac{1}{2}(2x)(\sqrt{3}\,x) = x^2\sqrt{3}.\; V = \int_0^2 x^2\sqrt{3}\; dy = \int_0^2 2y\sqrt{3}\; dy = 4\sqrt{3}.$

3. $V = \int_a^b \pi y\, dy = 24\pi.$

4. (a) $V = \int_a^b \pi[f(x)]^2\, dx.$

   (b) $V = \int_a^b 2\pi(b - x)f(x)\, dx.$ (Use cylindrical shells.)

   (c) $V = \int_a^b 2\pi(c - x)f(x)\, dx.$

5. (a) $V(a) = 0.$

   (b) $V(x + \Delta x) - V(x) = \pi[f(x)]^2\, \Delta x.$

   (c) $[V(x + \Delta x) - V(x)]/\Delta x = \pi[f(x)]^2.\; dV/dx = \pi[f(x)]^2.\; \int dV = \int \pi[f(x)]^2\, dx.\; V(x) = F(x) + C.$ $V(x) = 0$ when $x = a.$ Solving for the arbitrary constant gives $C = -F(a).$ Therefore $V(x) = F(x) - F(a).$ The total volume of the solid will be $V = F(b) - F(a).$

6. $V = \int_{r/2}^r \pi(r^2 - y^2)\, dy.\; V = (5/24)\,\pi r^3.$

7. $V = \int_{-r}^r \pi(r^2 - y^2)\, dy.\; V = \pi 2r^3 - \pi(2/3)r^3 = (4/3)\pi r^3.$

8. Divide the pyramid into little squares of side $x.$ Let z be the distance from the top of the pyramid to each square. Let h be the height of the pyramid, and s^2 be the area of the base. $z/x = h/s$ (by similar triangles).

$$V = \int_{-h}^0 x^2\, dz = \frac{1}{3}s^2 h.$$

9. volume $= \int_{-1}^1 \pi(2 - x)^2\, dy$

   $= \pi \int_{-1}^1 (4 - 4x + x^2)\, dy$

   $= \pi \int_{-1}^1 [4 - 4\sqrt{1 - y^2} + (1 - y^2)]\, dy$

   $= \pi \left[5y \Big|_{-1}^1 - 4 \int_{-1}^1 \sqrt{1 - y^2}\, dy - 1/3\, y^3 \Big|_{-1}^1 \right]$

   $= \pi(10 - 2\pi + 0)$

   $= 10\pi - 2\pi^2$

10. The calculations are almost the same as the previous exercise, except the radius of each pancake is now $(2 + x).$ The resulting volume is $10\pi + 2\pi^2.$

11. volume $= \int_0^r 2\pi x(2y)\, dx$

    $= 2\pi \int_0^r \sqrt{r^2 - x^2}\, 2x\, dx$

Let $u = r^2 - x^2$. Then the integral becomes:

$$-2\pi \int_{r^2}^{0} u^{1/2} \, du = (-2\pi) \, 2/3 u^{3/2} \Big|_{r^2}^{0} = 4/3\pi \, r^3$$

12.   volume $= \int_{0}^{\sqrt{200}} 2\pi x (100 - x^2/2) \, dx$

$$= 2\pi \left(50 x^2 \Big|_{0}^{\sqrt{200}} - 1/8 x^4 \Big|_{0}^{\sqrt{200}} \right)$$

$$= 2\pi (50 \times 200 - 40.000/8) = 10.000\pi$$

13. $\int_{x=-r}^{x=r} \int_{y=-\sqrt{r^2-x^2}}^{y=\sqrt{r^2-x^2}} \sqrt{r^2 - x^2 - y^2} \, dy \, dx$

First, evaluate the inner integral involving y (treating x as a constant while evaluating this integral):

$$\int_{y=-\sqrt{r^2-x^2}}^{y=\sqrt{r^2-x^2}} \sqrt{r^2 - x^2 - y^2} \, dy$$

Define $A = \sqrt{r^2 - x^2}$, then use the trigonometric substitution $y = A \sin \theta$; $dy = A \cos \theta \, d\theta$; $\theta = \arcsin (y/A)$.

$$= \int_{y=-A}^{y=A} \sqrt{A^2 - y^2} \, dy$$

$$= \int_{y=-A}^{y=A} \sqrt{A^2 [1 - (y/A)^2]} \, dy$$

$$= \int_{\theta=\arcsin \, (-A/A)}^{\theta=\arcsin \, (A/A)} \sqrt{1 - \sin^2 \theta} \; A \cos \theta \, d\theta$$

$$= A^2 \int_{\arcsin \, (-1)}^{\arcsin \, (1)} \cos^2 \theta d\theta$$

$$= A^2 \int_{-\pi/2}^{\pi/2} [\tfrac{1}{2}(1 + \cos 2\theta] \, d\theta$$

$$= \frac{A^2}{2} \int_{-\pi/2}^{\pi/2} d\theta + \frac{A^2}{2} \int_{-\pi/2}^{\pi/2} \cos 2\theta \, d\theta$$

$$= \frac{A^2}{2} \theta \Big|_{-\pi/2}^{\pi/2} - \frac{A^2}{4} \sin 2\theta \Big|_{-\pi/2}^{\pi/2}$$

Since $A^2 = r^2 - x^2$:

$$= \frac{r^2 - x^2}{2} \left(\frac{\pi}{2} + \frac{\pi}{2} \right) - \frac{r^2 - x^2}{4} [\sin \pi - \sin (-\pi)]$$

The second term in the integral is zero, so the result for the inner integral is:

$$\int_{y=-\sqrt{r^2-x^2}}^{y=\sqrt{r^2-x^2}} \sqrt{r^2 - x^2 - y^2} \, dy = \frac{\pi}{2} (r^2 - x^2)$$

After you've evaluated the inner integral, you'll need to evaluate the outer integral involving x, which becomes:

$$\int_{-r}^{r} \frac{\pi}{2}(r^2 - x^2)\, dx = \frac{\pi}{2}(r^2 x - \frac{x^3}{3})\Big|_{x=-r}^{r}$$

$$= \frac{\pi}{2}[(r^2 r - \frac{r^3}{3}) - (r^2(-r) - \frac{-r^3}{3})]$$

$$= \frac{\pi}{2}(r^3 - \frac{r^3}{3} + r^3 - \frac{r^3}{3})$$

$$= \frac{\pi}{2}(2r^3 - \frac{2r^3}{3})$$

$$= \frac{\pi}{2}(\frac{6r^3}{3} - \frac{2r^3}{3})$$

$$= \frac{\pi}{2}(\frac{4r^3}{3})$$

$$= \frac{2}{3}\pi r^3$$

(Note that this is half of the volume of a complete sphere.)

Chapter 14

EXERCISES

1. $1 + (dy/dx)^2 = 1 + 9x/4.\ L = 2.09.$

2. $1 + (dy/dx)^2 = 1 + x^2.\ L = \int_0^a \sqrt{1 + x^2}\, dx.\ L = \frac{1}{2}a\sqrt{1 + a^2} + \frac{1}{2}\ln(a + \sqrt{1 + a^2}).$ When $a = 2$, $L = 2.96.$

3. $1 + (dy/dx)^2 = 1 + \tan^2 x = \sec^2 x.\ L = 0.88.$ The length of the curve from $x = 0$ to $x = \pi/2$ is infinity.

4. $1 + (dy/dx)^2 = 1 + a^2.\ L = b\sqrt{1 + a^2}.$

5. $L = \int_0^a \sqrt{1 + \frac{1}{4}x^{-1}}\, dx.$

6. $L = \int_1^a \sqrt{1 + x^{-2}}\, dx.$

7. $L = \int_0^\pi \sqrt{1 + \cos^2 x}\, dx.$

8. $1 + (dx/dy)^2 = 1 + 1/4y$. $A = 2\pi \int_{3/4}^{15/4} \sqrt{y}\sqrt{1 + 1/4y}\ dy$. $A = \pi \int_{3/4}^{15/4} \sqrt{4y + 1}\ dy = 28\pi/3$.

9. $1 + (dy/dx)^2 = 1 + a^2$. $A = 2\pi a \sqrt{1 + a^2} \int_0^b x\ dx$. $A = \pi ab^2 \sqrt{1 + a^2}$.

10. $1 + (dy/dx)^2 = 1 + \cos^2 x$. $A = 2\pi \int_0^\pi \sin x \sqrt{1 + \cos^2 x}\ dx$. Let $u = \cos x$.

    $A = \pi[2\sqrt{2} + \ln(1 + \sqrt{2}) - \ln(\sqrt{2} - 1)] = 14.42$.

11. $1 + (dy/dx)^2 = 1 + x^4$. $A = 2\pi \int_0^a \frac{1}{3}x^3 \sqrt{1 + x^4}\ dx$. Let $u = 1 + x^4$. $A = (\pi/6)\int_1^{1+a^4} u^{1/2}\ du$.

    $A = (\pi/9)[(1 + a^4)^{3/2} - 1]$. When $a = 2$, the surface area is 24.12.

12. $1 + (dy/dx)^2 = 1 + e^{2x}$. $A = 2\pi \int_0^a e^x \sqrt{1 + e^{2a}}\ dx$. Let $u = e^x$. $A = 2\pi \int_1^{a^a} \sqrt{1 + u^2}\ du$.

    $A = \pi[e^a \sqrt{1 + e^{2a}} + \ln(e^a + \sqrt{1 + e^{2a}}) - \sqrt{2} - \ln(1 + \sqrt{2})]$.

13. $A = \pi r^2$.

14. $dm = \rho\ dv$. $dV = 2xh\ dy$. $y_{\text{com}} = \int_0^a y\ dm / \int_0^a dm$. $y_{\text{com}} = 3a/5$.

15. $dm = \rho\pi x^2\ dy$. $y_{\text{com}} = 3b/4$.

16. $x_{\text{com}} = \dfrac{\int_a^b x[f(x)]^2\ dx}{\int_a^b [f(x)]^2\ dx}$.

17. $dm = \rho\ dV$. $\rho = \sqrt{x} + 2$. $dV = A\ dx$.

    $x_{\text{com}} = \dfrac{\int_0^L x\ dm}{\int_0^L dm} = \dfrac{\int_0^L (x^{3/2} + 2x)\ dx}{\int_0^L (x^{1/2} + 2)\ dx} = \dfrac{0.4L^{2.5} + L^2}{\frac{2}{3}L^{1.5} + 2L}$.

18. $I = \int_0^L x^2 \rho A\ dx = \rho A \frac{1}{3}x^3 \Big|_0^L = ML^2/3$.

19. $I = 2\pi h\rho R^4/4 = MR^2/2$.

20. $dm = 2\pi r\rho(2L)\ dr$. $L = (R^2 - r^2)^{1/2}$.

    $I = 4\pi\rho \int_0^R r^3 \sqrt{R^2 - r^2}\ dr$. (Remember that R is constant and r is a variable.) Let $r = R\sin\theta$.

    $I = 4\pi\rho R^5 \int_0^{\pi/2} \sin^3\theta \cos^2\theta\ d\theta$. $I = 4\pi\rho R^5 \int_0^{\pi/2} (\sin\theta \cos^2\theta - \sin\theta \cos^4\theta)\ d\theta$.

    Let $u = \cos\theta$. $du = -\sin\theta\ d\theta$.

$$I = 4\pi\rho R^5 \int_1^0 (-u^2 + u^4)\, du$$

$$= 4\pi\rho R^5 (\tfrac{1}{3} - \tfrac{1}{5})$$

$$= \tfrac{4}{3}\pi R^3 \rho R^2 (\tfrac{2}{5})$$

$$= \frac{2MR^2}{5}.$$

21. $I = 2\pi\rho \int_0^R x^3(h - hx/R)\, dx.\, I = 3MR^2/10.$

Chapter 15

EXERCISES

1. Nonlinear (because of the y^2 term).

2. Linear and nonhomogeneous.

3. Nonlinear (because of $(dy/dx)^2$).

4. Nonlinear (unless $f(y)$ is a linear function of y).

5. Linear and nonhomogeneous.

6. Nonlinear.

7. Nonlinear.

8. $T(ax_1 + bx_2)$

$$= \frac{d^n(ax_1 + bx_2)}{dt^n} + \cdots + f_1(t)\frac{d}{dt}(ax_1 + bx_2) + f_0(t)(ax_1 + bx_2)$$

$$= a\frac{d^n}{dt^n}x_1 + b\frac{d^n}{dt^n}x_2 + \cdots + af_1(t)\frac{dx_1}{dt} + bf_1(t)\frac{dx_2}{dt} + af_0(t)x_1 + bf_0(t)x_2$$

$$= a\left[\frac{d^n}{dt^n} + \cdots + f_1(t)\frac{d}{dt} + f_0(t)\right]x_1 + b\left[\frac{d^n}{dt^n} + \cdots + f_1(t)\frac{d}{dt} + f_0(t)\right]x_2$$

   $T(ax_1 + bx_2) = aTx_1 + bTx_2.$

9. Set up the characteristic equation: $r^2 + r - 6 = 0.\, r = 2, -3.$

   $x = Ae^{2t} + Be^{-3t}.\, A = 12, B = 9.$

10. $r^2 + r + 1 = 0.\, r = \tfrac{1}{2}(-1 \pm i\sqrt{3}).\, x = e^{-t/2}\,(A\,\sin\,\sqrt{3}\,t/2 + B\,\cos\,\sqrt{3}\,t/2).\, A = 2/3, B = 4/3.$

11. $r^2 + 9 = 0.\, r = \pm 3i.\, x = A\,\sin\,3t + B\,\cos\,3t.\, A = 160, B = 20.$

12. $r^2 + 2r + 1 = 0.\, r = -1.\, x = Ae^{-t} + Bte^{-t}.\, A = -4, B = -7.$

13. $r^2 - 4r = 0.\, r = 0, 4.\, x = A + Be^{4t}.\, A = 17, B = 3.$

14. In order for the characteristic equation to have one real root, the differential equation must be of the form:

$$\frac{d^2x}{dt^2} + b\frac{dx}{dt} + \frac{b^2x}{4} = 0.$$

When $x = Ae^{rt} + Bte^{rt}$, $dx/dt = (Ar + B)e^{rt} + Brte^{rt}$, and $d^2x/dt^2 = (Ar^2 + 2Br)e^{rt} + Br^2te^{rt}$. (Note that $b = -2r$.) Therefore:

$$\frac{d^2x}{dt^2} + b\frac{dx}{dt} + \frac{b^2x}{4} = (Ar^2 + 2Br)e^{rt} + Br^2te^{rt} - 2(Ar^2 + Br)e^{rt} - 2Br^2te^{rt} + r^2Ae^{rt} + Btr^2e^{rt} = 0.$$

15. $x = 0.658 \sin(2t - 1.4) + e^{-1.5t} A_1 \sin(1.66t + A_2)$.

16. $-0.6484 = 0.658 \sin(-1.4) + A_1 \sin A_2$. $1.884 = 2(0.658)\cos(-1.4) + 1.66A_1 \cos A_2 + 1.5A_1 \sin A_2$. $0 = A_1 \sin A_2$. $1.66 = 1.66A_1 \cos A_2 + 1.5A_1 \sin A_2$. $A_2 = 0$, $A_1 = 1$. $x = 0.658 \sin(2t - 1.4) + e^{-1.5t} \sin(1.66 t)$.

17. The equation of motion in this case is: $(d^2x/dt^2) + 4(dx/dt) + 4x = 0$. $r^2 + 4r + 4 = 0$. $r = -2$. $x = Ae^{-2t} + Bte^{-2t}$. Notice that, if the friction is strong enough, the ride will not oscillate at all. This situation is known as *critical damping*.

18. The characteristic equation is: $r^2 + 5r + 4 = 0$. $r = -1$, or $r = -4$. $x = Ae^{-t} + Be^{-4t}$. This situation is known as *overdamping*.

19.
$$y' = 1/2(e^{x/a} - e^{-x/a})$$
$$y'^2 = 1/4(e^{2x/a} - 2 + e^{-2x/a})$$
$$1 + y'^2 = 1/4(e^{2x/a} + 2 + e^{-2x/a})$$
$$= [1/2(e^{x/a} + e^{-x/a})]^2$$
$$y'' = 1/(2a)(e^{x/a} + e^{-x/a})$$

Therefore, $ay'' = \sqrt{1 + y'^2}$

20. Substitute into the formula given in the chapter.

Chapter 16

EXERCISES

1. 3,650; 3,649.97; 3,649.97; 3,649.92; 3,649.92

2. $\dfrac{\partial z / \partial x}{1}$ $\qquad\qquad$ $\dfrac{\partial z / \partial y}{1}$

3. $2x$ $\qquad\qquad$ $2y$

4. y $\qquad\qquad$ x

5. $2xy^2$ $\qquad\qquad$ $2x^2y$

6. $ax^{a-1}y^b$ $\qquad\qquad$ bx^ay^{b-1}

7. $x/\sqrt{x^2 + y^2}$ $\qquad\qquad\qquad$ $y/\sqrt{x^2 + y^2}$

8. ae^{ax+by} $\qquad\qquad\qquad\qquad$ $be^{ax + by}$

9. $2ax + by + d$ $\qquad\qquad\qquad$ $bx + 2cy + e$

10. $n(x - a)^{n-1}$ $\qquad\qquad\qquad$ $m(y - b)^{m-1}$

11. $a \cos (ax + by)$ $\qquad\qquad\quad$ $b \cos (ax + by)$

12. $\dfrac{\partial z}{\partial x} = 6x + y + 9$

$\dfrac{\partial z}{\partial y} = x + 4y - 7$

Set both of these equal to zero, and solve the resulting two-variable, two-equation system to find $x = -1.87, y = 2.217$. Now calculate the second derivatives:

$$z_{xx} = \frac{\partial^2 z}{\partial x^2} = \frac{\partial}{\partial x}[6x + y + 9] = 6$$

$$z_{yy} = \frac{\partial^2 z}{\partial y^2} = \frac{\partial}{\partial y}[x + 4y - 7] = 4$$

$$z_{xy} = \frac{\partial^2 z}{\partial x\, \partial y} = \frac{\partial}{\partial x}[x + 4y - 7] = 1$$

Since $6 \times 4 > 1^2$, therefore $z_{xx} z_{yy} > (f_{xy})^2$, and this point is either a maximum or a minimum. Since z_{xx} and z_{yy} are both positive, the point $(x = -1.87, y = 2.217)$ is a minimum.

13. $x = 0.5, y = 0$, saddle point

14. $x = 2.5, y = 1.5$, saddle point

15. $x = 1.739, y = -0.304$, maximum

16. $x = 1, y = -2/3$, saddle point

17. $\dfrac{\partial Z}{\partial L} = -\dfrac{1}{2}P(L^{-2} + K^{-2})^{-3/2}(-2L^{-3}) - w$

$\dfrac{\partial Z}{\partial K} = -\dfrac{1}{2}P(L^{-2} + K^{-2})^{-3/2}(-2K^{-3}) - r$

Set the partial derivatives equal to zero and divide the top equation by the bottom equation:

$$\left(\frac{L}{K}\right)^{-3} = \frac{w}{r}$$

$$\frac{L}{K} = \left(\frac{w}{r}\right)^{-1/3}$$

18. $x = 1, y = 3, z = 2$

19. $\mathbf{a} + \mathbf{b} = (2 + 8, 6 + 3) = (10, 9)$; $\mathbf{a} \cdot \mathbf{b} = 2 \times 8 + 6 \times 3 = 34$

20. $\mathbf{a} + \mathbf{b} = (4 + 7, 3 + 1) = (11, 4)$; $\mathbf{a} \cdot \mathbf{b} = 4 \times 7 + 3 \times 1 = 31$

21. $\mathbf{a} + \mathbf{b} = (8 + 3, 5 + 9) = (11, 14)$; $\mathbf{a} \cdot \mathbf{b} = 8 \times 3 + 5 \times 9 = 69$

22. $\mathbf{a} + \mathbf{b} = (3 + 2, 7 + 4) = (5, 11)$; $\mathbf{a} \cdot \mathbf{b} = 3 \times 2 + 7 \times 4 = 34$

23. $\mathbf{a} + \mathbf{b} = (20 - 20, 12 - 12) = (0, 0)$; $\mathbf{a} \cdot \mathbf{b} = 20 \times (-20) + 12 \times (-12) = -544$

24. $\mathbf{a} + \mathbf{b} = (60 - 59, 10 - 10) = (1, 0)$; $\mathbf{a} \cdot \mathbf{b} = 60 \times (-59) + 10 \times (-10) = -3{,}640$

25. $\mathbf{a} + \mathbf{b} = (4 + 0, 0 + 6) = (4, 6)$; $\mathbf{a} \cdot \mathbf{b} = 4 \times 0 + 0 \times 6 = 0$

26. $\mathbf{a} + \mathbf{b} = (1 - 1, 1 + 1) = (0, 2)$; $\mathbf{a} \cdot \mathbf{b} = 1 \times (-1) + 1 \times 1 = 0$

27. $\mathbf{a} + \mathbf{b} = (0.5000 + 0.8660, 0.8660 + 0.5000) = (1.3660, 1.3660)$;
$\mathbf{a} \cdot \mathbf{b} = 0.5000 \times 0.8660 + 0.8660 \times 0.5000 = 0.8660$

28. $\mathbf{a} \cdot \mathbf{b} = 0$; $\theta = \pi/2 = 90$ degrees

29. $\mathbf{a} \cdot \mathbf{b} = 18 \times 54 + 25 \times 75 = 2{,}847$

$$\|\mathbf{a}\| = 30.8058; \|\mathbf{b}\| = 92.4175; \cos\theta = \frac{\mathbf{a} \cdot \mathbf{b}}{\|\mathbf{a}\| \times \|\mathbf{b}\|} = \frac{2{,}847}{30.8058 \times 92.4175} = 1; \theta = 0$$

30. $\mathbf{a} \cdot \mathbf{b} = \dfrac{\sqrt{3}}{2\sqrt{2}} + \dfrac{1}{2\sqrt{2}} = \dfrac{\sqrt{3} + 1}{2\sqrt{2}}$; $\|\mathbf{a}\| = \sqrt{\dfrac{3}{4} + \dfrac{1}{4}} = 1$; $\|\mathbf{b}\| = \sqrt{\dfrac{1}{2} + \dfrac{1}{2}} = 1$; $\cos\theta = \dfrac{\frac{\sqrt{3} + 1}{2\sqrt{2}}}{1 \times 1}$
$= 0.9659$; $\theta = \pi/12 = 15$ degrees.

31. $\mathbf{a} \cdot \mathbf{b} = \dfrac{\sqrt{3}}{2}$; $\theta = \pi/6 = 30$ degrees.

32. $\mathbf{a} \cdot \mathbf{b} = \frac{1}{2}$; $\theta = \pi/3 = 60$ degrees.

33. $a_x b_x + a_y b_y = b_x a_x + b_y a_y$

$\mathbf{a} \cdot \mathbf{b} = \mathbf{b} \cdot \mathbf{a}$

34. $n a_x b_x + n a_y b_y = a_x n b_x + a_y n b_y$

$n\mathbf{a} \cdot \mathbf{b} = \mathbf{a} \cdot (n\mathbf{b})$

35. $(a_x + b_x)c_x + (a_y + b_y)c_y = a_x c_x + b_x c_x + a_y c_y + b_y c_y = (a_x c_x + a_y c_y) + (b_x c_x + b_y c_y)$

$(\mathbf{a} + \mathbf{b}) \cdot \mathbf{c} = \mathbf{a} \cdot \mathbf{c} + \mathbf{b} \cdot \mathbf{c}$

36. $\mathbf{a} = (a_x, a_y)$; $\mathbf{b} = (b_x, b_y)$ $\mathbf{a} \cdot \mathbf{b} = a_x b_x + a_y b_y$

$r_a = \sqrt{a_x^2 + a_y^2}, r_b = \sqrt{b_x^2 + b_y^2}$

$\mathbf{a} \cdot \mathbf{b} = r_a r_b \left(\dfrac{a_x b_x}{r_a r_b} + \dfrac{a_y b_y}{r_a r_b} \right)$

Let θ_a be the angle between vector $\mathbf{a}$ and the x axis, θ_b be the angle between vector $\mathbf{b}$ and the x-axis, and $\theta = \theta_a - \theta_b$ be the angle between the two vectors. Then:

$$\frac{a_x}{r_a} = \cos\theta_a; \quad \frac{b_x}{r_b} = \cos\theta_b$$

$$\frac{a_y}{r_a} = \sin\theta_a; \quad \frac{b_y}{r_b} = \sin\theta_b$$

We can rewrite the dot product formula: $\mathbf{a}\cdot\mathbf{b} = r_a r_b (\cos\theta_a \cos\theta_b + \sin\theta_a \sin\theta_b)$
Using the formula for the cosine of the difference between two angles gives

$$\mathbf{a}\cdot\mathbf{b} = r_a r_b \cos\theta$$

37. The basket hits the ground when $z = 0$:

$$0 = h - \tfrac{1}{2}gt^2; \; h = \tfrac{1}{2}gt^2; \; \frac{2h}{g} = t^2; \; t = \sqrt{\frac{2h}{g}}$$

The speed at impact will be $gt = \sqrt{2gh}$.

38. Let D be the square of the distance. The value of t that minimizes D will also minimize the distance.

$$D = (a + tu - h)^2 + (b + tv - j)^2 + (c + tw - k)^2$$

$$dD/dt = 2(a + tu - h)u + 2(b + tv - j)v + 2(c + tw - k)w = 0$$

$$t(u^2 + v^2 + w^2) = hu + jv + kw - ua - vb - wc$$

$$t = (hu + jv + kw - ua - vb - wc)/(u^2 + v^2 + w^2)$$

39. (a) $z = (a^2 + b^2 + c^2 - ax - by)/c$.

  (b) Let D be the square of the distance. The values of $x, y,$ and z that minimize D will also minimize the distance.

$$D = x^2 + y^2 + z^2$$

$$\partial D/\partial x = 2x + 2z\,\frac{\partial z}{\partial x} = 2x - 2za/c$$

$$\partial D/\partial y = 2y + 2z\,\frac{\partial z}{\partial y} = 2y - 2zb/c$$

Set both partial derivatives equal to zero:

$$2x - 2za/c = 0$$

$$2y - 2zb/c = 0$$

Solve the first equation for x and the second for y:

$$x = za/c; \; y = zb/c$$

Put these back into the original equation $ax + by + cz = a^2 + b^2 + c^2$:

$$a(za/c) + b(zb/c) + cz = a^2 + b^2 + c^2$$

$$za^2/c + zb^2/c + cz = a^2 + b^2 + c^2$$

$$z(a^2 + b^2 + c^2)/c = a^2 + b^2 + c^2$$

$z = c$; then find $x = a$ and $y = b$.

40. $\mathbf{F} = \left(-\dfrac{\partial U}{\partial x}, -\dfrac{\partial U}{\partial y} \right) = (g \sin \theta \sin \phi, -g \cos \theta \sin \phi)$

41. $\mathbf{F} = \left(-\dfrac{\partial U}{\partial x}, -\dfrac{\partial U}{\partial y} \right) = (-2x, -2y)$. Note: the marble's motion can be described using the same formulas used for the motion of the spring in Chapter 15, except now separate formulas are needed for the x and y dimensions.

42. (a) $\int_0^{t_2}(-gv_{up})\, dt = -gv_{up}t_2 = -gs$ (where $s = v_{up}t_2$ is the distance moved)

(b) 0 (a force acting at right angles to the motion does no work)

(c) $-gv \sin \phi$

43. Use the chain rule. First, find the gradient of T:

$$\frac{\partial T}{\partial x} = z^b a(x^2 + y^2)^{a-1} 2x$$

$$\frac{\partial T}{\partial y} = z^b a(x^2 + y^2)^{a-1} 2y$$

$$\frac{\partial T}{\partial z} = b(x^2 + y^2)^a z^{b-1}$$

Next, find $d\mathbf{r}/dt = (v_x, v_y, v_z)$. From the chain rule:

$$\frac{dT}{dt} = (\text{grad } T) \cdot \frac{d\mathbf{r}}{dt} = 2xv_x z^b a(x^2 + y^2)^{a-1} + 2\, yv_y z^b a(x^2 + y^2)^{a-1} + bv_z(x^2 + y^2)^a z^{b-1}$$

44. $$0 = -\frac{GMm}{r_0} + \frac{1}{2}mv^2$$

$$\frac{GMm}{r_0} = \frac{1}{2}mv^2$$

$$\frac{2GM}{r_0} = v^2$$

$$v = \sqrt{\frac{2GM}{r_0}}$$

For Earth, the escape velocity is:

$$\sqrt{\frac{2 \times 6.67 \times 10^{-11} \times 5.97 \times 10^{24}}{6.38 \times 10^6}} = \text{about } 11{,}200 \text{ meters per second}$$

45. Let Δ = the change in potential energy.

$$\Delta = -\frac{GMm}{R+h} - \left(-\frac{GMm}{R}\right)$$

$$= -\frac{GMm}{R+h} + \frac{GMm}{R}$$

$$= -\frac{GMmR}{R(R+h)} + \frac{GMm(R+h)}{R(R+h)}$$

$$= \frac{-GMmR + GMm(R+h)}{R(R+h)}$$

$$= \frac{-GMmR + GMmR + GMmh}{R(R+h)}$$

$$= \frac{GMmh}{R(R+h)}$$

Since R is much bigger than h, $R(R+h)$ is just about the same as R^2. Let $g = GM/R^2$. Then $\Delta = mgh$, which is why it works to use this formula for the gravitational potential energy near the surface of Earth.

46. (a) $\mathbf{v} = (-R\omega \sin(\omega t), R\omega \cos(\omega t), 0)$

   $v = R\omega$.

(b) $\mathbf{a} = (-R\omega^2 \cos(\omega t), -R\omega^2 \sin(\omega t), 0) = -\omega^2 \mathbf{r}$

   $\|\mathbf{a}\| = R\omega^2$

(c) $\mathbf{F} = -\dfrac{GMm}{R^3}\mathbf{r}$

(d) Since acceleration = force divided by mass:

$$-\omega^2 \mathbf{r} = -\frac{GM}{R^3}\mathbf{r}$$

$$\omega^2 = \frac{GM}{R^3}$$

$$\omega = \sqrt{\frac{GM}{R^3}}$$

$$v = R\omega = \sqrt{\frac{GM}{R}}$$

(e) If an orbit starts at time 0, it will be completed when $\omega t_{orbit} = 2\pi$, where $t_{orbit} = \dfrac{2\pi}{\omega}$ is the time to complete one orbit.

$$(f)\text{ potential energy} = U = -\frac{GMm}{R}$$

$$\text{kinetic energy} = KE = \frac{1}{2}m\mathbf{v}\cdot\mathbf{v} = \frac{1}{2}mR^2\omega^2$$

$$= \frac{1}{2}mR^2\left(\frac{GM}{R^3}\right)$$

$$= \frac{GMm}{2R}$$

$$\text{total energy } E = -\frac{GMm}{R} + \frac{GMm}{2R} = -\frac{GMm}{2R}$$

For the moon in circular orbit, the potential energy is constant (since the distance to Earth does not change), and the kinetic energy is constant (because it moves with constant speed).

47. (a) Since θ does not appear in the Lagrangian expression, $\partial L/\partial\theta = 0$. $\partial L/\partial\theta' = mr^2\,\theta'$. The Euler equation then states that

$$0 = \frac{d}{dt}\frac{\partial L}{\partial\theta'}$$

Therefore, $\partial L/\partial\theta'$ does not change as time changes. So $mr^2\,\theta'$ is a constant (call it A), which represents the angular momentum of the planet. This result is used in the next part of the problem.

(b) $E = 1/2m\left(\dfrac{A^2}{m^2r^2} + r'^2\right) - \dfrac{k}{r}$

(c) $r' = \sqrt{\dfrac{2E}{m} - \dfrac{A^2}{m^2r^2} + \dfrac{2k}{mr}}$

(d) From the chain rule, $\dfrac{dr}{d\theta} = \dfrac{r'}{\theta'}$, or $d\theta = \dfrac{\theta'}{r'}\,dr$

Therefore, since $\theta' = A/mr^2$:

$$\theta = \int\frac{A/mr^2}{\sqrt{\dfrac{2E}{m} - \dfrac{A^2}{m^2r^2} + \dfrac{2k}{mr}}}\,dr$$

$$\theta = \int\frac{1/r^2\,dr}{\sqrt{\dfrac{2mE}{A^2} + \dfrac{2mk}{A^2r} - \dfrac{1}{r^2}}}$$

(e) $v = 1/r$; $dr = -dv/v^2$

$$\theta = \int\frac{-1}{\sqrt{\dfrac{2mE}{A^2} + \dfrac{2mkv}{A^2} - v^2}}\,dv$$

Use the formula from Chapter 12, Exercise 40, with $a = -1$, $b = 2mk/A^2$, and $c = 2mE/A^2$:

$$\theta = \frac{1}{\sqrt{1}} \arccos\left(\frac{-2r + 2mk/A^2}{\sqrt{4m^2k^2/A^4 + 8mE/A^2}}\right)$$

$$\cos\theta = \frac{(2mk/A^2)(1 - A^2/rmk)}{\sqrt{(4m^2k^2/A^4)(1 + 2A^2E/mk^2)}}$$

$$\cos\theta = \frac{1 - A^2/rmk}{\sqrt{1 + 2A^2E/mk^2}}$$

(f) $\cos\theta = \dfrac{1 - B/r}{e}$

$$r = \frac{B}{1 - e\cos\theta}$$

This is the polar coordinate equation for an ellipse with eccentricity e with one focus at the origin (which represents the location of the sun). This problem establishes that planets move in orbits shaped like ellipses, which is a fact discovered by Johannes Kepler in the early 1600s. However, Kepler was not able to derive that fact from the principle of gravitation, as we have done here. Instead, he needed to rely on many observations of planetary positions, which is much harder than solving this problem has been. For more information on polar coordinates, see a book on trigonometry such as *E-Z Trigonometry*. For more information on the physics of planetary motion, see a book on mechanics such as Marion, *Classical Dynamics of Particles and Systems* (Academic Press, 1970).

48.   Note that L does not depend on y, so $\partial L/\partial y = 0$. This means that the right-hand side of the Euler equation must also be zero, or that $\partial L/\partial y'$ must be a constant (call it K). Solve for $\partial L/\partial y'$:

$$\frac{\partial L}{dy'} = \frac{1}{2}\frac{(1+y'^2)^{-1/2}(2y')}{\sqrt{2gx}}$$

$$K = \frac{y'}{\sqrt{2gx(1 + y'^2)}}$$

Square both sides:

$2K^2gx(1 + y'^2) = y'^2$

$y'^2(1 - 2K^2gx) = 2K^2gx$

$$y' = \frac{dy}{dx} = \sqrt{\frac{2K^2gx}{1 - 2K^2gx}}$$

$$y = \int\sqrt{\frac{x^2}{x/2K^2g - x^2}}\,dx$$

Let $a = 1/4K^2g$, and then use the formula from Chapter 12, Exercise 41. The resulting curve is known as a *cycloid*.

49. (All distances are expressed in meters; all velocities, in meters per second.)

(a) $v = \sqrt{\dfrac{GM}{r_1}} = \sqrt{\dfrac{6.67 \times 10^{-11} \times 5.97 \times 10^{24}}{6.54 \times 10^6}} = 7{,}802.99$.

(b) $v = \sqrt{\dfrac{GM}{r_2}} = \sqrt{\dfrac{6.67 \times 10^{-11} \times 5.97 \times 10^{24}}{7.18 \times 10^6}} = 7{,}447.11$.

(c) The eccentricity of the orbit is:

$$e = \frac{r_2 - r_1}{r_2 + r_1} = \frac{7.18 \times 10^6 - 6.54 \times 10^6}{7.18 \times 10^6 + 6.54 \times 10^6} = 0.0466$$

The semimajor axis is:

$$a = \frac{r_1}{1-e} = \frac{6.54 \times 10^6}{1 - 0.0466} = 6.86 \times 10^6$$

The velocity at perigee is:

$$v = \sqrt{GM\left(\frac{2}{r_1} - \frac{1}{a}\right)}$$

$$v = \sqrt{6.67 \times 10^{-11} \times 5.97 \times 10^{24}\left(\frac{2}{6.54 \times 10^6} - \frac{1}{6.86 \times 10^6}\right)} = 7{,}982.91$$

(d) Change in velocity $= 7{,}982.91 - 7{,}802.99 = 179.92$.

(e) The velocity at apogee is:

$$v = \sqrt{GM\left(\frac{2}{r_2} - \frac{1}{a}\right)}$$

$$v = \sqrt{6.67 \times 10^{-11} \times 5.97 \times 10^{24}\left(\frac{2}{7.18 \times 10^6} - \frac{1}{6.86 \times 10^6}\right)} = 7{,}271.34$$

(f) Change in velocity $= 7{,}447.11 - 7{,}271.34 = 175.77$.

See Figure A–13.

50. For convenience, we'll find the absolute value of the potential first.

Step 1: Integrate around the ring:

$$|PE_{\text{ring}}| \int_{\theta=0}^{\theta=2\pi} = Gm\rho u^{-1} r \, d\theta$$

As we integrate around the ring, θ varies from 0 to 2π. Every point on the ring is the same distance from our viewpoint; consequently, the other letters don't depend on θ so we can pull them outside the integral:

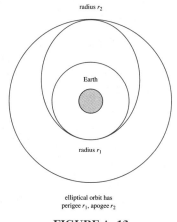

radius r_2

Earth

radius r_1

elliptical orbit has
perigee r_1, apogee r_2

FIGURE A–13.

$$|PE_{ring}| = Gm\rho u^{-1}r \int_{\theta=0}^{\theta=2\pi} d\theta$$

$$|PE_{ring}| = Gm\rho u^{-1}r\theta \Big|_{\theta=0}^{\theta=2\pi}$$

$$|PE_{ring}| = Gm\rho u^{-1}r(2\pi - 0)$$

Step 2: Integrate across rings to form a plate:

$$|PE_{plate}| = \int_{r=0}^{r=y} Gm\rho u^{-1}r(2\pi)\, dr$$

Recall $u = \sqrt{x^2 + r^2}$. Find du/dr:

$$du = 0.5(x^2 + r^2)^{(-0.5)}2r\, dr = \frac{r}{u}\, dr$$

$$u\, du = r\, dr$$

$$|PE_{plate}| = 2\pi Gm\rho \int_{u=x}^{u=\sqrt{x^2+y^2}} u^{-1}u\, du$$

Cancel u^{-1} and u:

$$|PE_{plate}| = 2\pi Gm\rho \int_{u=x}^{u=\sqrt{x^2+y^2}} du$$

$$|PE_{plate}| = 2\pi Gm\rho\, u \Big|_{u=x}^{u=\sqrt{x^2+y^2}}$$

$$|PE_{plate}| = 2\pi Gm\rho(\sqrt{x^2 + y^2} - x)$$

Step 3: Integrate across all the plates to form a sphere:

$$|PE_{sphere}| = \int_{x=D-R}^{x=D+R} 2\pi Gm\rho(\sqrt{x^2 + y^2} - x)\, dx$$

$$|PE_{sphere}| = 2\pi Gm\rho \int_{x=D-R}^{x=D+R} (\sqrt{x^2 + y^2} - x)\, dx \quad (1)$$

From Figure 16–7, we can see that:

$$(D - x)^2 + y^2 = R^2$$
$$D^2 - 2Dx + x^2 + y^2 = R^2$$
$$x^2 + y^2 = R^2 - D^2 + 2Dx$$

Put this formula for $x^2 + y^2$ into Equation (1):

$$|PE_{sphere}| = 2\pi Gm\rho \int_{x=D-R}^{x=D+R} (\sqrt{R^2 - D^2 + 2Dx} - x)\, dx \quad (2)$$

For convenience, define I_1 and I_2 as follows:

$$I_1 = \int_{x=D-R}^{x=D+R} (\sqrt{R^2 - D^2 + 2Dx}\, dx$$

$$I_2 = \int_{x=D-R}^{x=D+R} (-x)\, dx$$

Then:

$$|PE_{\text{sphere}}| = 2\pi Gm\rho(I_1 + I_2)$$

Work out I_1:

$$I_1 = \int_{x=D-R}^{x=D+R} \sqrt{R^2 - D^2 + 2Dx} \; dx$$

Solve by substitution. Let $v = R^2 - D^2 + 2Dx$, then $dv = 2D \; dx$ and $dx = \frac{dv}{2D}$.

$$I_1 = \int_{v=R^2 - D^2 + 2D(D-R)}^{v=R^2 - D^2 + 2D(D+R)} \left(\frac{v^{1/2}}{2D} \right) dv$$

$$I_1 = \frac{1}{2D} \int_{v=R^2 - D^2 + 2D^2 - 2DR}^{v=R^2 - D^2 + 2D^2 + 2DR} v^{1/2} \; dv$$

The limits of integration can be simplified:

$$I_1 = \frac{1}{2D} \left(\frac{2}{3} v^{3/2} \right) \Bigg|_{v=R^2 + D^2 - 2DR}^{v=R^2 + D^2 + 2DR}$$

$$I_1 = \frac{1}{2D} \left(\frac{2}{3} v^{3/2} \right) \Bigg|_{v=(D-R)^2}^{v=(D+R)^2}$$

$$I_1 = \frac{1}{3D} [(D + R)^3 - (D - R)^3]$$

$$I_1 = \frac{1}{3D} [(D^3 + 3D^2R + 3DR^2 + R^3 - (D^3 - 3D^2R + 3DR^2 - R^3)]$$

$$I_1 = \frac{1}{3D} [(D^3 + 3D^2R + 3DR^2 + R^3 - D^3 + 3D^2R - 3DR^2 + R^3)]$$

$$I_1 = \frac{1}{3D} [6D^2r + 2R^3]$$

$$I_1 = 2DR + \frac{2R^3}{3D}$$

Now work out I_2:

$$I_2 = \int_{x=D-R}^{x=D+R} (-x) \; dx$$

$$I_2 = -0.5x^2 \Big|_{x=D-R}^{x=D+R}$$

$$I_2 = -0.5[(D + R)^2 - (D - R)^2]$$

$$I_2 = -0.5[D^2 + 2DR + R^2 - (D^2 - 2DR + R^2]$$

$$I_2 = -0.5[D^2 + 2DR + R^2 - D^2 + 2DR - R^2]$$

$$I_2 = -0.5[4DR]$$

$$I_2 = -2DR$$

Now combine the results for I_1 and I_2:

$$|PE_{sphere}| = 2\pi Gm\rho(I_1 + I_2)$$

$$|PE_{sphere}| = 2\pi Gm\rho\left(2DR + \frac{2R^3}{3D} - 2DR\right)$$

$$|PE_{sphere}| = 2\pi Gm\rho\left(\frac{2R^3}{3D}\right)$$

$$|PE_{sphere}| = \frac{4\pi R^3 Gm\rho}{3D}$$

The volume (V) of the sphere is $4\pi R^3/3$:

$$|PE_{sphere}| = \frac{VGm\rho}{D}$$

The mass (M) of the sphere is volume times density ($V\rho$):

$$|PE_{sphere}| = \frac{GMm}{D}$$

This final answer is the negative of this expression:

$$PE_{sphere} = -\frac{GMm}{D}$$

In words: the gravitational potential of a sphere acts as if all of the mass was located at the center of the sphere (which is a distance D away from your position). If you take the negative of the gradient of the potential, you will get Newton's law—the force of gravity is inversely proportional to the distance squared.

Chapter 17

EXERCISES

The answers for Exercises 1 to 12 are at *http://myhome.spu.edu/ddowning/easycalc.html*

13. $\int y\,dy = -\int x\,dx$; $y^2/2 = -x^2/2 + C$; this can be rewritten as $x^2 + y^2 = r^2$, which is the equation of a circle.

14. $\int y^{-1}\,dy = \int x\,dx$; $\ln y = x^2/2 + C$; $y = C_2 e^{x^2/2}$

15. $\int y^{-1}\,dy = -\int x\,dx$; $\ln y = -x^2/2 + C$; $y = C_2 e^{-x^2/2}$

16. $f(x) = \ln(1 + x)$; $f'(x) = (1 + x)^{-1}$; $f''(x) = -(1 + x)^{-2}$; $f'''(x) = 2(1 + x)^{-3}$; $f''''(x) = -6(1 + x)^{-4}$; $f'''''(x) = 24(1 + x)^{-5}$. The ith derivative is $f^i(x) = (-1)^{i-1}(i - 1)!(1 + x)^{-i}$. The ith derivative evaluated at $x = 0$ is $f^i(0) = (-1)^{i-1}(i - 1)!(1 + 0)^{-i} = (-1)^{i-1}(i - 1)!$. (The factor $(-1)^{i-1}$ is included because the signs of the derivatives alternate.) The Taylor series is:

$$f(0 + t) = f(0) + f'(0)t + f''(0)t^2/2 + \cdots + f^i(0)t^i/i! + \cdots$$

$$\ln(1 + 0 + t) = \ln(1 + 0) + (1 + 0)^{-1}t - (1 + 0)^{-2}t^2/2 + \cdots + (-1)^{i-1}(i-1)!t^i/i! + \cdots$$

$$= \ln(1 + t) = \ln(1) + (1)^{-1}t - (1)^{-2}t^2/2 + \cdots + (-1)^{i-1}t^i/i + \cdots$$

$$\ln(1 + t) = t - t^2/2 + t^3/3 - t^4/4 + t^5/5 - \cdots$$

17. (a) $(1 + z + z^2 + z^3 + z^4 + \cdots + z^n) - (z + z^2 + z^3 + z^4 + \cdots + z^{n+1}) = 1 - z^{n+1}$

(b) As n approaches infinity:

$$1 + z + z^2 + z^3 + z^4 + \cdots = \frac{1}{1 - z}$$

Let $z = -x^2$:

$$1 - x^2 + x^4 - x^6 + x^8 + \cdots = \frac{1}{1 + x^2}$$

$$\int (1 - x^2 + x^4 - x^6 + x^8 + \cdots) \, dx = \int \frac{1}{1 + x^2} \, dx$$

$$x - \frac{x^3}{3} + \frac{x^5}{5} - \frac{x^7}{7} + \frac{x^9}{9} + \cdots = \arctan x$$

This formula provides a series approximation for the arctangent function.

18. $\displaystyle \lim_{i \to \infty} \frac{a_i + 1}{a_i} = \lim_{i \to \infty} \frac{x^{i+1}/(i+1)!}{x^i i!}$

$$= \lim_{i \to \infty} \frac{x^{i+1} i!}{x^i (i+1)!}$$

$$= \lim_{i \to \infty} \frac{x}{i + 1} = 0$$

The series converges.

19. $\displaystyle \lim_{x \to 0} \frac{\cos x}{1} = 1$

20. Let $L = \displaystyle \lim_{x \to 0} (1 + x)^{1/x}$.

$$\ln L = \frac{\ln(1 + x)}{x}; \text{ let } f(x) = \ln(1 + x); \text{ let } g(x) = x. \lim_{x \to 0} \ln L = \lim_{x \to 0} \frac{1/(1 + x)}{1} = 1$$

21. 10/19

22. The ratio of the derivatives is nx^{n-1}/e^x. This still leads to ∞/∞. Keep taking the derivative. Eventually the numerator will become $n!$. The indestructible function e^x won't be changed no matter how many times you take its derivative. Then $\lim\limits_{x\to\infty} n!/e^x$ becomes zero.

23. $\lim\limits_{x\to 8} \dfrac{3x^2 - 38x + 118}{2x - 11} = \dfrac{6}{5} = 1.2000$

24. $\lim\limits_{x\to 7} \dfrac{3x^2 - 38x + 111}{2x - 8} = \dfrac{-8}{6} = -1.3333$

25. $\lim\limits_{x\to 3} \dfrac{3x^2 - 32x + 79}{2x - 12} = \dfrac{10}{-6} = -1.6667$

26. $\lim\limits_{x\to 2} \dfrac{3x^2 - 24x + 41}{2x - 6} = \dfrac{5}{-2} = -2.5000$

27. $\lim\limits_{x\to -9} \dfrac{3x^2 - 46x - 48}{2x + 14} = \dfrac{609}{-4} = -152.2500$

28. $\lim\limits_{x\to 9} \dfrac{3x^2 + 2x - 66}{2x - 6} = \dfrac{195}{12} = 16.2500$

29. $\lim\limits_{x\to -10} \dfrac{3x^2 - 4x - 184}{2x - 8} = \dfrac{156}{-28} = -5.5714$

30. $\lim\limits_{x\to -24} \dfrac{3x^2 - 52x - 51}{2x - 18} = \dfrac{429}{30} = 14.3000$

31. $\lim\limits_{x\to -14} \dfrac{3x^2 - 88x + 321}{2x - 22} = \dfrac{-323}{6} = 53.83333$

32. $\lim\limits_{x\to 5} \dfrac{x^5 - 51x^4 + 951x^3 - 8{,}201x^2 + 33{,}060x - 50{,}400}{x^4 - 20x^3 + 146x^2 - 460x + 525}$

$$\lim\limits_{x\to 5} \frac{f'(x)}{g'(x)} = \frac{5x^4 - 204x^3 + 2{,}853x^2 - 16{,}402x + 33{,}060}{4x^3 - 60x^2 + 292x - 460}$$

This still is 0/0. Use L'Hôpital's rule again.

$$\lim\limits_{x\to 5} \frac{f''(x)}{g''(x)} = \frac{20x^3 - 612x^2 + 5{,}706x - 16{,}402}{12x^3 - 120x^2 + 292} = \frac{-672}{-8} = 84$$

Appendix 2: Summary of Trigonometric Formulas

TRIGONOMETRIC FUNCTIONS FOR RIGHT ANGLES

Let A be one of the acute angles in a right triangle. Then,

$$\sin A = \frac{\text{opposite side}}{\text{hypotenuse}}$$

$$\cos A = \frac{\text{adjacent side}}{\text{hypotenuse}}$$

$$\tan A = \frac{\text{opposite side}}{\text{adjacent side}}$$

TRIGONOMETRIC FUNCTIONS: GENERAL DEFINITION

Consider a point (x, y) in a Cartesian coordinate system. Let r be the distance from that point to the origin, and let A be the angle between the x axis and the line connecting the origin to that point. Then,

$$\sin A = \frac{y}{r}$$

$$\cos A = \frac{x}{r}$$

$$\tan A = \frac{y}{x}$$

RADIAN MEASURE

$$\pi \text{ rad} = 180°$$

SPECIAL VALUES

Degrees	Radians	sin	cos	tan
0°	0	0	1	0
30°	$\dfrac{\pi}{6}$	$\dfrac{1}{2}$	$\dfrac{\sqrt{3}}{2}$	$\dfrac{1}{\sqrt{3}}$
45°	$\dfrac{\pi}{4}$	$\dfrac{1}{\sqrt{2}}$	$\dfrac{1}{\sqrt{2}}$	1
60°	$\dfrac{\pi}{3}$	$\dfrac{\sqrt{3}}{2}$	$\dfrac{1}{2}$	$\sqrt{3}$
90°	$\dfrac{\pi}{2}$	1	0	Undefined (infinite)

TRIGONOMETRIC IDENTITIES
These equations are true for every allowable value of A and B.

Reciprocal functions

$$\sin A = \frac{1}{\csc A} \qquad \csc A = \frac{1}{\sin A}$$

$$\cos A = \frac{1}{\sec A} \qquad \sec A = \frac{1}{\cos A}$$

$$\tan A = \frac{1}{\operatorname{ctn} A} \qquad \operatorname{ctn} A = \frac{1}{\tan A}$$

Cofunctions (radian form)

$$\sin A = \cos\left(\frac{\pi}{2} - A\right) \qquad \cos A = \sin\left(\frac{\pi}{2} - A\right)$$

$$\tan A = \operatorname{ctn}\left(\frac{\pi}{2} - A\right) \qquad \operatorname{ctn} A = \tan\left(\frac{\pi}{2} - A\right)$$

$$\sec A = \csc\left(\frac{\pi}{2} - A\right) \qquad \csc A = \sec\left(\frac{\pi}{2} - A\right)$$

Negative angle relations

$$\sin(-A) = -\sin A$$
$$\cos(-A) = \cos A$$
$$\tan(-A) = -\tan A$$

Quotient relations

$$\tan A = \frac{\sin A}{\cos A}$$

$$\text{ctn } A = \frac{\cos A}{\sin A}$$

Supplementary angle relations

The angles A and B are supplementary angles if $A + B = \pi$.

$$\sin(\pi - A) = \sin A$$
$$\cos(\pi - A) = -\cos A$$
$$\tan(\pi - A) = -\tan A$$

Pythagorean identities

$$\sin^2 A + \cos^2 A = 1$$
$$\tan^2 A + 1 = \sec^2 A$$
$$\text{ctn}^2 A + 1 = \csc^2 A$$

Functions of the sum of two angles

$$\sin(A + B) = \sin A \cos B + \sin B \cos A$$
$$\cos(A + B) = \cos A \cos B - \sin A \sin B$$
$$\tan(A + B) = \frac{\tan A + \tan B}{1 - \tan A \tan B}$$

Functions of the difference of two angles

$$\sin(A - B) = \sin A \cos B - \sin B \cos A$$
$$\cos(A - B) = \cos A \cos B + \sin A \sin B$$
$$\tan(A - B) = \frac{\tan A - \tan B}{1 + \tan A \tan B}$$

Double-angle formulas

$$\sin(2A) = 2 \sin A \cos A$$
$$\cos(2A) = \cos^2 A - \sin^2 A$$
$$= 1 - 2 \sin^2 A$$
$$= 2 \cos^2 A - 1$$
$$\tan(2A) = \frac{2 \tan A}{1 - \tan^2 A}$$

Squared formulas

$$\sin^2 A = \frac{1}{2}(1 - \cos 2A)$$

$$\cos^2 A = \frac{1}{2}(1 + \cos 2A)$$

Half-angle formulas

$$\sin \frac{A}{2} = \pm \sqrt{\frac{1 - \cos A}{2}}$$

$$\cos \frac{A}{2} = \pm \sqrt{\frac{1 + \cos A}{2}}$$

$$\tan \frac{A}{2} = \pm \sqrt{\frac{1 - \cos A}{1 + \cos A}}$$

Product formulas

$$\sin A \cos B = \frac{1}{2}[\sin (A + B) + \sin (A - B)]$$

$$\cos A \sin B = \frac{1}{2}[\sin (A + B) - \sin (A - B)]$$

$$\cos A \cos B = \frac{1}{2}[\cos (A + B) + \cos (A - B)]$$

$$\sin A \sin B = -\frac{1}{2}[\cos (A + B) - \cos (A - B)]$$

Sum formulas

$$\sin A + \sin B = 2 \sin \frac{A + B}{2} \cos \frac{A - B}{2}$$

$$\cos A + \cos B = 2 \cos \frac{A + B}{2} \cos \frac{A - B}{2}$$

Difference formulas

$$\sin A - \sin B = 2 \cos \frac{A + B}{2} \sin \frac{A - B}{2}$$

$$\cos A - \cos B = -2 \sin \frac{A + B}{2} \sin \frac{A - B}{2}$$

FORMULAS FOR TRIANGLES

Let a be the side of a triangle opposite angle A, let b be the side opposite angle B, and let c be the side opposite angle C.

Law of cosines

$$c^2 = a^2 + b^2 - 2ab \cos C$$

Law of sines

$$\frac{a}{\sin A} = \frac{b}{\sin B} = \frac{c}{\sin C}$$

Appendix 3:
Brief Table of Integrals

a, b, c, m, n represent constants: C represents the arbitrary constant of integration.

PERFECT INTEGRAL

$$\int dx = x + C$$

MULTIPLICATION BY CONSTANT

$$\int n\, dx = nx + C$$

$$\int nf(x)\, dx = n \int f(x)\, dx$$

$$\int f(nx)\, dx = \frac{1}{n} \int f(u)\, du \qquad \text{where } u = nx$$

Addition

$$\int [\, f(x) + g(x)]\, dx = \int f(x)\, dx + \int g(x)\, dx$$

Powers

$$\int x\, dx = \frac{x^2}{2} + C$$

$$\int x^n\, dx = \frac{x^{n+1}}{n+1} + C \text{ if } n \neq -1$$

$$\int x^{-1}\, dx = \ln |x| + C$$

POLYNOMIALS

$$\int (a_n x^n + a_{n-1} x^{n-1} + \cdots + a_2 x^2 + a_1 x + a_0)\, dx =$$

$$\frac{a_n x^{n+1}}{n+1} + \frac{a_{n-1} x^n}{n} + \cdots + \frac{a_2 x^3}{3} + \frac{a_1 x^2}{2} + a_0 x + C$$

SUBSTITUTION

$$\int f[u(x)]\, dx = \int f(u)\, \frac{dx}{du}\, du$$

For example:

$$\int x f(x^2 + a)\, dx = \int x f(u) \left(\frac{1}{2x}\right) du = \frac{1}{2} \int f(u)\, du$$

where $u = x^2 + a$.

INTEGRATION BY PARTS

$$\int u\, dv = uv - \int v\, du$$

Note: The arbitrary constant of integration C will not be explicitly listed in the integrals that follow, but it must always be remembered.

TRIGONOMETRY

$$\int \sin x\, dx \quad = \quad -\cos x$$

$$\int \cos x\, dx \quad = \quad \sin x$$

$$\int \tan x\, dx \quad = \quad \ln |\sec x|$$

$$\int \sec x\, dx \quad = \quad \ln |\sec x + \tan x|$$

$$\int \sin^2 x\, dx \quad = \quad \frac{x}{2} - \frac{\sin 2x}{4}$$

$$\int x \sin x\, dx \quad = \quad \sin x - x \cos x$$

$$\int x^2 \sin x\, dx \quad = \quad -x^2 \cos x + 2x \sin x + 2 \cos x$$

$$\int \cos^2 x \, dx \quad = \quad \frac{x}{2} + \frac{\sin 2x}{4}$$

$$\int x \cos x \, dx \quad = \quad \cos x + x \sin x$$

$$\int \sin x \cos x \, dx \quad = \quad \frac{\sin^2 x}{2}$$

$$\int \sin^m x \, dx \quad = \quad -\frac{\sin^{m-1} x \cos x}{m} + \frac{m-1}{m} \int \sin^{m-2} x \, dx$$

$$\int \arcsin x \, dx \quad = \quad x \arcsin x + \sqrt{1-x^2}$$

$$\int \arctan x \, dx \quad = \quad x \arctan x - \frac{\ln (1+x^2)}{2}$$

EXPONENTIAL FUNCTIONS AND LOGARITHMS

$$\int e^x \, dx \quad = \quad e^x$$

$$\int x e^x \, dx \quad = \quad x e^x - e^x$$

$$\int x^2 e^x \, dx \quad = \quad x^2 e^x - 2x e^x + 2 e^x$$

$$\int a^x \, dx \quad = \quad \frac{a^x}{\ln a}$$

$$\int e^x \cos x \, dx \quad = \quad \frac{e^x (\sin x + \cos x)}{2}$$

$$\int \ln x \, dx \quad = \quad x \ln x - x$$

$$\int x \ln x \, dx \quad = \quad \frac{x^2 \ln x}{2} - \frac{x^2}{4}$$

$$\int x^2 \ln x \, dx \quad = \quad \frac{x^3 \ln x}{3} - \frac{x^3}{9}$$

$$\int_{-\infty}^{\infty} e^{-x^2/2} \, dx \quad = \quad \frac{1}{\sqrt{2\pi}}$$

Integrals Involving $ax^2 + bx + c$

For this section, let $D = b^2 - 4ac$. These integrals can be simplified by substituting $u = x + b/2a$:

$$ax^2 + bx + c = \frac{4a^2u^2 - D}{4a}$$

(1) Let y

$$= \int \frac{1}{ax^2 + bx + c}\, dx$$

If $D < 0 : y$

$$= \frac{2}{\sqrt{-D}} \arctan\left(\frac{2ax+b}{\sqrt{-D}}\right)$$

If $D > 0 : y$

$$= \frac{1}{\sqrt{D}} \ln\left|\frac{2ax+b-\sqrt{D}}{2ax+b+\sqrt{D}}\right|$$

If $D = 0 : y$

$$= -\frac{2}{2ax+b}$$

Specific examples of form (1) include:

$$\int \frac{1}{1+x^2}\, dx = \arctan x$$

$$\int \frac{1}{1+x^2}\, dx = \frac{1}{2} \ln\left|\frac{1+x}{1-x}\right|$$

$$\int \frac{1}{m^2+n^2x^2}\, dx = \frac{1}{mn} \arctan\left(\frac{nx}{m}\right)$$

$$\int \frac{1}{m^2-n^2x^2}\, dx = \frac{1}{2mn} \ln\left|\frac{m+nx}{m-nx}\right|$$

(2) Let y

$$= \int \frac{1}{\sqrt{ax^2 + bx + c}}\, dx$$

If $a > 0 : y$

$$= \frac{1}{\sqrt{a}} \ln\left|2\sqrt{a(ax^2+bx+c)} + 2ax + b\right|$$

If $a < 0$ and $D > 0 : y$

$$= \frac{-1}{\sqrt{-a}} \arcsin\left(\frac{2ax+b}{\sqrt{D}}\right)$$

(provided $|2ax + b| < \sqrt{D}$).

Specific examples of form (2) include:

$$\int \frac{1}{\sqrt{1-x^2}}\, dx \quad = \arcsin x$$

$$\int \frac{1}{\sqrt{1+x^2}}\, dx \quad = \ln\left(x + \sqrt{1+x^2}\right)$$

$$\int \frac{1}{\sqrt{x^2-1}}\, dx \quad = \ln\left(x + \sqrt{x^2-1}\right)$$

$$\int \frac{1}{\sqrt{m^2-n^2 x^2}}\, dx \quad = \frac{1}{n}\arcsin\left(\frac{nx}{m}\right)$$

$$\int \frac{1}{\sqrt{n^2 x^2 + m^2}}\, dx \quad = \frac{1}{n}\ln\left|\frac{nx}{m} + \sqrt{1 + \frac{n^2 x^2}{m^2}}\right|$$

(3) Let $y \quad = \int \sqrt{ax^2 + bx + c}\, dx$

$$y \quad = \frac{2ax+b}{4a}\sqrt{ax^2 + bx + c} + \left(\frac{4ac-b^2}{8a}\right)\int \frac{1}{\sqrt{ax^2 + bx + c}}\, dx$$

Specific examples of form (3) include:

$$\int \sqrt{1-x^2}\, dx \quad = \frac{\arcsin x + x\sqrt{1-x^2}}{2}$$

$$\int \sqrt{1+x^2}\, dx \quad = \frac{x\sqrt{1+x^2} + \ln\left|x+\sqrt{1+x^2}\right|}{2}$$

$$\int \sqrt{x^2-1}\, dx \quad = \frac{x\sqrt{x^2-1} - \ln\left|x+\sqrt{x^2-1}\right|}{2}$$

$$\int \sqrt{m^2-n^2 x^2}\, dx \quad = \frac{m^2}{2n}\left[\arcsin\left(\frac{nx}{m}\right) + \frac{nx}{m}\sqrt{1-\left(\frac{nx}{m}\right)^2}\right]$$

$$\int \sqrt{m^2+n^2 x^2}\, dx \quad = \frac{m^2}{2n}\left[\left(\frac{nx}{m}\right)\sqrt{1+\left(\frac{nx}{m}\right)^2} + \ln\left|\frac{nx}{m} + \sqrt{1+\left(\frac{nx}{m}\right)^2}\right|\right]$$

Glossary

acceleration — The rate of change of the velocity of an object; it is equal to the second derivative of the function giving the position of the object as a function of time.

antiderivative — If the derivative of a function $F(x)$ is equal to $f(x)$, then $F(x)$ is the antiderivative of $f(x)$. (See integral.)

chain rule — Rule that says the derivative of the function $f(g(x))$ is equal to $(df/dg)(dg/dx)$.

concave downward curve — Curve oriented so that its concave side faces downward; this occurs in places where the second derivative of the function representing that curve is negative.

concave upward curve — Curve oriented so that its concave side faces upward; this occurs in places where the second derivative of the function representing that curve is positive.

constant of integration — An arbitrary constant that must be added to an indefinite integral.

definite integral — If $f(x)$ represents a function of x that is always nonnegative, then the definite integral of $f(x)$ between a and b (symbolized by $\int_b^a f(x)\,dx$) represents the area under the curve $y = f(x)$, above the x axis, to the right of the line $x = a$, and to the left of the line $x = b$. If $F(x)$ is an antiderivative function for $f(x)$, then the value of the definite integral is $F(b) - F(a)$.

derivative — The derivative of a function is the rate of change of that function; on the graph of the curve $y = f(x)$ the value of the derivative at a point x represents the slope of the tangent line to the curve at that point; if $f(t)$ is a function giving the position of an object as a function of time, then the value of the derivative at time t represents the velocity of the object at that time.

differential equation — An equation involving derivatives.

differentiate — The process of taking the derivative of a function.

e — The symbol e represents a number approximately equal to 2.71828 . . . ; it is used as the base of the natural logarithm function.

indefinite integral — If $f(x)$ is a function of x, then the indefinite integral of a function (symbolized by $\int f(x)\, dx$) is equal to $F(x) + C$, where $F(x)$ is a function whose derivative is equal to $f(x)$, and C is an arbitrary constant that may have any value. (However, the value of C may be determined by using an initial condition.)

inflection, point of — A point such that a curve is concave downward on one side of the point and concave upward on the other side; the second derivative is zero at points of inflection.

integral — See definite integral and indefinite integral.

integrand — In the integral expression $\int f(x)\, dx$, the function $f(x)$ is called the integrand.

integration by partial fractions — Integration method for finding an integral of a function consisting of several factors multiplied together in the denominator of a fraction; the method works by writing the fraction as a sum of several fractions with simpler denominators.

integration by parts — Integration method based on the formula:

$$\int u\, dv = uv - \int v\, du.$$

natural logarithm — The logarithm of x to the base e, which is symbolized by $\ln x$; if $y = \ln x$, then $x = e^y$; the natural logarithm can also be defined by the integral:

$$\ln x = \int_1^x q^{-1}\, dq.$$

numerical integration — Procedure for finding the definite integral of a function by approximating the area under the curve by a series of rectangles or other simple shapes; this procedure is needed when it is not possible to find a formula for the antiderivative of a function.

partial derivative — The derivative of a multivariable function with respect to one of the variables which is found by treating all of the other variables in the function as constant.

power rule for derivatives — Rule that says the derivative of the function x^n is nx^{n-1}.

power rule for integrals — Rule that says the indefinite integral of the function x^n is $(x^{n+1})/(n + 1) + C$.

product rule — Rule that says the derivative of $u(x) \times v(x)$ is equal to $u(x) \times v'(x) + u'(x) \times v(x)$.

secant — A line that crosses a curve at two points.

tangent — A line that just touches a curve at one point without crossing it at that point.

velocity — The rate of change of the position of an object; for an object moving in one dimension it is equal to the derivative of the function that gives the position of the object as a function of time.

Index

MOVE TO THE HEAD OF YOUR CLASS
THE EASY WAY!

Barron's presents **THE E-Z SERIES** (formerly THE EASY WAY SERIES)—specially prepared by top educators, it maximizes effective learning while minimizing the time and effort it takes to raise your grades, brush up on the basics, and build your confidence. Comprehensive and full of clear review examples, **THE E-Z SERIES** is your best bet for better grades, quickly!

ISBN 978-0-7641-4256-7 **E-Z Accounting**—$16.99, *Can$19.99*
ISBN 978-0-7641-4257-4 **E-Z Algebra**—$16.99, *Can$19.99*
ISBN 978-1-4380-0039-8 **E-Z Algebra 2**—$16.99, *Can$19.50*
ISBN 978-0-7641-4258-1 **E-Z American History**—$18.99, *Can$22.99*
ISBN 978-0-7641-4458-5 **E-Z American Sign Language**—$16.99, *Can$19.99*
ISBN 978-0-7641-4468-4 **E-Z Anatomy and Physiology**—$16.99, *Can$19.99*
ISBN 978-0-7641-4466-0 **E-Z Arithmetic**—$16.99, *Can$19.99*
ISBN 978-0-7641-4134-8 **E-Z Biology**—$18.99, *Can$21.99*
ISBN 978-0-7641-4133-1 **E-Z Bookkeeping**—$14.99, *Can$17.99*
ISBN 978-0-7641-4259-8 **E-Z Business Math**—$14.99, *Can$17.99*
ISBN 978-0-7641-4461-5 **E-Z Calculus**—$16.99, *Can$19.99*
ISBN 978-0-7641-4128-7 **E-Z Chemistry**—$16.99, *Can$19.99*
ISBN 978-0-7641-2579-9 **Creative Writing the Easy Way**—$14.99, *Can$17.99*
ISBN 978-0-7641-4464-6 **E-Z Earth Science**—$16.99, *Can$19.99*
ISBN 978-0-7641-3736-5 **English for Foreign Language Speakers the Easy Way**—$18.99, *Can$21.99*
ISBN 978-0-7641-4260-4 **E-Z English**—$14.99, *Can$17.99*
ISBN 978-0-7641-3050-2 **Forensics the Easy Way**—$14.99, *Can$17.99*
ISBN 978-0-7641-4455-4 **E-Z French**—$16.99, *Can$19.99*
ISBN 978-0-7641-2435-8 **French Grammar the Easy Way**—$16.99, *Can$19.99*
ISBN 978-0-7641-3918-5 **E-Z Geometry**—$16.99, *Can$19.99*
ISBN 978-0-7641-4261-1 **E-Z Grammar**—$14.99, *Can$16.99*
ISBN 978-0-7641-4454-7 **E-Z Italian**—$14.99, *Can$17.99*
ISBN 978-0-8120-9627-9 **Japanese the Easy Way**—$18.99, *Can$22.99*
ISBN 978-0-7641-3237-7 **Macroeconomics the Easy Way**—$14.99, *Can$17.99*
ISBN 978-0-7641-9369-9 **Mandarin Chinese the Easy Way**—$21.99, *Can$26.50*
ISBN 978-0-7641-4132-4 **E-Z Math**—$14.99, *Can$17.99*
ISBN 978-0-7641-1871-5 **Math Word Problems the Easy Way**—$16.99, *Can$19.99*
ISBN 978-0-7641-4456-1 **E-Z Microbiology**—$18.99, *Can$22.99*
ISBN 978-0-8120-9601-9 **Microeconomics the Easy Way**—$16.99, *Can$19.99*
ISBN 978-0-7641-4467-7 **E-Z Organic Chemistry**—$16.99, *Can$19.99*
ISBN 978-0-7641-4126-3 **E-Z Physics**—$16.99, *Can$19.99*
ISBN 978-1-4380-0011-4 **E-Z Pre-Algebra**—$14.99, *Can$16.99*
ISBN 978-0-7641-4465-3 **E-Z Precalculus**—$16.99, *Can$19.99*
ISBN 978-0-7641-4462-2 **E-Z Psychology**—$16.99, *Can$19.99*
ISBN 978-0-7641-4129-4 **E-Z Spanish**—$16.99, *Can$19.99*
ISBN 978-0-7641-4249-9 **E-Z Spanish Grammar**—$16.99, *Can$19.99*
ISBN 978-0-7641-4459-2 **E-Z Spelling**—$14.99, *Can$17.99*
ISBN 978-0-7641-3978-9 **E-Z Statistics**—$16.99, *Can$19.99*
ISBN 978-0-7641-4251-2 **E-Z Trigonometry**—$16.99, *Can$19.99*
ISBN 978-0-8120-9765-8 **World History the Easy Way, Vol. One**—$18.99, *Can$22.99*
ISBN 978-0-8120-9766-5 **World History the Easy Way, Vol. Two**—$21.99, *Can$26.50*